Photocatalysts based on Carbon Scaffolds

Environmental and Industrial Applications

Edited by

Ramesh Oraon, Pardeep Singh, Mika Sillanpää, and Ganesh Chandra Nayak

WILEY VCH

Ramesh Oraon
Department of Chemistry, Central University of Jharkhand, Ranchi, Jharkhand, India

Pardeep Singh
Assistant Professor
Department of Environmental Studies,
PGDAV College, University of Delhi,
New Delhi, India

Mika Sillanpää
Professor
Department of Chemical Engineering,
School of Mining, Metallurgy and
Chemical Engineering,
University of Johannesburg,
P. O. Box 17011, Doornfontein 2028,
South Africa

Ganesh Chandra Nayak
Associate Professor
Chemistry and Chemical Biology,
IIT (ISM) Dhanbad, Jharkhand, India

Cover Design: Wiley
Cover Image: © remotevfx.com/Shuttertock

Typesetting: Lumina Datamatics

Library of Congress Card No.: applied for

British Library Cataloguing-in-Publication Data A catalogue record for this book is available from the British Library.

Bibliographic information published by the Deutsche Nationalbibliothek The Deutsche Nationalbibliothek lists this publication in the Deutsche Nationalbibliografie; detailed bibliographic data are available on the Internet at <http://dnb.d-nb.de>.

Print ISBN: 9783527353972
ePDF ISBN: 9783527847389
ePub ISBN: 9783527847372
oBook ISBN: 9783527847365

Printed and bound by CPI Group (UK) Ltd, Croydon, CR0 4YY

C9783527353972_050326

Contents

Preface

Ever since the inception of the Industrial Revolution, the global demand for advanced materials has triggered the need for sustainable energy and environmental protection. Central to the advancement of this field, carbon scaffold-based nanophotocatalysts have emerged as a transformative class of materials, offering remarkable potential to address key issues such as water purification, air remediation, and solar-to-fuel conversion, etc. Recently, due to their inherent properties viz. high surface area to volume ratio, tunable electronic structures, and exceptional stability, positioned them toward functional platforms for constructing next-generation nanophotocatalyst. The combination of nanoscale photocatalysts with carbon-based structures has opened innovative pathways for utilizing light energy to power chemical reactions, including hydrogen generation through water splitting, pollutant breakdown, and carbon dioxide reduction. These developments pave the way for a future where clean energy and environmental stewardship go hand in hand. Advances in nanotechnology have also enabled nano-accessibility of these photocatalytic materials, allowing to tailor properties efficiently according to end application. The purpose of this book is to feature such ongoing trends, developments, challenges, and prospects in conventional energy sectors as well as remediate technologies that involve nanophotocatalysis. This book also aims to publish several promising strategies in conventional energy sectors involving nanophotocatalysis. This book (consisting of nine chapters) aims to provide a comprehensive exploration of carbon scaffold-based nanophotocatalysts, blending fundamental principles with cutting-edge advancements. It is designed for a diverse audience, including graduate students, researchers, and professionals in materials science, chemistry, and environmental engineering. Special attention is given to the challenges and opportunities in scaling up these technologies for industrial use, as well as the emerging trends that promise to shape the future of the field.

Editing this book (with editors Dr Ramesh Oraon, Central University of Jharkhand; Dr Pardeep Singh, Delhi University; Prof Ganesh Chandra Nayak, IIT (ISM) Dhanbad; Prof Mika Sillanpää, University of Johannesburg) has been a collaborative effort, drawing on the expertise of leading researchers and the latest developments in the literature. We have endeavored to present a balanced perspective, highlighting both the remarkable achievements and the unresolved questions that continue to drive innovation. Our hope is that this work will serve as both a valuable resource and an inspiration for those seeking to push the boundaries of photocatalysis and contribute to a sustainable world. The first chapter highlights the fundamental mechanism of nanophotocatalysis and its implications for environmental remediation. The second chapter deals with the computation estimation of supplemented photocatalytic activity of carbon scaffolds involved in the degradation process. The third chapter augments the industrial synthesis of carbon scaffold-based photocatalyst and their degradation mechanism toward sustainability. The fourth chapter provides an overview of the development of carbon scaffold-based nanophotocatalyst and challenges associated remediation of persistent pollutants in the biosphere. Fabrication, performance, and prospects of the carbon scaffold have been demonstrated in Chapter 5. Design and properties evaluation for water splitting and fuel production by carbon scaffold are extensively dealt with in Chapters 6 and 7, respectively. Chapters 8 and 9 elaborate on industry economics and eco-friendly strategies for energy environment conservation, respectively. We give deep sense of gratitude to all the contributing authors for sharing their expertise, making the book clear and understandable. Their significant efforts and knowledge were essential to its success. These contributors include esteemed academics and scientists from renowned institutions, universities, and organizations worldwide. The publisher and their administrative team also merit sincere appreciation for their unwavering support throughout the project.

Dr Ramesh Oraon
Dr Pardeep Singh
Prof Mika Sillanpää
Prof Ganesh Chandra Nayak

Section 1

Fundamental Insight and Novelty

1

Nanophotocatalysts: Fundamentals, Mechanisms, and Advances in Remediation Technologies

*Ehiaghe Agbovhimen Elimian[1,2], Claude Kiki[3], Andrew Nosakhare Amenaghawon[4], and Fidelis Odedishemi Ajibade[5]**

[1] *Department of Chemical and Environmental Engineering, University of Nottingham Ningbo China, Ningbo, China*
[2] *Department of Plant Biology and Biotechnology, Faculty of Life Sciences, University of Benin, Benin City, Nigeria*
[3] *CAS Key Laboratory of Urban Pollutant Conversion, Fujian Key Laboratory of Watershed Ecology, Institute of Urban Environment, Xiamen, China*
[4] *Department of Chemical Engineering, Faculty of Engineering, University of Benin, Benin City, Nigeria*
[5] *Department of Civil and Environmental Engineering, Federal University of Technology Akure, Akure, Nigeria*

*Corresponding author: foajibade@futa.edu.ng

1.1 Introduction

Sustainable and nontoxic technologies are being developed globally in response to the growing environmental pollution and the increasing energy demand. Photocatalysis is a highly effective technology for breaking down organic pollutants under light exposure in ambient conditions (Bianchi et al., 2015; Wu et al., 2021; Xu et al., 2019). In semiconducting materials, light energy excites electrons, resulting in the formation of electron-hole pairs. These pairs then engage in secondary reactions that produce radicals, which help detoxify contaminants. In "photocatalysis," light and a semiconducting substance initiate the reaction, which combines photochemical and catalytic processes (Su et al., 2018a; Maeda, 2011; Sano et al., 2006). In a nanophotocatalyst, electrons in the valence band are excited by incident photons and move to the conduction band. These charge carriers recombine to release energy and reactive species, which facilitate organic pollutant degradation (Fujishima et al., 2000; Carp et al., 2004; Duan et al., 2021). The crucial aspect of the photocatalysis reaction is the separation of charge carriers (Luo et al., 2020; Yang et al., 2011; Zhao et al., 2018). Nanocatalysts are derived from nanomaterials capable of catalyzing in both homogeneous and heterogeneous reactions (Moon et al., 2016; Tong et al., 2012).

Photocatalysts based on Carbon Scaffolds: Environmental and Industrial Applications, First Edition. Edited by Ramesh Oraon et al.
Published 2026 by Wiley-VCH GmbH.

Nanomaterials, falling within the size range of 1–100 nm, exhibit a notable surface-to-volume ratio and exceptional reactivity. These nanomaterials can be broadly categorized into three groups based on their composition: carbon materials (carbon nanotubes, fullerene C60 and C540, and graphene) (Bhattacharjee et al., 2022; Liang et al., 2019; Hendriks et al., 2014; Wang et al., 2019), polymeric materials and inorganic (Long et al., 2012; Zhou et al., 2019; Deng et al., 2023) and metal and metal oxide materials (Guerra et al., 2018; Fattakhova-Rohlfing et al., 2014; Xu et al., 2012; Jemec et al., 2015). Among these, carbon nanoparticles are deemed the most effective in eliminating organic pollutants.

Over the last decade, nanocatalysts have gained significant importance due to their diverse applications in wastewater treatment, drug delivery, energy production, and environmental remediation (Kumar et al., 2020; Moore & Wei, 2021; Zhang et al., 2021). By employing condition-based synthesis of nanoparticles (NPs) and their corresponding composites, the activity of the nanophotocatalysts can be precisely adjusted and controlled at the nanoscale, influencing their physical, chemical, optical, electronic, and magnetic characteristics. The increase in surface area and the quantum effects are the primary factors contributing to the distinct properties observed in comparison to bulk materials (Baig et al., 2021). The variations in the properties of bulk materials primarily stem from two factors: the increase in surface area and quantum effects. Also, reduced catalyst particle sizes yield a larger surface area, resulting in heightened reactivity and improved catalytic performance. Additionally, quantum effects at the nanoscale influence electrical, optical, and magnetic properties, causing distinct behavior compared to bulk materials (Wu et al., 2021; Baig et al., 2021). These unique properties at the nanoscale can be optimized through various characterizations, owing to their intermediate relationship between macroscopic and nanoscopic levels. Given their numerous benefits and promising prospects, nanophotocatalysts have become essential assets in the remediation of environmental pollutants (Yu et al., 2018; Zhang et al., 2017). These catalysts play a critical role in pollution control, particularly in the treatment of hazardous waste, the purification of air and water, and the overall sustenance of a healthy ecosystem. Their importance lies in their ability to efficiently break down various contaminants, including organic dyes, antibiotics, and phenolic compounds, through photochemical reactions (Singh & Dutta, 2018; Kumar et al., 2021; Dinh et al., 2018; Araya et al., 2017; Badvi & Javanbakht, 2021). Carbon scaffolds are a kind of carbon nanostructure that contains open, interconnected microporous structures. Their properties may be tuned to efficiently break down various pollutants. Additionally, their capability to harness solar energy for photocatalytic processes presents a viable and environmentally friendly approach to remediating pollutants. Furthermore, their nontoxic nature and ability to function under ambient conditions make them a promising solution for addressing pollution issues without further impacting the environment. As nanotechnology continues to advance, nanophotocatalysts hold significant promise for transforming pollution control and environmental remediation efforts, providing a sustainable and effective pathway toward a cleaner and healthier environment.

1.2 Fundamental Principles of Semiconductor Photocatalysis

In the presence of irradiated semiconductors, photocatalysis plays the function of starting or accelerating certain reduction and oxidation processes (redox). Light is absorbed, and electron-hole pairs are subsequently photoexcited when the energy of the input photons is ≥ the bandgap. The electrons (e_{CB}^-) play a reduction role with a chemical potential of +0.5 to −1.5 V per normal hydrogen electrode (NHE), whereas holes (h_{vb}^+) play an oxidative role with a chemical potential of +1.0 to +3.5 V (Rosman et al., 2021; Liu et al., 2020). The semiconductor stores photoexcitation-induced energy from incoming photons, which is then transformed into chemical form via surface/interface interactions and electronic processes. In contrast to traditional catalysis's thermodynamics, photocatalysis may facilitate both spontaneous ($\Delta G < 0$) and non-spontaneous ($\Delta G > 0$) processes (Liu et al., 2020; Yang & Wang, 2018). The activation barrier is broken down in spontaneous reactions by the input energy, which enables photocatalysis to occur more quickly or in more hospitable environments. Chemical energy that builds up in the reaction products during non-spontaneous reactions is partially transformed from the input energy. The typical sequence of a photocatalytic semiconductor cycle involves three main stages. Generally, heterogeneous photocatalysis involves the absorption of photons, excitation, separation, migration, transport, and possibly recombination of charges. The visual representation of the photocatalytic process is depicted in Figure 1.1.

Certain charge carriers have the ability to travel to the surface and instigate a sequence of surface reactions encompassing reduction and oxidation processes, ultimately resulting in the breakdown of pollutants. Consequently, the collective impact of these four successive steps significantly influences the overall efficacy of

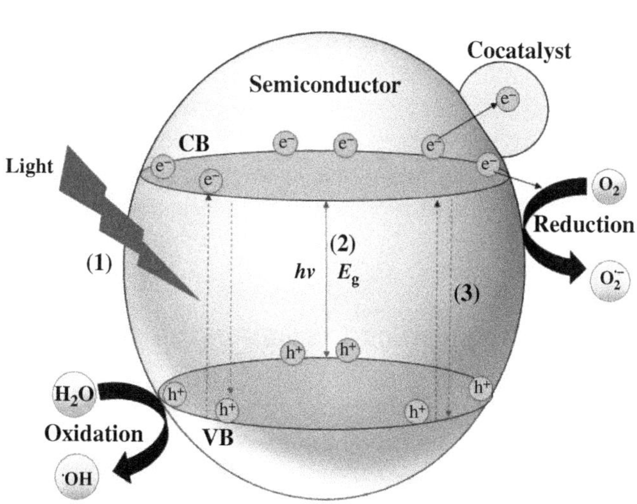

Figure 1.1 The schematic diagram for the photocatalytic process.

photocatalysis, a relationship that can be mathematically represented by Eq. (1.1) (Li et al., 2016).

$$\eta_c = \eta_{abs} \times \eta_{cs} \times \eta_{cmt} \times \eta_{cu} \tag{1.1}$$

Where solar energy conversion efficiency, light absorption efficiency, electron-hole pair excitation, migration and transportation, and utilization efficiency are denoted as η_c, η_{abs}, η_{cs}, η_{cmt}, and η_{cu} in Eq. (1.1) respectively.

The interplay between the redox potential of the substrate and the band edges of the semiconductor photocatalyst is crucial for the feasibility of reductive, oxidative, or nonreactive pathways (Fu et al., 2019). Specifically, if the redox potential of the substrate falls below the conduction band potential (CB) edge of the semiconductor photocatalyst, it enables reductive reactions. Conversely, when the redox potential exceeds the valence band potential (VB) edge, the substrate is primed for oxidative reactions. However, if the redox potential straddles the CB and VB edges, the substrate is precluded from participating in either reductive or oxidative reactions. Notably, if the redox potential simultaneously falls below the CB edge and exceeds the VB of the semiconductor photocatalyst, the substrate can engage in either reductive or oxidative reactions, illustrating the delicate balance that governs the outcome of photocatalytic processes (Fu et al., 2019; Liu et al., 2014a).

According to thermodynamics, the photogenerated e^- or h^+ are only necessary for the reduction and oxidation processes on the surface if their potentials lie between the semiconductor's VB and CB. For the effective photocatalytic degradation of a contaminant, the photogenerated charge carriers should possess adequate redox capacity that facilitates the generation of free radicals, which can further degrade the adsorbed pollutant (Yang & Wang, 2018; Li et al., 2015). The schematics in Figure 1.1 and the processes they illustrate can be used to elucidate the overall mechanism of carbon scaffolds as charge carriers or electron transport/migration in GO-metal oxide nanocomposites. Upon stimulation by light, photogenerated electrons are initially excited from the metal oxide semiconductor's VB to its CB. In this scenario, the carbon backbone acts as an electron trap, enabling the photoexcited electrons to be transported to the nanocomposite and preventing their recombination. When these electrons interact with oxygen on the surface, reactive oxygen species (ROS) are formed. In contrast, hydroxyl ROS are produced when water molecules react with photogenerated holes in the VB of metal oxide semiconductors.

1.3 General Mechanism on the Generation and Utilization of ROS

ROS such as hydroxyl ($^{\bullet}OH$) species, hydrogen peroxide (H_2O_2), superoxide ($^{\bullet}O_2^-$), and singlet oxygen ($^1O_2$) are generated when adsorbed oxygen and water molecules interact with the photoexcited charge carriers during the photocatalysis process (Li et al., 2011; Nosaka & Nosaka, 2016; Li et al., 2021). Among these ROS, $^1O_2$, an excited state of O_2, has a brief lifespan and may revert to its stable O_2 state without engaging in chemical processes or electron transfer (Li et al., 2021). In contrast to the $^{\bullet}O_2^-$ radical exhibits one of the longest lifespans among the primary ROS and demonstrates moderate reactivity (Yuan et al., 2021). The $^{\bullet}OH$ radical is acknowledged as the most potent ROS for degradation during the photocatalysis of semiconductors owing to its substantial reactivity, although it frequently induces nonselective oxidation reactions. Following multiple radical chain reactions in the presence of oxygen molecules (O_2), these intermediate radicals ultimately decompose into carbon dioxide (CO_2) and water. The reduction process in the conduction band is carried out by the photoelectrons (Ribao et al., 2019). The electrons from the reduction process act on dissolved oxygen species (O_2) in the air, leading to the generation of superoxide anions. These superoxide anions then react with the intermediate radicals formed during oxidation to produce H_2O_2, resulting in the formation of water molecules as the end product of the reduction process (Nosaka & Nosaka, 2016). Studies show that the $^{\bullet}OH$ radical can adsorb onto the photocatalyst's surface, trapping photogenerated holes. These trapped charge carriers are more reactive than bulk excitons, making them primary sites for adsorbing aromatic molecules, while freely available $^{\bullet}OH$ primarily reacts with non-adsorbed organic pollutants. For example, under visible light, the role of ROS radicals in the photocatalytic degradation of NAP over biochar-supported graphitic carbon nitride functionalized with potassium (biochar@K@g-C_3N_4) was investigated using scavengers such as ascorbic acid (AA), ethylenediaminetetraacetic acid (EDTA), and isopropanol (IPA) (Li & Lin, 2020). The nanocomposite achieved a photocatalytic performance of 82.19%. However, when AA, EDTA, and IPA were added, the performance decreased to 32%, 44%, and 65%, respectively. These findings suggest that the primary ROS generated during photocatalysis are h^+ and superoxide radicals (Li & Lin, 2020). The hypothesis arising from these findings suggests that g-C_3N_4 photoactivation facilitates e^-/h^+ separation, allowing their subsequent transport to the surface of the biochar. This proposed reaction mechanism extends the lifespan of photogenerated charge carriers, enabling h^+ to combine with H_2O, thereby producing hydroxyl radicals. Subsequently, these hydroxyl radicals lead to the production of active oxygen species through a series of chemical processes.

1.3.1 Generation of H_2O_2 Oxidative Molecules

Theoretically, both oxygen reduction, which involves reduction with two electrons ($O_2 \rightarrow O_2^- \rightarrow H_2O_2$), and the water oxidation reaction, which involves a one-step two-electron process ($O_2 \rightarrow H_2O_2$), can potentially produce H_2O_2 by photocatalysis (Chen et al., 2023; Cheng et al., 2022). The electrons (e^-) in the CB would interact with adsorbed O_2 to generate H_2O_2, while the holes (h^+) in the VB would oxidize H_2O to generate O_2 and H^+ (Eq. 1.2) (Hou et al., 2020).

$$2H_2O + 4h^+ \rightarrow O_2 + 4H^+ \tag{1.2}$$

For the two-electron oxygen reduction route to produce H_2O_2, the photocatalyst's CB must be more negative than 0.68 V. When the CB falls between + 0.68 and −0.33 V, the indirect $2e^-$ is thermodynamically disallowed. It is common for the direct and indirect $2e^-$ to coexist when the CB is less than −0.33 V (Chen et al., 2023; Fukuzumi et al., 2018). The indirect $2e^-$, requiring only one electron in each step, offers kinetic benefits, while the direct $2e^-$, with a higher positive reaction potential, is thermodynamically favorable. Enhancing the selectivity of the $2e^-$ for H_2O_2 synthesis is crucial, particularly because the $4e^-$, which has a more favorable reaction potential, can occur across all oxygen reduction processes (Chen et al., 2023; Xu et al., 2024). The production of H_2O_2 through the indirect, sequential, two-step reduction pathway, each step involving a single electron, is depicted in Eqs. (1.3–1.6). Initially, O_2 undergoes a one-electron reduction, resulting in the formation of the superoxide radical (Eq. 1.3), which subsequently reacts with H+ to generate the hydroxyl radical (Eq. 1.4). Following this, the hydroxyl radical undergoes an additional single-electron reduction (Eq. 1.5), leading to the generation of hydroxyl anions. Ultimately, as outlined in (Eq. 1.6), the hydroxyl anions react with H^+ to culminate in the formation of the final product, H_2O_2 (Hou et al., 2020).

$$O_2 + e^- \rightarrow O_2^- \tag{1.3}$$

$$O_2^{\bullet -} + H^+ \rightarrow OH_2^- \tag{1.4}$$

$$OH_{2^2}^{\bullet -} + e^- \rightarrow OH_2^- \tag{1.5}$$

$$OH_2^- + H^+ \rightarrow H_2O_2 \tag{1.6}$$

A positive VB of over 1.76 V is necessary for the oxidative reaction, attributed to the one-step two-electron pathway (Chen et al., 2023). The direct two-electron pathways are thermodynamically more advantageous (Badvi & Javanbakht, 2021; Chen et al., 2023). Equation (1.7) illustrates a straightforward one-step two-electron reduction for the creation of H_2O_2. In this process, O_2 directly engages with two H^+ ions to generate H_2O_2 through a two-electron photoreduction (Hou et al., 2020).

$$O_2 + 2H^+ + 2e^- \rightarrow H_2O_2 \tag{1.7}$$

The double H_2O_2 evolution route, which incorporates both reduction-oxidation routes, is preferred because it produces H_2O_2 from O_2 and H_2O without the usage of sacrificial materials. The photocatalytic synthesis of H_2O_2 from H_2O and O_2 is a growing process, and the two routes may be characterized as a single photocatalytic reaction (Eq. 1.8) (Hou et al., 2020).

$$2H_2O + O_2 \rightarrow 2H_2O_2 \tag{1.8}$$

Although H_2O_2 is synthesized on the photocatalyst surface, it tends to disintegrate after a few seconds, and methods are used to stabilize and recover it. Therefore, the redox potential of photocatalytic materials must meet the requirements for thermodynamic oxygen reduction and water oxidation processes (Xu et al., 2024). If the CB and VB of these materials fall within the single-electron oxygen reduction/water oxidation potential range, the single-electron route is chosen. Furthermore, the superoxide and hydroxyl radical produced in the one-electron pathway may be converted to H_2O_2, making it more beneficial than the four-electron pathway for photocatalytic H_2O_2 generation (Chen et al., 2023). To ensure efficient H_2O_2 production, photocatalytic materials must not only have a good band structure, but also effectively limit competition with the four-electron pathway and increase the selectivity of the H_2O_2 production pathway. Recently, the research focus has shifted to the realization of the redox double pathway for H_2O_2 production by considering the single-electron water oxidation with only one electron to replace the traditional four-electron O_2 production pathway.

1.3.2 Generation of Superoxide Species

The superoxide ion ($O_2^{\bullet-}$) plays a crucial role as a radical species in numerous photochemical processes. When O_2 undergoes monovalent reduction, it generates $O_2^{\bullet-}$, which possesses both radical and anionic properties, indicated by the radical sign (•) and a charge of −1 (Eq. 1.9) (Hayyan et al., 2016). The $O_2^{\bullet-}$ with 17 electrons is a detectable paramagnetic ion belonging to the main group, and it serves as an intermediate species in the oxygen reduction reaction.

$$O_2 + e^- \rightarrow O_{2^-} \tag{1.9}$$

Superoxide is highly reactive, serving as an effective oxidizing agent and radical starter. Free radicals, characterized by their unpaired electrons, exhibit strong reactivity, enabling them to oxidize various organic contaminants. However, the significant reactivity and brief lifespan of free radicals create challenges for accurate quantitative measurement. Consequently, only a limited number of measurement methods, such as chemical reactions, spin trapping, and direct detection, are viable for assessing superoxide and other free radicals (Nosaka & Nosaka, 2017).

1.3.3 Generation of Hydroxyl Radicals

The essential role of hydroxyl radicals ($^{\bullet}OH$) in photocatalysis is due to their high reactivity, making them the most potent ROS capable of completely oxidizing organic pollutants. With a high redox capacity (E0 = 2.7 VNHE) and nonselectivity, hydroxyl radicals can effectively eliminate organic contaminants in water (Wang et al., 2022; Xiao et al., 2017).

In typical photocatalysis, the involvement of three electrons will result in the generation of only one hydroxyl radical. The conversion of photoinduced electrons into highly oxidative hydroxyl radicals is hindered by a three-electron pathway of oxygen activation. This pathway begins with the reaction between electrons and dissolved oxygen (DO), leading to the formation of hydrogen peroxide. Subsequently, hydrogen peroxide is converted into hydroxyl radical with the involvement of an additional electron (Eq. 1.10) (Nosaka & Nosaka, 2017; Xiao et al., 2017).

$$O_2 + CB - e^- \rightarrow O_2^{\bullet-} + 2H^+ + CB - e^- \rightarrow H_2O_2 + CB - e^- \rightarrow HO + OH^- \tag{1.10}$$

The production of hydroxyl radicals during heterogeneous catalytic processes has not been intensely reviewed since these the radical production quantum yields cannot be established with certainty (Gligorovski et al., 2015).

1.3.4 Formation of Singlet Oxygen Species

There has been an observation that $^1O_2$ is produced by energy transfer from excited photocatalysts in TiO_2 photocatalysis through the transfer of electrons. However, due to its excited state, the reactivity of $^1O_2$ can be quenched by energy transfer without participating in chemical reactions. Even though $^1O_2$ has an inherent lifetime, the actual lifetime of the compound in aqueous solution is reduced to a few microseconds (Nosaka & Nosaka, 2017). By oxidizing superoxide with holes, the resulting $^1O_2$ accumulates on the surface of the photocatalyst, where it is subsequently converted into oxygen (Buchalska et al., 2013). Additionally, the generation of $^1O_2$ through the oxidation of hydroxyl radical and superoxide species is characterized by a change in Gibbs energy of −28 kJ/mol attributed to the respective standard reduction potentials at pH 7. The reaction to form $^1O_2$, given by Eq. (1.11) (Nosaka & Nosaka, 2017). However, if the singlet oxygen cannot participate in a reaction within a short timeframe, it cannot be involved in a solution reaction.

$$O_2^{\bullet-} + OH_2\bullet + H^+ \rightarrow {}^1O_2 + H_2O_2 \qquad (1.11)$$

Thus, singlet oxygen is generated by superoxide oxidation with a strong oxidant such as the photo-induced holes and oxidized ligands.

1.4 Environmental Application of Carbon Scaffolds

Nanoscale materials have gained prominence in recent years due to their exceptional physiochemical characteristics. These materials are more reactive than bulk materials due to their higher surface-to-volume ratios, making them more efficient at remediation. It is possible to functionalize these surfaces with specific molecules in order to optimize the removal of contaminants. Furthermore, nanomaterials can be tailored in terms of their size, shape, porosity, and composition in order to offer better performance. Hybrids and composites of different materials can offer advantages in terms of selectivity and stability over single-material approaches. It should also be noted that nanoparticles can enhance the stability of a scaffold, and the addition of specific chemicals can significantly increase its selectivity and efficiency against a particular contaminant.

1.4.1 Bacteria Inactivation and Antibiotic Removal

When the surface of the metal oxide is exposed to photons equal to or > the band, electron-hole pairs are generated (Fagan et al., 2016). The charge carriers generated by the light are then trapped on the semiconductor surface leading to the formation of highly ROS. These ROS serve as powerful oxidizing agents for organic contaminants and break down stable organic compounds into intermediates that can be further degraded to less harmful intermediates (Kumar et al., 2023). They also have the potential to inactivate bacteria and denature proteins, resulting in a variety of end products (Eq. 1.12) (Baaloudj et al., 2021).

$$\frac{\text{Organic pollutions}}{\text{Bacteria membrane}} + HO^{\bullet} + O_2^{\bullet -} \rightarrow \frac{\text{water } a + \text{ carbon dioxide}}{\text{disruption } + \text{ inactivation}} \tag{1.12}$$

The carbon scaffold, g-C_3N_4, has been employed for visible-light disinfection (Teng et al., 2019). A disinfection rate exceeding 99.9% was achieved within 30 minutes using g-C_3N_4, outperforming other metal oxides. Its catalytic performance is comparable to that of noble metal-based photocatalysts. Further analysis indicated that the presence of carboxyl and carbonyl groups at the edges of g-C_3N_4 nanosheets enhanced charge separation and facilitated the formation of reactive species (Teng et al., 2019). In a recent study reported by Zeng et al. (2020), C_3N_4 nanosheets were modified with polyethylenimine (PEI) for the inactivation of microbes. It was demonstrated that the PEI/C_3N_4 nanocomposite was capable of rapid photocatalytic activity against bacteria, resulting in *Escherichia coli* and *Enterococcus faecalis* levels being reduced within 45 and 60 minutes, respectively, under simulated solar light. It has been shown that PEI is an effective cell adhesion substance due to its electrostatic interaction with negatively charged surfaces when the amino groups of the protein are protonated in an aqueous environment. Furthermore, PEI is capable of interacting with C_3N_4 nanosheets in order to influence the electronic structure of those nanosheets. Combining PEI with C_3N_4 improves the separation of charge carriers and the conversion of superoxide anions. Through electrostatic adhesion, PEI increases the zeta potential of C_3N_4 nanosheets, thereby enhancing indirect ROS production during photocatalysis. This study demonstrates a facile method for enhancing the disinfection performance of g-C_3N_4 by introducing PEI on the nanosheet surface (Zeng et al., 2020).

Antibiotics, widely used to treat bacterial infections, contribute to water pollution when their waste is discharged into water sources from various origins (Wang et al., 2020; Godoy & Sánchez, 2020). Carbon scaffold nanomaterials, such as the graphene subgroups (graphene oxide, GO, and its reduced form, rGO), possess properties similar to graphene and have been studied for their photocatalytic activity in degrading organic pollutants under light. Graphene-based materials exhibit a large surface area, high electron mobility, and extensive π-bond conjugation (Yang et al., 2012). Recently, SnO_2/GO and ZnO/GO nanocomposites have shown promising results in enhancing the photodegradation of organic pollutants (Kumar et al., 2023). Furthermore, the efficient removal of organic contaminants such as ofloxacin, bisphenol A, and *p*-bromophenol was studied by using *n*-ZnO/*n*-SnO_2 heterojunctions combined with rGO nanosheets (Kumar et al., 2023). The nanocomposite with 5 wt% rGO exhibited good photocatalytic efficiency, achieving high removal rates for the pollutants after specific durations of UV light exposure. It was found that the combined effects of ZnO/SnO_2/rGO significantly enhanced the photodegradation of contaminants owing to their synergistic effect. In the study, superoxide and hydroxyl radical anions were identified as the main active oxygen species for photocatalytic degradation.

Wastewater treatment plants have the potential to aid the formation of extracellular antibiotic resistance genes (eARGs) and antibiotic-resistant bacteria (ARBs), which might result in environmental resistance. Moreover, the release of ARBs and eARGs is not inhibited by traditional secondary disinfection techniques including chlorination and ultraviolet (UV) light (Ju et al., 2016). Moreover, residual disinfectants may encourage the horizontal transfer of genes associated with antibiotic resistance (ARGs), which would increase the overall level of resistance (He et al., 2019). As a result, technical innovation is desperately needed to create reliable and long-lasting methods for degrading eARGs and inactivating ARBs to successfully manage the growth of antimicrobial resistance. Deyi et al. (Li et al., 2020) reported that modified GO could be used to inactivate antibiotic-resistant bacteria and genes. To enhance the inactivation of plasmid-encoded antibiotic resistance genes and multidrug-resistant *E. coli*, N_2-rGO was coated onto hierarchically structured microspheres of $Bi_2O_2CO_3$. In addition to the enhanced light absorption and shortened electron transfer time, N_2-rGO coatings increased reactive oxygen species threefold (Li et al., 2020). When compared to a commercial TiO_2 photocatalyst, the coating greatly outperformed it due to lower ROS scavenging by background components and higher E. coli adsorption around the photocatalytic sites. The discussion above suggests that carbon scaffolds could enhance secondary effluent photocatalytic treatment by inactivating bacteria and reducing antibiotic resistance. Under light exposure, these photocatalysts can cause charge carriers to interact with oxygen and water molecules, resulting in the formation of a variety of strong reactive oxygen species. These species deactivate infections by altering the cell membrane structure. As a result, nanophotocatalysts provide an efficient and cost-effective method for eliminating pollution.

1.4.2 Phenolic Compounds

Phenolic compounds are frequently detected in surface water discharged by the chemical, refinery, and pharmaceutical sectors. Phenolic compounds include phenol, aminophenols, chlorophenols, and nitrophenols. These compounds pose a major worldwide issue due to their high level of toxicity toward species, especially when introduced into the natural environment. The efficient decomposition of organic contaminants using photocatalytic oxidation has garnered increasing attention in recent years. According to several studies, carbon scaffold-based photocatalysts are very effective at oxidizing and getting rid of phenolic compounds from wastewater when exposed to light (Hu et al., 2018; Guo et al., 2016; Ma et al., 2021; Chen et al., 2017a), as shown in Table 1.1. Hui et al. (Zhang et al., 2015) reported the performance of supported Cu^{2+} clusters on TiO_2/rGO nanocomposite for phenol degradation. These clusters formed strong connections with O-containing groups on rGO, enhancing interfacial charge transfer between TiO_2 and copper ions. A significant finding was the discovery of a two-electron transfer mechanism that generated highly oxidative $^{\bullet}OH$ for effective phenol degradation. The improved photocatalytic activity of this ternary system was demonstrated through electron spin resonance (ESR) spectra and electrochemical measurements. Through the two-electron reduction of oxygen, RGO and copper ions facilitate interfacial charge transfer, enabling in situ hydrogen peroxide generation (Zhang et al., 2015). The performance of reduced TiO_{2-x} with rGO nanocomposite (TiO_{2-x}/rGO) for the degradation of bisphenol

Table 1.1 Performance of carbon scaffolds for phenolic compounds degradation.

Nanophotocatalyst	Pollutant	Initial concentration (mg/L)	Degradation efficiency	Duration (min)	Reference
g-C_3N_4/GO/rGO	Phenol	10.0	28%	4,320	Aleksandrzak et al. (2017)
$BiVO_4$/GR		20.0	95%	80	Wang et al. (2014)
Fe3O4/rGO		50.0	76%	120	Padhi et al. (2017)
Fe_3O_4@rGO@Ag		50.0	99%	540	Rehman et al. (2019)
Ag-Cu_2O/rGO		10.0	97%	210	Wei et al. (2018)
SnO_2/rGO	*p*-Nitrophenol	20.0	96	90	Chen et al. (2017b)
Bi_2WO_6/rGO	Bisphenol	10.0	95	80	Yang et al. (2015)
α-Fe_2O_3/rGO		10.0	51%	180	Zhang et al. (2016b)
rGO/SiO_2/ZrO_2		10.0	87	90	Hassan et al. (2019)
Bi_2S_3 NP/rGO	2,4-dichlorophenol	60.0	92	120	Chen et al. (2016)
BiOI/Fe_3O_4@GO	2,4-dintirophenol	129.0	100	100	Singh et al. (2019)

(BPA) under visible light has been reported (Xu et al., 2018). TiO_2 nanoparticles were produced via the hydrothermal-calcination synthesis method and dispersed uniformly over rGO sheets by means of Ti—O—C linkages. After calcination in an inert environment, Ti^{3+} and O vacancies were then added to the composites concurrently. The study found that the pH of the solution affected the amount of BPA removed, with a pH of 4.0 producing the fastest rate of breakdown. In comparison to pristine TiO_2, TiO_2/rGO, and $TiO_{2-x,}$ TiO_{2-x}/rGO demonstrated much quicker response rates for BPA degradation, with enhancement factors of 6.16, 2.92, and 2.55, respectively. The study also investigated the effects of inorganic anions, catalyst dosage, solution pH, and initial substrate concentration on BPA removal. According to EPR data, superoxide ions (O_2^-) were the main oxidizing species that breaks down BPA. Solvent thermal methods were utilized to mix ZnS-MoS_2 nanoparticles with rGO to form a new ternary composite (rGO/ZnS-MoS_2) that exhibited efficient photocatalytic activity (Hu et al., 2018). The PL intensities of the rGO/ZnS-MoS_2 composites saw a considerable drop upon the addition of rGO. In 180 minutes, under visible light, the 11% rGO/ZnS-MoS_2 sample effectively removed 90.57% of 2-nitrophenol demonstrating optimum catalytic activity. With the advancement of the reactions, the 2-NP was completely oxidized as evidenced by the decrease in intensity of its distinctive absorption peaks and the absence of any new absorption peaks. Unlike binary metal sulfides, ternary compounds have a greater visible light response [72]. The performance of a ternary GO/Ag_2CrO_4/g-C3N4 (GO/ACR/CN) composite was reported for the photocatalytic degradation of phenol, oxytetracycline, MB, and RhB under visible light [75]. The heterojunction formed by the well-incorporated GO, Ag_2CrO_4, and g-C_3N_4 can be viewed as a combination of the ACR/GO and ACR/CN binary heterostructures.

1.4.3 Organic Dye

In the past two decades, the escalation of organic pollutants in natural resources can be attributed to technological advancements and rapid industrial expansion. Notably, common pollutants such as organic synthetic dyes primarily originate from industries like textiles, food, cosmetics, and leather. These dyes disrupt aquatic photosynthesis, deplete dissolved oxygen, and pose risks to the health of plants, animals, and humans. Their intricate structure results in substantial biological, chemical, and photocatalytic stability, rendering them resistant to degradation (Martínez-Huitle & Brillas, 2009). To address these challenges, a semiconductor photocatalyst is utilized to adsorb and chemically transform these pollutants through light-induced redox reactions (Ye et al., 2020; Azhar et al., 2018). This semiconductor serves as a dependable catalyst for photolysis and is resilient to the toxicity of organic dyes. The primary objective is to develop an affordable, nontoxic catalyst with a tunable bandgap, abundant active sites, and pores to remove pollutants (Ye et al., 2020). A summary of the dye degradation process utilizing enhanced carbon scaffold-based photocatalysts is provided in Table 1.2.

Table 1.2 Performance of photocatalytic carbon scaffolds for some phenolic compounds.

Nanophotocatalyst	Weight (g/L)	Dye	Initial conc (mg/L)	Degradation efficiency	Duration (min)	Reference
$g\text{-}C_3N_4/RGO/Bi_2MoO_6$	1.0	Rhodamine B (RhB)	10.0	97%	60	Ma et al. (2017)
$FeWO_4$-RGO	0.8		5.0	24%	240	He et al. (2015)
RGO/In_2TiO_5	0.2		10.0	98%	120	Zhang et al. (2019)
Cu_2O-rGO	1.3		/	65%	80	Sun et al. (2015)
MoS_2/Graphene	0.1		5.0	95%	80	Zhao et al. (2017)
CdS-rGO heterojunction	0.4		20.0	92%	60	Wei et al. (2019)
CdS@RGO microspheres	1.0		5.0	95%	60	Liu et al. (2014b)
Ag_3PO_4@GO composites	1.0		6.0	100	40	Liu et al. (2018)
CN/rGO@BPQDs	0.1		10.0	97%	20	Xiong et al. (2020)
Cu_3Se_2/rGO	0.33	Methyl blue (MB)	10.0	95%	60	Baghchesara et al. (2019)
ZnSe/graphene	0.17		35.0	99%	360	Hsieh et al. (2015)
CoS-graphene sheet	0.2		10.0	92%	180	Kong et al. (2014)
$CoFe_2O_4$/graphene nanocomposites	0.25		10.0	100%	60	Gan et al. (2015)
N/Ti^{3+}-TiO_2(GNTT)	0.5		/	61%	80	Lin et al. (2019)
BiOI-rGO photocatalyst	1.0	Methyl orange (MO)	10.0	85%	60	Vinoth et al. (2017)
BiOBr/RGO aerogel	1.0		10.0	95%	120	Yu et al. (2017)
CeO_2/RGO nanocomposites	0.5		/	88%	60	Verma and Samdarshi (2016)
TiO_2@rGO	0.2	Rhodamine 6G (Rh 6G)	20.0	100%	70	Pu et al. (2017)
BiOCOOH/g-C_3N_4 composite	0.02	Amido black 10B	10.0	89.2%	120	Cui et al. (2019)

Due to its rapid electron mobility, abundant reactive sites, and large specific surface area, graphene oxide is an excellent catalyst support and photocatalyst (Su et al., 2018b; Zhai et al., 2015; Xu et al., 2015). For example, Verma and Dutta (2017) reported the photodegradation of Trypan blue dye (TB) using a nanocomposite consisting of ZnO NPs and NH_3-reduced GO. Under irradiation, the nanocomposites made with ZnO NPs and 5 wt% ammonia-modified graphene oxide showed a 7–10-times increase in photocatalytic activity in comparison to pure ZnO NPs. The conductive reduced graphene oxide's ability to effectively separate photoexcited electrons and holes is responsible for the notable increase in photocatalytic activity. This process produces higher concentrations of reactive species as demonstrated by ROS scavengers. Additionally, the GC–MS analysis of the degraded sample revealed no intermediates, confirming the complete mineralization of the dye. Similarly, the rGO-modified Ni–Mn–S nanocomposite (Ni–Mn–S/rGO) effectively catalyzed the degradation of AY and RB dyes, owing to its small bandgap (Kandhasamy et al., 2024). Within 150 minutes of visible light exposure, the Ni–Mn–S/rGO nanocomposite achieved a 96% degradation of AY dye and a 93% degradation of RB dye. These results affirm the viability of Ni–Mn–S/rGO for textile dye degradation, supported by its structural and optical characteristics. Using a hydrothermal technique, ZnO-Al_2O_3-rGO/MWCNT was synthesized for organic contaminant degradation. This study showed that the rGO/MWCNT matrix and ZnO-Al_2O_3 nanoparticles together improved the degradability of Congo red and RhB dyes under visible light by 90% and 80%, respectively. The synergistic interaction decreased the electron-hole pair recombination, facilitated charge transfer, and increased the surface area. Several studies have shown that Cu_2MoS_4 nanosheets supported on rGO are more efficient at degrading methyl orange (MO) (Zhang et al., 2016a; Rameshbabu et al., 2017). Rameshbabu et al. (2017) observed a 99% photocatalytic degradation efficiency for the Cu_2MoS_4/rGO nanocomposite compared to Cu_2MoS_4 (64%). Compared to bare Cu_2MoS_4 nanosheets, the interaction between rGO sheets and Cu_2MoS_4 nanosheets expanded the absorption to the visible light spectrum. The Cu_2MoS_4 and rGO interfacial contact enhanced photocatalytic methyl orange efficiency due to the separation of the charge carriers (Zhang et al., 2016a; Rameshbabu et al., 2017).

Since the valence and conduction bands of g-C_3N_4 are located at +1.6 eV and −1.1 eV, respectively, and it has a bandgap of 2.7 eV, it is more effective in suppressing organic pollutants when exposed to visible light than pure TiO_2, which is only active in the UV range. The flexible tunability of the bandgap with other cocatalysts has increased the application for various organic dyes (Ahmaruzzaman & Mishra, 2021; Guo et al., 2022). The hydrothermal process was utilized to generate an efficient Z-scheme composite driven by visible light. When compared to Bi_2MoO_6, g-C_3N_4, and g-C_3N_4/Bi_2MoO_6 composites, the g-C_3N_4/RGO/Bi_2MoO_6 composite exhibited superior photocatalytic activity in degrading RhB (Ma et al., 2017). The charge transfer is enhanced by the matching Fermi levels in the Z-scheme system. It provides a solid scientific basis for planning, designing, and implementing Z-scheme photocatalysts since the Fermi level is the key to the Z-scheme system.

1.5 Challenges and Prospects

Carbon-based materials have recently seen significant advancements in doping, immobilizing photocatalysts, designing photonic crystals, optimizing light harvesting, selective cocatalyst attachment, sensitization, and nanostructuring to increase surface area. These diverse tactics aim to maximize the efficiency of photocatalysis, leading to groundbreaking discoveries across various sectors and fields (Baaloudj et al., 2021). The common approach for using carbon-based materials to degrade organic pollutants involves combining semiconductor materials and their composites with carbon counterparts (Ahmaruzzaman & Mishra, 2021; Singh et al., 2020). However, achieving good performance becomes increasingly challenging due to diminishing returns when loading exceeds a certain threshold. This necessitates careful control of surface area, pore distribution, and particle size. Furthermore, morphological changes, electronic properties, and stability across different operational conditions. In order to accelerate the photocatalytic breakdown of environmental contaminants, scientists are investigating the possibility of combining alone using several materials. This might lead to the development of future techniques that are both more effective and sustainable. These developments pave the way for novel approaches to environmental problems and have the potential to significantly enhance the use of customized carbon-based materials in photocatalytic pollutant degradation.

1.6 Conclusion

The chapter emphasizes the potential of carbon scaffolds to enhance photocatalytic degradation of organic pollutants. Although carbon-based photocatalysis is recognized as a sustainable solution, challenges in material synthesis, scalability, and stability remain, requiring collaboration among academia, industry, and policymakers. In environmental remediation, novel carbon materials represent sustainability. To combat pollution in an eco-friendly manner, researchers can use these materials to develop more effective strategies for degrading photo-induced pollutants.

References

Ahmaruzzaman, M. & Mishra, S. R. (2021). Photocatalytic performance of g-C_3N_4 based nanocomposites for effective degradation/removal of dyes from water and wastewater. *Materials Research Bulletin*, 143, 111417. https://doi.org/10.1016/j.materresbull.2021.111417

Aleksandrzak, M., Kukulka, W., & Mijowska, E. (2017). Graphitic carbon nitride/graphene oxide/reduced graphene oxide nanocomposites for photoluminescence and photocatalysis. *Applied Surface Science*, 398, 56–62. https://doi.org/10.1016/j.apsusc.2016.12.023

Araya, T., cheng Chen, C., ke Jia, M., Johnson, D., Li, R., & ping Huang, Y. (2017). Selective degradation of organic dyes by a resin modified Fe-based metal-organic framework under visible light irradiation. *Optical Materials*, 64, 512–523. https://doi.org/10.1016/j.optmat.2016.11.047

Azhar, M. R., Vijay, P., Tadé, M. O., Sun, H., & Wang, S. (2018). Submicron sized water-stable metal organic framework (bio-MOF-11) for catalytic degradation of pharmaceuticals and personal care products. *Chemosphere*, 196, 105–114.

Baaloudj, O., Assadi, I., Nasrallah, N., El Jery, A., Khezami, L., & Assadi, A. A. (2021). Simultaneous removal of antibiotics and inactivation of antibiotic-resistant bacteria by photocatalysis: A review. *Journal of Water Process Engineering*, 42, 102089. https://doi.org/10.1016/j.jwpe.2021.102089

Badvi, K. & Javanbakht, V. (2021). Enhanced photocatalytic degradation of dye contaminants with TiO_2 immobilized on ZSM-5 zeolite modified with nickel nanoparticles. *Journal of Cleaner Production*, 280, 124518. https://doi.org/10.1016/j.jclepro.2020.124518

Baghchesara, M. A., Azimi, H. R., Ghorban Shiravizadeh, A., Asri Mat Teridi, M., & Yousefi, R. (2019). Improving the intrinsic properties of rGO sheets by S-doping and the effects of rGO improvements on the photocatalytic performance of Cu_3Se_2/rGO nanocomposites. *Applied of Surface Sciences*, 466, 401–410. https://doi.org/10.1016/j.apsusc.2018.10.082

Baig, N., Kammakakam, I., & Falath, W. (2021). Nanomaterials: A review of synthesis methods, properties, recent progress, and challenges. *Materials Advances*, 2, 1821–1871. https://doi.org/10.1039/D0MA00807A

Bhattacharjee, B., Ahmaruzzaman, M., Djellabi, R., Elimian, E., & Rtimi, S. (2022). Advances in 2D MXenes-based materials for water purification and disinfection: Synthesis approaches and photocatalytic mechanistic pathways. *Journal of Environmental Management*, 324, 116387. https://doi.org/10.1016/j.jenvman.2022.116387

Bianchi, C. L., Pirola, C., Galli, F., Cerrato, G., Morandi, S., & Capucci, V. (2015). Pigmentary TiO_2: A challenge for its use as photocatalyst in NO_x air purification. *Chemical Engineering Journal*, 261, 76–82. https://doi.org/10.1016/j.cej.2014.03.078

Buchalska, M., Łabuz, P. Ł. B., Szewczyk, G., Sarna, T., Maćkowski, S., & Macyk, W. (2013). New insight into singlet oxygen generation at surface modified nanocrystalline TiO_2– the effect of near-infrared irradiation. *Dalton Transactions*, 42, 9468–9475. https://doi.org/10.1039/C3DT50399B

Carp, O., Huisman, C. L., & Reller, A. (2004). Photoinduced reactivity of titanium dioxide. *Progress in Solid State Chemistry*, 32, 33–177. https://doi.org/10.1016/j.progsolidstchem.2004.08.001

Chen, F., Yang, Q., Wang, S. et al. (2017a). Graphene oxide and carbon nitride nanosheets co-modified silver chromate nanoparticles with enhanced visible-light photoactivity and anti-photocorrosion properties towards multiple refractory pollutants degradation. *Applied Catalysis B: Environmental*, 209, 493–505. https://doi.org/10.1016/j.apcatb.2017.03.026

Chen, Y., Sun, F., Huang, Z. et al. (2017b). Photochemical fabrication of SnO_2 dense layers on reduced graphene oxide sheets for application in photocatalytic degradation of p-Nitrophenol. *Applied Catalysis B: Environmental*, 215, 8–17. https://doi.org/10.1016/j.apcatb.2017.03.082

Chen, Y., Tian, G., Mao, G., Li, R., Xiao, Y., & Han, T. (2016). Facile synthesis of well-dispersed Bi_2S_3 nanoparticles on reduced graphene oxide and enhanced photocatalytic activity. *Applied Surface Science*, 378, 231–238. https://doi.org/10.1016/j.apsusc.2016.03.194

Chen, Z., Yao, D., Chu, C., & Mao, S. (2023). Photocatalytic H_2O_2 production systems: Design strategies and environmental applications. *Chemical Engineering Journal*, 451, 138489. https://doi.org/10.1016/j.cej.2022.138489

Cheng, H., Cheng, J., Wang, L., & Xu, H. (2022). Reaction pathways toward sustainable photosynthesis of hydrogen peroxide by polymer photocatalysts. *Chemistry of Materials*, 34, 4259–4273. https://doi.org/10.1021/acs.chemmater.2c00936

Cui, Y., Zhang, X., Zhang, H., Cheng, Q., & Cheng, X. (2019). Construction of BiOCOOH/g-C_3N_4 composite photocatalyst and its enhanced visible light photocatalytic degradation of amido black 10B. *Separation and Purification Technology*, 210, 125–134. https://doi.org/10.1016/j.seppur.2018.07.059

Deng, Y., Yang, G., Lens, P. N. L. et al. (2023). Enhanced removal of mixed VOCs with different hydrophobicities by Tween 20 in a biotrickling filter: Kinetic analysis and biofilm characteristics. *Journal of Hazardous Materials*, 450, 131063. https://doi.org/10.1016/j.jhazmat.2023.131063

Dinh, V. T., Thu, P. A., An, N. T., Nhan, D. N. T., & Long, N. Q. (2018). Toluene removal under humid conditions by synergistic adsorption–photocatalysis using nano TiO_2 supported on ZSM-5 synthesized from rice-husk without structure-directing agent. *Reaction Kinetics, Mechanisms and Catalysis*, 125, 1039–1054. https://doi.org/10.1007/s11144-018-1452-7

Duan, X., Yang, J., Hu, G. et al. (2021). Optimization of TiO_2/ZSM-5 photocatalysts: Energy band engineering by solid state diffusion method with calcination. *Journal of Environmental Chemical Engineering*, 9, 105563. https://doi.org/10.1016/j.jece.2021.105563

Fagan, R., McCormack, D. E., Dionysiou, D. D., & Pillai, S. C. (2016). A review of solar and visible light active TiO_2 photocatalysis for treating bacteria, cyanotoxins and contaminants of emerging concern. *Materials Science in Semiconductor Processing*, 42, 2–14. https://doi.org/10.1016/j.mssp.2015.07.052

Fattakhova-Rohlfing, D., Zaleska, A., & Bein, T. (2014). Three-dimensional titanium dioxide nanomaterials. *Chemical Reviews*, 114, 9487–9558. https://doi.org/10.1021/cr500201c

Fu, Y., Li, J., & Li, J. (2019). Metal/semiconductor nanocomposites for photocatalysis: Fundamentals, structures, applications and properties. *Nanomaterials*, 9, 359. https://doi.org/10.3390/nano9030359

Fujishima, A., Rao, T. N., & Tryk, D. A. (2000). Titanium dioxide photocatalysis. *Journal of Photochemistry and Photobiology C: Photochemistry Reviews*, 1(1), 1–21. https://doi.org/10.1016/S1389-5567(00)00002-2

Fukuzumi, S., Lee, Y. M., & Nam, W. (2018). Solar-driven production of hydrogen peroxide from water and dioxygen. *Chemistry - A European Journal*, 24, 5016–5031. https://doi.org/10.1002/chem.201704512

Gan, L., Shang, S., Yuen, C. W. M., Jiang, S., & Hu, E. (2015). Hydrothermal synthesis of magnetic $CoFe_2O_4$/graphene nanocomposites with improved photocatalytic activity. *Applied of Surface Sciences*, 351, 140–147. https://doi.org/10.1016/j.apsusc.2015.05.130

Gligorovski, S., Strekowski, R., Barbati, S., & Vione, D. (2015). Environmental implications of hydroxyl radicals (•OH). *Chemical Reviews*, 115, 13051–13092. https://doi.org/10.1021/cr500310b

Godoy, M. & Sánchez, J. (2020). Chapter 12 - Antibiotics as emerging pollutants in water and its treatment. In V. Kokkarachedu, V. Kanikireddy, & R. Sadiku (Eds.), *Antibiotic Materials in Healthcare* (pp. 221–230). Academic Press. https://doi.org/10.1016/B978-0-12-820054-4

Guerra, F. D., Attia, M. F., Whitehead, D. C., & Alexis, F. (2018). Nanotechnology for environmental remediation: Materials and applications. *Molecules*, 23, 1–23. https://doi.org/10.3390/molecules23071760

Guo, R., Wang, J., Bi, Z., Chen, X., Hu, X., & Pan, W. (2022). Recent advances and perspectives of g–C_3N_4–based materials for photocatalytic dyes degradation. *Chemosphere*, 295, 133834. https://doi.org/10.1016/j.chemosphere.2022.133834

Guo, Y., Gong, Z., Li, P., Zhang, W., & Gao, B. (2016). Preparation, characterization and enhancement of the visible-light photocatalytic activity of In_2O_3/Na-bentonite composite. *Ceramics International*, 42, 8850–8855. https://doi.org/10.1016/j.ceramint.2016.02.132

Hassan, N. S., Jalil, A. A., Khusnun, N. F., Ali, M. W., & Haron, S. (2019). Role of reduced graphene oxide in improving interfacial charge transfer of hybridized rGO/silica/zirconia for enhanced Bisphenol A photodegradation. *Journal of Alloys and Compounds*, 789, 221–230. https://doi.org/10.1016/j.jallcom.2019.03.105

Hayyan, M., Hashim, M. A., & AlNashef, I. M. (2016). Superoxide ion: Generation and chemical implications. *Chemical Reviews*, 116, 3029–3085. https://doi.org/10.1021/acs.chemrev.5b00407

He, G. L., Chen, M. J., Liu, Y. Q., Li, X., Liu, Y. J., & Xu, Y. H. (2015). Hydrothermal synthesis of $FeWO_4$-graphene composites and their photocatalytic activities under visible light. *Applied of Surface Sciences*, 351, 474–479. https://doi.org/10.1016/j.apsusc.2015.05.159

He, H., Zhou, P., Shimabuku, K. K. et al. (2019). Degradation and deactivation of bacterial antibiotic resistance genes during exposure to free chlorine, monochloramine, chlorine dioxide, ozone, ultraviolet light, and hydroxyl radical. *Environmental Science & Technology*, 53, 2013–2026. https://doi.org/10.1021/acs.est.8b04393

Hendriks, K. H., Li, W., Wienk, M. M., & Janssen, R. A. J. (2014). Small-bandgap semiconducting polymers with high near-infrared photoresponse. *Journal of the American Chemical Society*, 136, 12130–12136. https://doi.org/10.1021/ja506265h

Hou, H., Zeng, X., & Zhang, X. (2020). Production of hydrogen peroxide by photocatalytic processes. *Angewandte Chemie International Edition*, 59, 17356–17376. https://doi.org/10.1002/anie.201911609

Hsieh, S. H., Chen, W. J., & Yeh, T. H. (2015). Effect of various amounts of graphene oxide on the degradation characteristics of the ZnSe/graphene nanocomposites. *Applied of Surface Sciences*, 358, 63–69. https://doi.org/10.1016/j.apsusc.2015.08.220

Hu, X., Deng, F., Huang, W., Zeng, G., Luo, X., & Dionysiou, D. D. (2018). The band structure control of visible-light-driven rGO/ZnS-MoS_2 for excellent photocatalytic degradation performance and long-term stability. *Chemical Engineering Journal*, 350, 248–256. https://doi.org/10.1016/j.cej.2018.05.182

Jemec, A., Djinović, P., Črnivec, I. G. O., & Pintar, A. (2015). The hazard assessment of nanostructured CeO_2-based mixed oxides on the zebrafish *Danio rerio* under environmentally relevant UV-A exposure. *Science of the Total Environment*, 506–507, 272–278. https://doi.org/10.1016/j.scitotenv.2014.10.120

Ju, F., Li, B., Ma, L., Wang, Y., Huang, D., & Zhang, T. (2016). Antibiotic resistance genes and human bacterial pathogens: Co-occurrence, removal, and enrichment in municipal sewage sludge digesters. *Water Research*, 91, 1–10. https://doi.org/10.1016/j.watres.2015.11.071

Kandhasamy, N., Murugadoss, G., Kannappan, T. et al. (2024). Synthesis of nickel–manganese sulfide decorated with reduced graphene oxide nanocomposite for ultra-fast photocatalytic degradation of organic dye molecules. *Carbon Letters*, 34, 827–840. https://doi.org/10.1007/s42823-023-00654-5

Kong, S., Jin, Z., Liu, H., & Wang, Y. (2014). Morphological effect of graphene nanosheets on ultrathin CoS nanosheets and their applications for high-performance Li-Ion batteries and photocatalysis. *The Journal of Physical Chemistry C*, 118, 25355–25364. https://doi.org/10.1021/jp508698q

Kumar, S., Kaushik, R. D., & Purohit, L. P. (2023). RGO supported ZnO/SnO_2 Z-scheme heterojunctions with enriched ROS production towards enhanced photocatalytic mineralization of phenolic compounds and antibiotics at low temperature. *Journal of Colloid and Interface Science*, 632, 196–215. https://doi.org/10.1016/j.jcis.2022.11.040

Kumar, V., Lee, Y., Shin, J., Kim, K., & Kukkar, D. (2020). Potential applications of graphene-based nanomaterials as adsorbent for removal of volatile organic compounds. *Environment International*, 135, 105356. https://doi.org/10.1016/j.envint.2019.105356

Kumar, V., Singh, V., Kim, K. H., Kwon, E. E., & Younis, S. A. (2021). Metal-organic frameworks for photocatalytic detoxification of chromium and uranium in water. *Coordination Chemistry Reviews*, 447, 214148. https://doi.org/10.1016/j.ccr.2021.214148

Li, D., Yu, P., Zhou, X., Kim, J. H., Zhang, Y., & Alvarez, P. J. J. (2020). Hierarchical $Bi_2O_2CO_3$ wrapped with modified graphene oxide for adsorption-enhanced photocatalytic inactivation of antibiotic resistant bacteria and resistance genes. *Water Research*, 184, 116157. https://doi.org/10.1016/j.watres.2020.116157

Li, F. & Lin, M. (2020). Synthesis of biochar-supported K-doped g-C_3N_4 photocatalyst for enhancing the polycyclic aromatic hydrocarbon degradation activity. *International Journal of Environmental Research and Public Health*, 17, 2065. https://doi.org/10.3390/ijerph17062065

Li, Q., Guo, B., Yu, J. et al. (2011). Highly efficient visible-light-driven photocatalytic hydrogen production of CdS-cluster-decorated graphene nanosheets. *Journal of the American Chemical Society*, 133, 10878–10884. https://doi.org/10.1021/ja2025454

Li, X., Lyu, S., & Lang, X. (2021). Superoxide generated by blue light photocatalysis of g-C_3N_4/TiO_2 for selective conversion of amines. *Environmental Research*, 195, 110851. https://doi.org/10.1016/j.envres.2021.110851

Li, X., Yu, J., & Jaroniec, M. (2016). Hierarchical photocatalysts. *Chemical Society Reviews*, 45, 2603–2636. https://doi.org/10.1039/C5CS00838G

Li, X., Yu, J., Low, J., Fang, Y., Xiao, J., & Chen, X. (2015). Engineering heterogeneous semiconductors for solar water splitting. *Journal of Materials Chemistry A*, 3, 2485–2534. https://doi.org/10.1039/C4TA04461D

Liang, Y., Shang, R., Lu, J. et al. (2019). 2D MOFs enriched g-C_3N_4 nanosheets for highly efficient charge separation and photocatalytic hydrogen evolution from water. *International Journal of Hydrogen Energy*, 44, 2797–2810. https://doi.org/10.1016/j.ijhydene.2018.12.036

Lin, X., Sun, M., Sun, S., & Zhang, Z. (2019). Graphene promoted triphasic N/Ti_3+-TiO_2 heterostructures: In-situ hydrothermal synthesis and enhanced photocatalytic performance. *Journal of Alloys and Compounds*, 785, 732–741. https://doi.org/10.1016/j.jallcom.2019.01.266

Liu, B., Wu, H., & Parkin, I. P. (2020). New insights into the fundamental principle of semiconductor photocatalysis. *ACS Omega*, 5, 14847–14856. https://doi.org/10.1021/acsomega.0c02145

Liu, B., Zhao, X., Terashima, C., Fujishima, A., & Nakata, K. (2014a). Thermodynamic and kinetic analysis of heterogeneous photocatalysis for semiconductor systems. *Physical Chemistry Chemical Physics*, 16, 8751–8760. https://doi.org/10.1039/C3CP55317E

Liu, H., Lv, T., Wu, X., Zhu, C., & Zhu, Z. (2014b). Preparation and enhanced photocatalytic activity of CdS@RGO core–shell structural microspheres. *Applied of Surface Sciences*, 305, 242–246. https://doi.org/10.1016/j.apsusc.2014.03.045

Liu, R., Li, H., Duan, L., Shen, H., Zhang, Q., & Zhao, X. (2018). The synergistic effect of graphene oxide and silver vacancy in Ag_3PO_4-based photocatalysts for Rhodamine B degradation under visible light. *Applied of Surface Sciences*, 462, 263–269. https://doi.org/10.1016/j.apsusc.2018.07.173

Long, C., Li, Y., Yu, W., & Li, A. (2012). Removal of benzene and methyl ethyl ketone vapor: Comparison of hypercrosslinked polymeric adsorbent with activated carbon. *Journal of Hazardous Materials*, 203–204, 251–256. https://doi.org/10.1016/j.jhazmat.2011.12.010

Luo, L., Zhang, T., Wang, M., Yun, R., & Xiang, X. (2020). Recent advances in heterogeneous photo-driven oxidation of organic molecules by reactive oxygen species. *ChemSusChem*, 13, 5173–5184. https://doi.org/10.1002/cssc.202001398

Ma, D., Wu, J., Gao, M., Xin, Y., Sun, Y., & Ma, T. (2017). Hydrothermal synthesis of an artificial Z-scheme visible light photocatalytic system using reduced graphene oxide as the electron mediator. *Chemical Engineering Journal*, 313, 1567–1576. https://doi.org/10.1016/j.cej.2016.11.036

Ma, H., Liu, J., Zuo, S. et al. (2021). One-step fabrication of 2D/2D Z-scheme BiOCl/g-C_3N_4 nanosheets heterojunction for efficient degradation of RhB and Cr(VI) ions reduction under visible-light illumination. *ChemistrySelect*, 6, 10097–10104. https://doi.org/10.1002/slct.202102668

Maeda, K. (2011). Photocatalytic water splitting using semiconductor particles: History and recent developments. *Journal of Photochemistry and Photobiology C: Photochemistry Reviews*, 12, 237–268. https://doi.org/10.1016/j.jphotochemrev.2011.07.001

Martínez-Huitle, C. A. & Brillas, E. (2009). Decontamination of wastewaters containing synthetic organic dyes by electrochemical methods: A general review. *Applied Catalysis B: Environmental*, 87, 105–145. https://doi.org/10.1016/j.apcatb.2008.09.017

Moon, S. Y., Naik, B., Jung, C. H., Qadir, K., & Park, J. Y. (2016). Tailoring metal-oxide interfaces of oxide-encapsulated Pt/silica hybrid nanocatalysts with enhanced thermal stability. *Catalysis Today*, 265, 245–253. https://doi.org/10.1016/j.cattod.2015.08.036

Moore, K. & Wei, W. (2021). Applications of carbon nanomaterials in perovskite solar cells for solar energy conversion. *Nano Materials Science*, 3, 276–290. https://doi.org/10.1016/j.nanoms.2021.03.005

Nosaka, Y. & Nosaka, A. (2016). Understanding hydroxyl radical (•OH) generation processes in photocatalysis. *ACS Energy Letters*, 1, 356–359. https://doi.org/10.1021/acsenergylett.6b00174

Nosaka, Y. & Nosaka, A. Y. (2017). Generation and detection of reactive oxygen species in photocatalysis. *Chemical Reviews*, 117, 11302–11336. https://doi.org/10.1021/acs.chemrev.7b00161

Padhi, D. K., Panigrahi, T. K., Parida, K., Singh, S. K., & Mishra, P. M. (2017). Green synthesis of Fe_3O_4/RGO nanocomposite with enhanced photocatalytic performance for Cr(VI) reduction, phenol degradation, and antibacterial activity. *ACS Sustainable Chemistry & Engineering*, 5, 10551–10562. https://doi.org/10.1021/acssuschemeng.7b02548

Pu, S., Zhu, R., Ma, H. et al. (2017). Facile in-situ design strategy to disperse TiO_2 nanoparticles on graphene for the enhanced photocatalytic degradation of rhodamine 6G. *Applied of Surface Sciences*, 218, 208–219. https://doi.org/10.1016/j.apcatb.2017.06.039

Rameshbabu, R., Vinoth, R., Navaneethan, M., Harish, S., Hayakawa, Y., & Neppolian, B. (2017). Visible light responsive Cu_2MoS_4 nanosheets incorporated reduced graphene oxide for efficient degradation of organic pollutant. *Applied of Surface Sciences*, 418, 128–137. https://doi.org/10.1016/j.apsusc.2017.02.126

Rehman, G. U., Tahir, M., Goh, P. S., Ismail, A. F., & Khan, I. U. (2019). Controlled synthesis of reduced graphene oxide supported magnetically separable Fe_3O_4@rGO@AgI ternary nanocomposite for enhanced photocatalytic degradation of phenol. *Powder Technology*, 356, 547–558. https://doi.org/10.1016/j.powtec.2019.08.026

Ribao, P., Corredor, J., Rivero, M. J., & Ortiz, I. (2019). Role of reactive oxygen species on the activity of noble metal-doped TiO_2 photocatalysts. *Journal of Hazardous Materials*, 372, 45–51. https://doi.org/10.1016/j.jhazmat.2018.05.026

Rosman, N., Salleh, W. N. W., & Razali, N. A. M. (2021). Chapter 4 - Heterojunction-based photocatalyst. In M. Sakar, R. G. Balakrishna, & T. O. Do (Eds.), *Photocatalytic Systems by Design* (pp. 85–130). Elsevier. https://doi.org/10.1016/B978-0-12-820532-7

Sano, T., Negishi, N., Sakai, E., & Matsuzawa, S. (2006). Contributions of photocatalytic/catalytic activities of TiO_2 and γ-Al_2O_3 in nonthermal plasma on oxidation of acetaldehyde and CO. *Journal of Molecular Catalysis A: Chemical*, 245, 235–241. https://doi.org/10.1016/j.molcata.2005.10.002

Singh, P., Shandilya, P., Raizada, P., Sudhaik, A., Rahmani-Sani, A., & Hosseini-Bandegharaei, A. (2020). Review on various strategies for enhancing photocatalytic activity of graphene based nanocomposites for water purification. *Arabian Journal of Chemistry*, 13, 3498–3520. https://doi.org/10.1016/j.arabjc.2018.12.001

Singh, P., Sudhaik, A., Raizada, P., Shandilya, P., Sharma, R., & Hosseini-Bandegharaei, A. (2019). Photocatalytic performance and quick recovery of BiOI/Fe_3O_4@graphene oxide ternary photocatalyst for photodegradation of 2, 4-dintirophenol under visible light. *Materials Today Chemistry*, 12, 85–95. https://doi.org/10.1016/j.mtchem.2018.12.006

Singh, R. & Dutta, S. (2018). Synthesis and characterization of solar photoactive TiO_2 nanoparticles with enhanced structural and optical properties. *Advanced Powder Technology*, 29, 211–219. https://doi.org/10.1016/j.apt.2017.11.005

Su, S., Han, Q., Wang, X., & Zhu, J. (2018a). Synthesis of nanosheet-based hierarchical BiO_2 microtubes and its photocatalytic performance. *Applied Surface Science*, 455, 616–621. https://doi.org/10.1016/j.apsusc.2018.05.164

Su, X., Yang, J., Yu, X., Zhu, Y., & Zhang, Y. (2018b). In situ grown hierarchical 50%BiOCl/BiOI hollow flowerlike microspheres on reduced graphene oxide nanosheets for enhanced visible-light photocatalytic degradation of Rhodamine B. *Applied of Surface Sciences*, 433, 502–512. https://doi.org/10.1016/j.apsusc.2017.09.258

Sun, L., Wang, G., Hao, R., Han, D., & Cao, S. (2015). Solvothermal fabrication and enhanced visible light photocatalytic activity of Cu_2O-reduced graphene oxide composite microspheres for photodegradation of Rhodamine B. *Applied of Surface Sciences*, 358, 91–99. https://doi.org/10.1016/j.apsusc.2015.08.128

Teng, Z., Yang, N., Lv, H. et al. (2019). Edge-functionalized g-C_3N_4 nanosheets as a highly efficient metal-free photocatalyst for safe drinking water. *Chem*, 5, 664–680.

Tong, H., Ouyang, S., Bi, Y., Umezawa, N., Oshikiri, M., & Ye, J. (2012). Nano-photocatalytic materials: Possibilities and challenges. *Advanced Materials*, 24, 229–251. https://doi.org/10.1002/adma.201102752

Verma, R. & Samdarshi, S. K. (2016). In situ decorated optimized CeO_2 on reduced graphene oxide with enhanced adsorptivity and visible light photocatalytic stability and reusability. *The Journal of Physical Chemistry C*, 120, 22281–22290. https://doi.org/10.1021/acs.jpcc.6b04493

Verma, S. & Dutta, R. K. (2017). Enhanced ROS generation by ZnO-ammonia modified graphene oxide nanocomposites for photocatalytic degradation of trypan blue dye and 4-nitrophenol. *Journal of Environmental Chemical Engineering*, 5, 4776–4787. https://doi.org/10.1016/j.jece.2017.08.026

Vinoth, R., Babu, S. G., Ramachandran, R., & Neppolian, B. (2017). Bismuth oxyiodide incorporated reduced graphene oxide nanocomposite material as an efficient photocatalyst for visible light assisted degradation of organic pollutants. *Applied of Surface Sciences*, 418, 163–170. https://doi.org/10.1016/j.apsusc.2017.01.278

Wang, S., Ma, X., Liu, Y., Yi, X., Du, G., & Li, J. (2020). Fate of antibiotics, antibiotic-resistant bacteria, and cell-free antibiotic-resistant genes in full-scale membrane bioreactor wastewater treatment plants. *Bioresource Technology*, 302, 122825. https://doi.org/10.1016/j.biortech.2020.122825

Wang, Y., Li, X., Liu, S. et al. (2022). Roles of catalyst structure and gas surface reaction in the generation of hydroxyl radicals for photocatalytic oxidation. *ACS Catalysis*, 12, 2770–2780. https://doi.org/10.1021/acscatal.1c05447

Wang, Y., Wang, W., Mao, H. et al. (2014). Electrostatic self-assembly of $BiVO_4$–reduced graphene oxide nanocomposites for highly efficient visible light photocatalytic activities. *ACS Applied Materials & Interfaces*, 6, 12698–12706. https://doi.org/10.1021/am502700p

Wang, Z., Xie, X., Wang, X., Mahmood, A., Qiu, H., & Sun, J. (2019). Difference of photodegradation characteristics between single and mixed VOC pollutants under simulated sunlight irradiation. *Journal of Photochemistry and Photobiology A: Chemistry*, 384, 112029. https://doi.org/10.1016/j.jphotochem.2019.112029

Wei, Q., Wang, Y., Qin, H. et al. (2018). Construction of rGO wrapping octahedral Ag-Cu_2O heterostructure for enhanced visible light photocatalytic activity. *Applied Catalysis B: Environmental*, 227, 132–144. https://doi.org/10.1016/j.apcatb.2018.01.003

Wei, X. N., Ou, C. L., Guan, X. X., Peng, Z. K., & Zheng, X. C. (2019). Facile assembly of CdS-reduced graphene oxide heterojunction with enhanced elimination performance for organic pollutants in wastewater. *Applied of Surface Sciences*, 469, 666–673. https://doi.org/10.1016/j.apsusc.2018.11.102

Wu, H., Sun, Q., Chen, J. et al. (2021). Citric acid-assisted ultrasmall CeO_2 nanoparticles for efficient photocatalytic degradation of glyphosate. *Chemical Engineering Journal*, 425, 130640. https://doi.org/10.1016/j.cej.2021.130640

Xiao, J., Rabeah, J., Yang, J., Xie, Y., Cao, H., & Brückner, A. (2017). Fast electron transfer and •OH formation: Key features for high activity in visible-light-driven ozonation with C_3N_4 catalysts. *ACS Catalysis*, 7, 6198–6206. https://doi.org/10.1021/acscatal.7b02180

Xiong, J., Li, X., Huang, J. et al. (2020). CN/rGO@BPQDs high-low junctions with stretching spatial charge separation ability for photocatalytic degradation and H_2O_2 production. *Applied Catalysis B: Environmental*, 266, 118602. https://doi.org/10.1016/j.apcatb.2020.118602

Xu, J., Zhang, J., Cai, Z. et al. (2019). Facile and large-scale synthesis of defective black TiO_{2-x} (B) nanosheets for efficient visible-light-driven photocatalytic hydrogen evolution. *Catalysts*, 9, 2–9. https://doi.org/10.3390/catal9121048

Xu, L., Wei, Y., Guo, W., Guo, Y., & Guo, Y. (2015). One-pot solvothermal preparation and enhanced photocatalytic activity of metallic silver and graphene co-doped $BiVO_4$ ternary systems. *Applied of Surface Sciences*, 332, 682–693. https://doi.org/10.1016/j.apsusc.2015.01.235

Xu, L., Yang, L., Johansson, E. M. J., Wang, Y., & Jin, P. (2018). Photocatalytic activity and mechanism of bisphenol a removal over TiO_{2-x}/rGO nanocomposite driven by visible light. *Chemical Engineering Journal*, 350, 1043–1055. https://doi.org/10.1016/j.cej.2018.06.046

Xu, P., Zeng, G. M., Huang, D. L. et al. (2012). Use of iron oxide nanomaterials in wastewater treatment: A review. *Science of the Total Environment*, 424, 1–10. https://doi.org/10.1016/j.scitotenv.2012.02.023

Xu, X., Sui, Y., Chen, W. et al. (2024). The photocatalytic H_2O_2 production by metal-free photocatalysts under visible-light irradiation. *Applied Catalysis B: Environmental*, 341, 123271. https://doi.org/10.1016/j.apcatb.2023.123271

Yang, C.-C., Vernimmen, J., Meynen, V., Cool, P., & Mul, G. (2011). Mechanistic study of hydrocarbon formation in photocatalytic CO_2 reduction over Ti-SBA-15. *Journal of Catalysis*, 284, 1–8. https://doi.org/10.1016/j.jcat.2011.08.005

Yang, J., Wang, X., Zhao, X., Dai, J., & Mo, S. (2015). Synthesis of uniform Bi_2WO_6-reduced graphene oxide nanocomposites with significantly enhanced photocatalytic reduction activity. *The Journal of Physical Chemistry C*, 119, 3068–3078. https://doi.org/10.1021/jp510041x

Yang, X., Chen, C., Li, J., Zhao, G., Ren, X., & Wang, X. (2012). Graphene oxide-iron oxide and reduced graphene oxide-iron oxide hybrid materials for the removal of organic and inorganic pollutants. *RSC Advance*, 2, 8821–8826. https://doi.org/10.1039/c2ra20885g

Yang, X. & Wang, D. (2018). Photocatalysis: From fundamental principles to materials and applications. *ACS Applied Energy Mater*, 1, 6657–6693. https://doi.org/10.1021/acsaem.8b01345

Ye, G., Luo, P., Zhao, Y. et al. (2020). Three-dimensional Co/Ni bimetallic organic frameworks for high-efficient catalytic ozonation of atrazine: Mechanism, effect parameters, and degradation pathways analysis. *Chemosphere*, 253, 126767.

Yu, L., Wang, L., Xu, W. et al. (2018). Adsorption of VOCs on reduced graphene oxide. *Journal of Environmental Sciences*, 67, 171–178. https://doi.org/10.1016/j.jes.2017.08.022

Yu, X., Shi, J., Feng, L., Li, C., & Wang, L. (2017). A three-dimensional BiOBr/RGO heterostructural aerogel with enhanced and selective photocatalytic properties under visible light. *Applied of Surface Sciences*, 396, 1775–1782. https://doi.org/10.1016/j.apsusc.2016.11.219

Yuan, L., Geng, Z., Xu, J., Guo, F., & Han, C. (2021). Metal-semiconductor heterostructures for photoredox catalysis: Where are we now and where do we go? *Advanced Functional Materials*, 31, 2101103. https://doi.org/10.1002/adfm.202101103

Zeng, X., Liu, Y., Xia, Y. et al. (2020). Cooperatively modulating reactive oxygen species generation and bacteria-photocatalyst contact over graphitic carbon nitride by polyethylenimine for rapid water disinfection. *Applied Catalysis B: Environmental*, 274, 119095. https://doi.org/10.1016/j.apcatb.2020.119095

Zhai, J., Yu, H., Li, H., Sun, L., Zhang, K., & Yang, H. (2015). Visible-light photocatalytic activity of graphene oxide-wrapped Bi_2WO_6 hierarchical microspheres. *Applied of Surface Sciences*, 344, 101–106. https://doi.org/10.1016/j.apsusc.2015.03.100

Zhang, C., Zhang, Y., & Xie, W. (2021). Plasmonic metal/semiconductor hybrid nanomaterials for solar to chemical energy conversion. *Journal of Energy Chemistry*, 63, 40–53. https://doi.org/10.1016/j.jechem.2021.08.036

Zhang, H., Guo, L. H., Wang, D., Zhao, L., & Wan, B. (2015). Light-induced efficient molecular oxygen activation on a Cu(II)-Grafted TiO_2/graphene photocatalyst for phenol degradation. *ACS Applied Material Interfaces*, 7, 1816–1823. https://doi.org/10.1021/am507483q

Zhang, K., Lin, Y., Wang, C. et al. (2016a). Facile synthesis of hierarchical Cu_2MoS_4 hollow sphere/reduced graphene oxide composites with enhanced photocatalytic performance. *The Journal of Physical Chemistry C*, 120, 13120–13125. https://doi.org/10.1021/acs.jpcc.6b03767

Zhang, L., Bao, Z., Yu, X. et al. (2016b). Rational design of α-Fe_2O_3/reduced graphene oxide composites: Rapid detection and effective removal of organic pollutants. *ACS Applied Material Interfaces*, 8, 6431–6438. https://doi.org/10.1021/acsami.5b11292

Zhang, Q., Han, F., Yan, Y. et al. (2019). Preparation and properties of visible light responsive RGO/In_2TiO_5 nanobelts for photocatalytic degradation of organic pollutants. *Applied of Surface Sciences*, 485, 547–553. https://doi.org/10.1016/j.apsusc.2019.04.185

Zhang, X., Gao, B., Creamer, A. E., Cao, C., & Li, Y. (2017). Adsorption of VOCs onto engineered carbon materials: A review. *Journal of Hazardous Materials*. https://doi.org/10.1016/j.jhazmat.2017.05.013

Zhao, X., Zhang, Y., Wen, P., Xu, G., Ma, D., & Qiu, P. (2018). NH_2-MIL-125(Ti)/TiO_2 composites as superior visible-light photocatalysts for selective oxidation of cyclohexane. *Molecular Catalysis*, 452, 175–183. https://doi.org/10.1016/j.mcat.2018.04.004

Zhao, Y., Zhang, X., Wang, C. et al. (2017). The synthesis of hierarchical nanostructured MoS_2/Graphene composites with enhanced visible-light photo-degradation property. *Applied of Surface Sciences*, 412, 207–213. https://doi.org/10.1016/j.apsusc.2017.03.181

Zhou, B., Sun, B., Qiu, W. et al. (2019). Adsorption/desorption of toluene on a hypercrosslinked polymeric resin in a highly humid gas stream. *Chinese Journal of Chemical Engineering*, 27, 863–868. https://doi.org/10.1016/j.cjche.2018.09.027

2

Photocatalytic Activity: Computational Estimation, Optimization, and Degradation Processes

*Arnab Sankar Bhattacharyya**

Department of Metallurgical and Materials Engineering, Central University of Jharkhand, Ranchi, India

*Corresponding author: 2006asb@gmail.com, arnab.bhattacharya@cuj.ac.in

2.1 Introduction

Self-assembled carbon scaffolds have emerged as a promising material for photocatalytic applications due to their unique structure, high surface area, and tunable properties. These scaffolds are composed of carbon-based building blocks that are assembled through non-covalent interactions, such as π–π stacking, hydrogen bonding, and van der Waals forces. The resulting materials exhibit high surface areas, porosity, and electrical conductivity, making them ideal for photocatalytic reactions (Tan et al., 2017; Li et al., 2016). The unique structure of self-assembled carbon scaffolds is attributed to the non-covalent interactions between the carbon-based building blocks. These interactions allow for the formation of complex networks and hierarchically porous structures, which enhance the surface area and porosity of the material. The high surface area and porosity of self-assembled carbon scaffolds provide many active sites for photocatalytic reactions, enabling efficient absorption of light and interaction with reactants, and used in energy storage (Weingarten et al., 2014; Ko et al., 2017).

The tunable properties of self-assembled carbon scaffolds are another advantage for photocatalytic applications. The scaffolds' morphology, surface chemistry, and electrical conductivity can be controlled by adjusting the building blocks' composition, size, and functional groups. This tunability enables the optimization of self-assembled carbon scaffolds for specific photocatalytic reactions, such as water splitting, carbon dioxide reduction, or organic pollutant degradation (Wei et al., 2020). The high electrical conductivity of self-assembled carbon scaffolds is also beneficial for photocatalytic applications. The scaffolds' conductivity

Photocatalysts based on Carbon Scaffolds: Environmental and Industrial Applications, First Edition. Edited by Ramesh Oraon et al.
Published 2026 by Wiley-VCH GmbH.

enables efficient charge transport and separation, reducing recombination losses and improving photocatalytic performance. Additionally, the electrical conductivity of self-assembled carbon scaffolds can be enhanced by introducing dopants or functional groups, further improving their photocatalytic efficiency. Self-assembled carbon scaffolds have been successfully applied in various photocatalytic reactions, including water splitting, carbon dioxide reduction, and organic pollutant degradation. These scaffolds have shown high photocatalytic efficiency, stability, and recyclability, making them a promising material for large-scale industrial applications (Sun et al., 2020; Wei et al., 2017).

Self-assembled carbon scaffolds can be synthesized through scalable methods, enabling large-scale production without compromising their unique properties. This scalability is crucial for industrial applications, where high-volume production is essential. Additionally, the scaffolds' cost-effectiveness stems from their synthesis using abundant and inexpensive carbon-based precursors, reducing material costs (Pavlenko et al., 2022; Li et al., 2019c; Yuan et al., 2018). The scalability and cost-effectiveness of self-assembled carbon scaffolds make them an attractive alternative to traditional materials, which often require complex and expensive synthesis methods. For instance, self-assembled carbon scaffolds can be synthesized through a facile and scalable solvothermal method, enabling large-scale production (Ran et al., 2018; Guo et al., 2023; Whitby, 2014).

The tunability and adaptability of self-assembled carbon scaffolds enable precise control over their structure, functionality, and properties. This tunability is achieved through control over the assembly process, allowing for the creation of scaffolds with tailored properties for specific applications. For example, self-assembled carbon scaffolds can be combined with metals to create hybrid materials with enhanced electrical conductivity (Khan et al., 2019; Yuan et al., 2018). The tunability and adaptability of self-assembled carbon scaffolds make them suitable for a wide range of applications, from energy storage to biomedical devices (Luo et al., 2019; Wagh et al., 2020; Serrano et al., 2014). For instance, self-assembled carbon scaffolds have been used as electrodes in supercapacitors, demonstrating high energy storage capacity. Additionally, self-assembled carbon scaffolds have been used as support for catalytic materials, enhancing their stability and activity (Nithiyanantham et al., 2015; Li et al., 2020a).

Self-assembled carbon scaffolds offer a sustainable and environmentally friendly alternative to traditional materials. Their synthesis from abundant carbon-based precursors reduces reliance on scarce resources, minimizing environmental impact. Additionally, the scaffolds' biodegradability and non-toxicity make them suitable for use in biomedical applications, reducing long-term toxicity and environmental harm (Zhang et al., 2023; Soufi & Iravani, 2020). The scaffolds' recyclability and reusability further enhance their sustainability, enabling repeated use and minimizing waste. Furthermore, self-assembled carbon scaffolds can be designed to degrade into harmless byproducts, reducing environmental pollution and promoting a circular economy (Venkateswaran et al., 2024; Gavilán et al., 2023).

Self-assembled carbon scaffolds can be synthesized using templating methods, which involve the use of a template or mold to direct the assembly of carbon-based building blocks. This approach enables the creation of scaffolds with precise control over their structure and morphology. Templating methods can be further divided into hard templating and soft templating. Hard templating involves the use of rigid templates, such as silica or metal oxides, to direct the assembly of carbon-based building blocks. Soft templating, on the other hand, involves the use of flexible templates, such as surfactants or polymers, to guide the assembly process (Islam et al., 2022). Sol–gel processing is another method used to synthesize self-assembled carbon scaffolds. This approach involves the hydrolysis and condensation of carbon-based precursors, such as resorcinol or phenol, to form a gel-like network. The sol–gel process can be controlled to produce scaffolds with tailored properties, such as porosity, surface area, and electrical conductivity. Hydrothermal treatment is a synthesis method that involves the use of high temperature and pressure to drive the assembly of carbon-based building blocks. This approach enables the creation of scaffolds with high crystallinity and purity, making them suitable for applications where high performance is essential (Adhikari, 2022; Titirici, 2013).

Sol–gel processing offers several advantages, including low synthesis temperatures, high purity, and ease of scalability. Additionally, the sol–gel process can be combined with other methods, such as templating or hydrothermal treatment, to create hybrid scaffolds with enhanced properties. Hydrothermal treatment is a synthesis method that involves the use of high temperature and pressure to drive the assembly of carbon-based building blocks. This approach enables the creation of scaffolds with high crystallinity and purity, making them suitable for applications where high performance is essential (Fagiolari & Bella, 2019). The hydrothermal process can be controlled to produce scaffolds with tailored properties, such as surface area, porosity, and electrical conductivity. Additionally, hydrothermal treatment can be combined with other methods, such as templating or sol–gel processing, to create hybrid scaffolds with enhanced properties. Self-assembled carbon scaffolds have been used in a wide range of applications, including energy storage, catalysis, and biomedical devices. Their unique properties, such as high surface area, electrical conductivity, and biocompatibility, make them an attractive material for these applications (Fagiolari & Bella, 2019; Sun et al., 2016; Zhou et al., 2020).

Photocatalytic activity is a crucial process for environmental remediation, renewable energy, and chemical synthesis. This involves a process where light energy is used to drive chemical reactions, typically in the presence of a catalyst (Guo et al., 2020; Zaleska-Medynska, 2018). In recent years, photocatalytic materials have gained significant attention due to their potential applications in environmental remediation, energy conversion, and chemical synthesis. The photocatalytic activity of a material is influenced by various factors, including its electronic structure, surface properties, and reaction conditions (Kumari et al., 2023a,b; Kumar et al., 2022; Neha et al., 2020; Nair et al., 2016). Computational methods have emerged as powerful tools for estimating and optimizing photocatalytic activity, allowing researchers to design and develop more efficient photocatalysts (Singh et al., 2015).

This chapter provides an overview of the computational estimation and optimization of photocatalytic activity, as well as the degradation processes involved in photocatalysis. Recent advances in photocatalytic materials and processes have been reviewed in this communication, which also includes self-assembled carbon scaffolds, with a focus on their potential applications and future directions.

2.2 Photocatalysis and Related Computational Studies

A photocatalyst is primarily a semiconductor that increases the rate of reaction in the presence of light. Photocatalysis offers a wide range of applications, including wastewater treatment (Ameta et al., 2018). The photocatalyst's job is to provide a substrate for adsorption of the water molecule and to create electron-hole pairs under photonic irradiation. Metal oxides have a significant role in photocatalysis (Zaleska-Medynska, 2018). Photocatalytic conversion is an essential approach for producing energy while being environmentally friendly. Transition metal oxides such as ZnO, TiO_2, WOx, along with Nb_2O_5 and CeO_2 have been employed as photocatalysts (Kumari et al., 2023a,b; Kumar et al., 2022; Neha et al., 2020; Nair et al., 2016). To boost efficiency and wavelength range, these materials are doped and nanocomposites are created. Composites comprising ZnO-ceria, titania-tungsten oxide, titania-Niobium pentoxide, and tungsten oxide-ceria have demonstrated good performance. The increased photoactivity in the composites was owing to the charged crystallite surface and the larger surface area provided by the smaller crystallite size, which allows organic molecules to adsorb more and be destroyed more easily. A large decrease in bandgap makes the photocatalyst useful in the visible range as well (Faisal et al., 2013; Zou et al., 2016). Enhanced photoactivity occurs in mixed phase transition metal oxide composites such as ZnO-CeO_2, TiO_2-WO_3, and WO_3-CeO_2. An increase in the calcination temperature induces oxygen deficit, increased crystallite size, reduced bandgap, and higher photoactivity in Nb_2O_5. Three factors could explain this increase in photoactivity: (i) increased adsorption of organic molecules, which are then destroyed, due to the charged crystallite surface; and (ii) increased surface area due to the small crystallite size. The smaller crystallite size and increased adsorption of MB molecules appear to permit slightly enhanced photoactivity. The considerable decrease in bandgap suggests that the composite might likewise be used to capture visible light (Sun et al., 2015; Verma et al., 2015; Kumari et al., 2024; Gaurav et al., 2019).

Zinc oxide (ZnO) and ceria-doped zinc oxide (Ce-ZnO) nanoparticles were synthesized utilizing a simple chemical precipitation approach with no additional catalyst. Ceria-doped ZnO showed increased photoactivity as compared to ZnO nanoparticles for MB breakdown in an aqueous solution under UV irradiance. The ensuing heterojunction between ZnO and ceria improves charge separation efficiency, demonstrating a substantial association between ZnO and CeO_2 heterojunctions on charge transfer mechanism across the interface (Kumari et al., 2023a; Cerrato et al., 2018). Recently, niobium oxide nanoparticles were synthesized using polyethyleneimine (PEI) as a capping agent and a template. The sample calcined at 800 °C demonstrated the greatest adsorption capability and photocatalytic activity, with a combined degradation efficiency of 98.8% of methylene blue dye. The adsorption qualities, stability of the hydrocolloid system, the presence of oxygen vacancies, and the peculiar morphology of the photocatalytic nano-rod/tripods are primarily responsible for this behavior (Neha et al., 2020; Kumari et al., 2023b). The doping element Cu in ZnO was found to be more suitable for photocatalysis compared to Ni, which influenced the magnetic properties more (Gurjar et al., 2023). Doping with more than one element creates a preferential growth, and the temperature at this point plays a significant role (Dash et al., 2023).

Future challenges in modelling complex nano-photocatalysts include combining different methods in a hierarchical manner to ensure that resources are spent wisely at each length scale, as well as accounting for excited state chemistry, which is important for photocatalysis, a path that will bring devices closer to the theoretical limit of photocatalytic efficiency (Prakash et al., 2023; Samanta et al., 2022).

2.2.1 Computational Estimation of Photocatalytic Activity

Computational methods have become essential tools for understanding and predicting photocatalytic activity. These methods can be broadly classified into three categories: density functional theory (DFT) calculations, time-dependent DFT (TD-DFT) calculations, and machine learning (ML) approaches (Prakash et al., 2023; Samanta et al., 2022; Yang et al., 2018; Chen et al., 2018a).

DFT Calculations: DFT calculations are widely used to study the electronic structure of photocatalytic materials. These calculations can provide insights into the material's bandgap, band edges, and electronic density of states, which are critical factors influencing photocatalytic activity.

TD-DFT Calculations: TD-DFT calculations are used to study the excited-state properties of photocatalytic materials. These calculations can provide insights into the material's optical absorption spectrum, exciton binding energy, and charge carrier dynamics.

ML Approaches: ML approaches are being increasingly used to predict photocatalytic activity. These approaches can learn from large datasets and identify patterns that may not be apparent through experimental or theoretical studies alone.

2.2.2 Optimization of Photocatalytic Activity

Optimizing photocatalytic activity involves modifying the material's electronic structure, surface properties, and reaction conditions to enhance its performance. Modifying the electronic structure of a photocatalyst can enhance its light absorption, charge carrier separation, and reaction kinetics. Strategies include bandgap engineering, doping, and heterojunction formation (Li et al., 2019a, 2020c; Wang et al., 2014).

2.2.3 Surface Modification and Functionalization

Surface modification and functionalization can improve the photocatalyst's surface area, adsorption capacity, and reaction selectivity. Strategies include nanostructuring, surface coating, and functional group attachment

2.2.4 Nanostructuring and Morphology Control

Controlling the photocatalyst's morphology and nanostructure can enhance its surface area, light absorption, and charge carrier dynamics. Strategies include nanoparticle synthesis nanowire growth, and mesoporous structure formation.

2.2.5 Degradation Processes

Photocatalytic degradation processes involve the breakdown of organic pollutants, water splitting, and CO_2 reduction. This section discusses the mechanisms and applications of these processes (Chen et al., 2020; Herrmann, 2019; Li et al., 2019b).

Photocatalytic Degradation of Organic Pollutants: Photocatalytic degradation is a promising approach for removing organic pollutants from water and air. The process involves the generation of reactive oxygen species (ROS) that break down organic molecules.

Photocatalytic Water Splitting and Hydrogen Production: Photocatalytic water splitting is a process that uses light energy to split water into hydrogen and oxygen. This process has potential applications in renewable energy and chemical synthesis.

Photocatalytic CO_2 Reduction and Utilization: Photocatalytic CO_2 reduction involves the conversion of CO_2 into valuable chemicals and fuels. This process has potential applications in carbon capture and utilization.

Mechanisms and Kinetics: The mechanisms and kinetics of photocatalytic degradation processes involve complex interactions between light, catalysts, and reactants. Understanding these mechanisms is crucial for optimizing photocatalytic performance.

2.2.6 DFT-based Computation

A feasible computational framework is proposed, concentrating on the static and dynamic features of the relevant excited states as well as the chemistry involved in the reactions of interest. It investigates the nature of charge carriers, the excited-state potential energy surface, and its temporal evolution. Excited-state descriptors are introduced to help find and characterize the electrons and holes produced during excitation. Nonadiabatic molecular dynamics simulations are offered as a useful technique for describing the time evolution of photogenerated species and their propagation through the crystalline structure of photoactive material, resulting in information regarding charge carrier lifetime (Xie et al., 2020).

A heterogeneously photocatalyzed process starts with an electron excited by light of appropriate wavelength/frequency, causing a transition from the highest occupied molecular orbital (HOMO) or valence band (VB) to the lowest unoccupied molecular orbital (LUMO) or conduction band (CB), creating an e-h pair (exciton). In materials having photocatalytic property, the charges get separated on being constantly radiated, which are responsible for the degradation process. Understanding the surface-active site through DFT and new, more accurate exchange-correlation functionals is critical for surface engineering. DFT at this stage is used for band structure determination. The Kohn–Sham implementation of DFT, known for its electronic self-interaction inaccuracy, causes the generalized gradient approximation (GGA) to significantly underestimate the bandgap (Morales-García et al., 2023; Sousa et al., 2013). The Materials Studio program was used for theoretical modeling and analysis of TiO_2 crystal samples, and the optical characterization was performed using the CASTEP module and DFT. The thin film samples' performance was evaluated using methylene blue (MB) dye degradation kinetics, with the sample demonstrating the highest rate constant in each cycle. This can be attributed to the nanosamples' acceptable bandgap and phase ratio, as well as the resulting reduction in recombination of photo-generated charge carriers. The rate constants of three continuous cycles of thin film samples were determined (de P. R. Moreira et al., 2002).

2.2.7 Photocatalytic Reactor Designs

The dissemination of important photocatalytic reactor design ideas within the scientific community and the water industry is still one of the most significant barriers to turning photocatalytic water treatment research into widespread real-world use. The photoreactor ideas and models are explored, along with fundamental design guidelines. Other engineering concerns, including operation, cost analysis, patents, and a variety of industrial uses for photocatalytic reactors used in water treatment, have been investigated (Harsh et al., 2023). The charge carrier generation and mass transfer through the porous structure as the two important aspects that need to be taken care of while designing the structure. Doping is an effective means of reducing the charge recombination. The recombination rate is defined as $\mathrm{r}_{\mathrm{recomb}} = k_{\mathrm{recomb}}\,[\mathrm{h}+][\mathrm{e}-]$ where k_{recomb} is the recombination rate constant. Doping will increase the concentration of either the positive or negative charge. The competition between electron-hole recombination and photocatalytic reactions causes changes in the exponent of the I (current) term in the charge carrier governing equation as r_{recomb} is proportional to I^n where $n = 0.5 - 1.0$. (Wang et al., 2021; Visan et al., 2019) The scalar radiative transfer equation (RTE) uses the light intensity distribution where the spectral volumetric absorption, scattering coefficients, and the scattering phase function were used to determine the spectra intensity distribution (Ripoll, 2011). The slurry reactors representing the dispersed phase, where mass transfer occurs through convection, are good scattering systems although having lower quantum efficiencies due to scattering, which can be avoided in micro-reactors (Visan et al., 2014). This internal mass transport in slurry systems is governed by diffusion and solved analytically for a first-order reaction rate $r = kc$, where k is the rate constant. The dominating mechanism can be evaluated using the Thiele modulus and the internal effectiveness factor η, which directly expresses the fraction of the catalyst that is being utilized for a first-order reaction rate (Rafieian et al., 2015).

2.2.8 Molecular Dynamics

Molecular dynamics can be used to tackle larger-scale dynamics at the catalyst/electrolyte interface, albeit greater generalizations of force fields are required. Monte Carlo and continuum modeling techniques are not yet the most popular methods for modelling water splitting, but interest is developing due to their decreased computational cost and the ability to compare modeling results directly to experimental data (Akimov et al., 2015). The electron transfer dynamics in electroneutral Ru(di-X-bpy) $(CO)_2Cl_2$ complexes with X = COOH and PO_3H_2 attached to the N–Ta_2O5 substrate were analyzed. Nonadiabatic molecular dynamics simulations indicate that the electron transfer is faster in complexes with COOH anchors than in complexes with PO_3H_2 groups, due to larger nonadiabatic coupling (Bhattacharyya & Dash, 2022).

2.2.9 Defect Energy Levels and Phonon-assisted Non-radiative Transitions

The discrete exciton energy states caused by quantum confinement and lattice imperfections affected the absorption spectra in the case of ZnO. The possibility of tuning the bandgap of ZnO by changing the temperature was explored by computational studies. Simulation studies showed the growth of a thermodynamically stable Wurtzite phase and a decrease in indirectness with the rise in temperature. A threshold temperature that initiates crystallization was also obtained (Varshni, 1967). The UV spectra of the ZnO synthesized by the sol–gel mechanism taking zinc nitrate as precursor are shown in Figure 2.1a. The adsorption of light takes place between 200 and 800 nm. The absorption peaks arise due to the surface plasma resonance effect of ZnO nanocrystals, and the redshift is due to the quantum size effect. The most probable transition takes place from the highest occupied molecular orbital (HOMO) to the lowest unoccupied molecular orbital (LUMO). An increase in absorption intensity was observed with the increase of temperature (Hyperchromic shift). Absorption of light occurs in a material due to electronic polarization or excitation. An increase in temperature causes the Zn and O atoms to vibrate along their bond axis (phonons). The higher the temperature, the more will be the vibration. During the vibration, there are instances when the Zn and O atoms are at a maximum distance apart. The atoms, while vibrating, are also creating a vibration in the electron distribution of the existing bond. Light being an electromagnetic wave

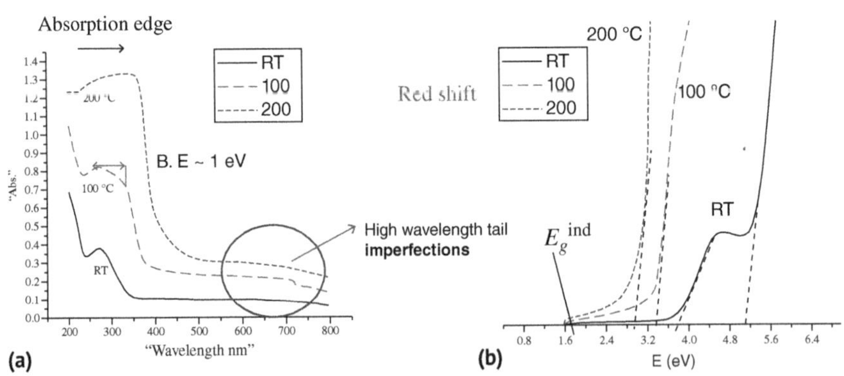

Figure 2.1 (a) UV spectra of synthesized ZnO at different temperatures. (b) Bandgap determination from the Tauc plot.

can induce electronic oscillations in the bonds, which may ultimately result in high light absorption. An increase in temperature also causes an increase in the number of electrons cutting loose from the valence band and excited to the conduction band which can also result in higher absorption intensity and broadening as observed. The increase in absorption is also due to an increase in crystallinity with temperature. However, it should be kept in mind that the frequency of electronic transition is much higher 10^{-15}/s, and hence in molecules, the vibrational levels are almost stationary when electronic transitions take place (Franck–Condon Principle). In semiconductors, however, due to the formation of excitons with temperature, the scenario is different. The excitons create discrete energy states in previously existing continuum states as there is a restriction imposed on the electrons and holes associated with excitons. A decrease in the bandgap was observed with an increase in temperature, as observed from the Tauc plot (Figure 2.1b), which indicates highly confined carriers to be present. The first absorption edge at a higher wavelength (lower energy) is associated with the transfer of electrons from the filled band to energy bands just below the conduction band, called the exciton band (Figure 2.2a). An exciton is a bound e-h pair. Since the electron excited is still bound to the hole produced by it by the Coulomb field, we call it an exciton. It results from the transfer of an electron from a negative ion, which happens to be O^{-2} in this case, to the nearest neighbor positive ion (Zn^{+2}). The bandgaps were obtained based on the relation (Eq. 2.1)

$$(\alpha h\nu)^{1/n} = A\left(h\nu - E_g\right) \tag{2.1}$$

where $n = 0.5$ for direct transition, showed a variation with a change in temperature. At room temperature two different bandgaps were obtained at 3.8 and 5.1 eV, respectively. The higher bandgap was probably due to the fact that full transformation to ZnO has not yet taken place, and a major part of the materials was still in the precursor state as seen in XRD. So, we take 3.8 eV as the bandgap

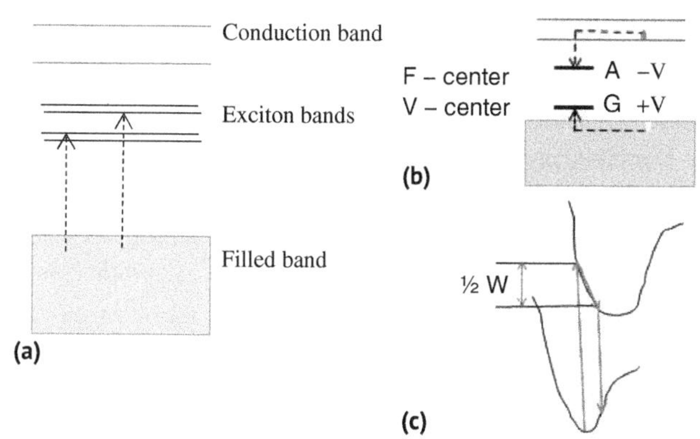

Figure 2.2 (a) Exciton band formation just below the conduction band. (b) Exciton interaction with lattice imperfection forming F and V centers. (c) Broadening of the absorption peak with temperature.

for this case. On increasing the temperature to 100 °C, a decrease in value to 3.4 eV was observed, which indicates that the precursor materials have been completely consumed to form ZnO phase. A further rise in temperature to 200 °C, however, caused a decrease in value to 2.9 eV due to the formation of defects. A lattice will expand and contract with rise in temperature, which will result in increased oscillations of the atoms around their mean position. The electron lattice interaction also changes with temperature. The impurity potential of donors and acceptors also get affected by rise in temperature. According to Varshini, the bandgap is related to temperature as Eq. (2.2), where A and B are constants (Zanatta, 2019).

$$E_g(T) = E_g(0) - \frac{AT^2}{T + B} \tag{2.2}$$

The initial wave vector k of the exciton is the same as that of the photon that was absorbed to produce it. However, as the exciton propagates through the lattice, it undergoes scattering from lattice imperfections. Lattice imperfections such as a positive ion vacancy act like a negative charge and make it easier to excite an electron from the negative ion surrounding it. Hence, the outer electrons do not stay in the filled bands but occupy a level above the filled band. The effect of positive ion vacancy is propagated further to the next nearest negative ions, which makes them closer to the filled bands. Similarly, a negative ion vacancy acts as a positive charge and attracts electrons that have been created by some means in the conduction band. This leads to an empty state of the positive ions lying below the conduction band. An electron in this state can get trapped between the positive ions, which are called F-centers in the case of alkali halides. A hole trapped surrounding negative ions due to a positive ion vacancy is called V-centers (Figure 2.2b). These new levels also create a new absorption band, which forms the tail of the first absorption peak. The peak height is proportional to the number of F-centers per unit volume and hence shows a higher value with an increase in temperature, indicating the formation of more lattice defects. The broadening of the peaks at high temperatures is due to the changed interatomic distance of an excited molecule and the Franck–Condon principle, which states that the vibrational levels are static during an electronic transition (Figure 2.2c).

For direct transitions between conduction and valence bands, it is considered that there is no change in the electron wave vector, and the absorption coefficient for α ($E = h\nu$) for direct transitions can be expressed as given in Eq. 2.3, where A is the constant for direct transition. The linear dependence of $\alpha(h\nu)^2$ with $h\nu$, therefore, indicates direct transition with the determination of E_g possible using a linear fit. Having a long tail is a sign of indirectness, which was found in all the cases. It is indicative of the fact that a fraction of indirectness is present in direct transitions. The absorption coefficient for indirect transitions can be expressed as Eq. (2.4), where $n = 2$ and $\hbar\Omega$ denotes a phonon being emitted ($+\hbar\Omega$) or absorbed ($-\hbar\Omega$). The equation shows that deviation from linearity is related to indirectness as the order index changes from ½ to 2 and can be used to find the indirect bandgap (Zanatta, 2019).

$$\alpha_{dir}\ (h\nu < E) = 0 \text{ and } \alpha_{dir}\left(h\nu \geq E_g\right) = A(h\nu - E_g)^{1/2} \tag{2.3}$$

$$\alpha_{ind}\ (E < E_g) = 0 \text{ and } \alpha_{ind}\ (E \geq E_g) \propto (E \pm \hbar\Omega - E_g)^n \tag{2.4}$$

In a degenerately n-doped semiconductor, photoexcited holes come to the top of the valence band and recombine with an electron, with a possible change in the k vector. The relaxation of the *k*-selection rule is caused by crystal imperfections introduced by dopants or by momentum taken up by the electron gas. A shift of the absorption band edge toward a higher wavelength (red shift) is observed, which corresponds to a change in bandgap (Davis, 2021). The sample prepared at RT showed a sharp edge at 275 nm. An increase in temperature to 100 °C, caused the formation of a secondary edge around 330 nm apart from 275 nm. A further increase in temperature to 200 °C caused the disappearance of the first edge, and the second edge shifted to 350 nm, which is in proximity to the values reported earlier.

The bandgap usually shows a slight decrease with the rise of temperature. The curve fitting of the Tauc plots at different temperatures is shown in Figure 2.3. The magnitude of the coefficients of E, E^2, and E^3, which had a major impact on the nature of the curve, is given in Table 2.1. The absorption coefficient α is related to the optical energy bandgap (E_g) for high photon energies. The coefficients a and b are associated with indirectness, while coefficient c is related to a direct transition between energy levels (Eq. 2.5). The coefficient a corresponds to $n = 1.5$, while b and c will correspond to $n = 1$ and 0.5, respectively. The ratio b/c and a/c will quantify the amount of indirectness present in the material, which is found to decrease with an increase in temperature.

For lower temperatures, where the crystallization is still in the nascent state, the contribution from the impurities gets reflected in the coefficients of E^2 and E, having higher values which are close to each other as observed for RT and 100 °C. For the 200 °C, the dominant coefficient was (for E), indicating a decrease in impurities that cause indirectness and growth of crystallites leading to the formation of the thermodynamically stable Wurtzite phase, where the valence band splits into three nondegenerate sub-bands referred to as A, B, and C bands, which also account for the decrease in the direct bandgap, E_g^{dir}. The indirect bandgap was, however, the same (E_g^{dir} ~1.6 eV) for all the cases, which shows the fact that it's not the decrease of defects or impurities that provides the directness in ZnO with the increase in temperature, but rather the crystallographic growth toward a stable phase formation.

$$(\alpha h\nu)^2 = aE^3 + bE^2 + cE \tag{2.5}$$

The coefficients corresponding to different Tauc exponent (n) were plotted with respect to temperature as shown in Figure 2.3d. A similar trend of an increase in value followed by a decrease reflects the synthesis where the precursors having their own crystallographic nomenclature underwent a chemical reaction and the process

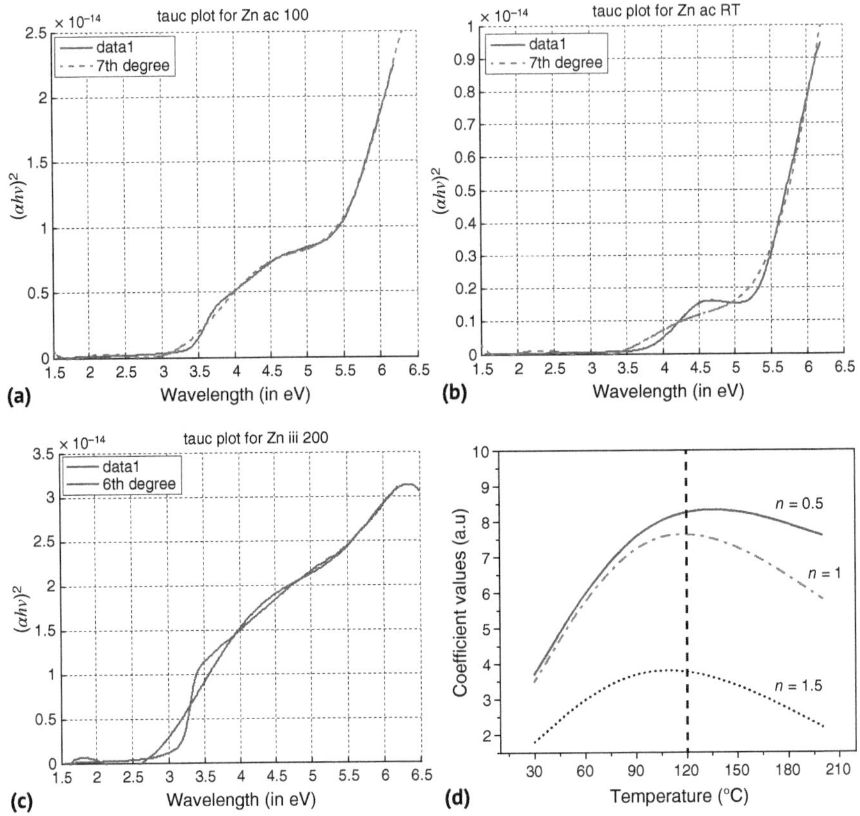

Figure 2.3 Curve fitting of Tauc plots of ZnO synthesized at (a) RT, (b) 100 °C, and (c) 200 °C (d) the variation of coefficients with temperature for the different Tauc exponents.

Table 2.1 The variation of coefficients E, E^2, and E^3 in the scale of 10^{-13}.

Temperature (°C)	**a (E^3) $n = 1.5$**	**b (E^2) $n = 1$**	**c (E) $n = 0.5$**	**Indirectness (b/c)**	**Indirectness (a/c)**	E_g^{dir} **(eV)**
30	1.8	3.5	3.7	0.95	0.49	3.8
100	4.7	9.0	9.2	0.98	0.51	3.4
200	2.2	5.8	7.6	0.76	0.29	2.9

of crystallization started once again, but this time into the final product, ZnO. A clear demarcation at 120 °C shows the initiation of the growth of ZnO crystals, as the two plots corresponding to $n = 1$ and $n = 0.5$ move downward, being deviated apart.

The key elements that contribute to optical absorption are related to the transition probability and are related to the optical absorption coefficient at the absorption onset by Eq. (2.6), where M is the matrix element and $\rho_{vc}(\hbar\omega)$ is the joint density of states

(JDOS). The constants λ_0, ε_0, and n_0 correspond to the photon wavelength and index of refraction at the bandgap, and the permittivity of free space (Rosencher & Vinter, 2002).

$$\alpha_0(\omega) = \frac{2\pi^2 q^2 \sqrt{\hbar}}{\lambda_0 \varepsilon_0 n_0} |M|^2 \rho_{vc}(\hbar\omega) \tag{2.6}$$

Drawing an analogy with Eq. (2.5), the physical significance of coefficients a, b, and c lies again in the JDOS. An increase in its value will therefore indicate a greater number of possible optical transitions from the filled states around the valence band to vacant states around the conduction band. The JDOS was therefore found to be lower in magnitude at higher temperatures as temperature rise brings increased molecular vibration, which may cause more phononic non-radiative transitions decreasing the directness. The experimental results, however, show a decrease in E_g^{dir}, which is most probably due to a separate set of JDOS associated with defects that trap electrons, and a rise in temperature sets them free, which eventually lowers the bandgap (Chen et al., 2018b; Spadavecchia et al., 2011).

2.3 Self-assembled Carbon Scaffolds for Photocatalytic Applications

Self-assembled carbon scaffolds have garnered significant attention in recent years for their potential in photocatalytic applications, attributed to their distinctive structure, high surface area, and adaptable properties (Fang et al., 2023). These scaffolds comprise carbon-based building blocks that are assembled through non-covalent interactions, such as π–π stacking, hydrogen bonding, and van der Waals forces (Dharmarajan et al., 2024). The resulting materials exhibit high surface areas, porosity, and electrical conductivity, making them ideal for photocatalytic reactions, facilitating efficient charge transfer and reaction kinetics (Choi et al., 2016; Ma & Cheng, 2023). The unique structure of self-assembled carbon scaffolds enables them to interact with light in a way that enhances their photocatalytic activity (Ma et al., 2021). Moreover, the electrical conductivity of self-assembled carbon scaffolds enables the efficient transfer of charge carriers, further enhancing their photocatalytic performance (Nam et al., 2010). By manipulating the assembly conditions and building block composition, researchers can tailor the structure and properties of these materials to suit specific photocatalytic reactions (Fang et al., 2023). For example, self-assembled carbon scaffolds can be designed to exhibit specific bandgap energies, enabling them to absorb light in the visible or ultraviolet spectrum (Ma et al., 2021).

Self-assembled carbon scaffolds have been employed in various photocatalytic applications, including water splitting (Weingarten et al., 2014), carbon dioxide reduction (Yuan et al., 2018), and organic pollutant degradation (Choi et al., 2016). In these reactions, the scaffolds serve as a platform for the adsorption and activation of reactants, facilitating the transfer of charge carriers and reaction intermediates

(Sahani et al., 2022). The high surface area and porosity of these materials also enable the efficient diffusion of reactants and products, further enhancing reaction rates and yields (Jian et al., 2016; Nam et al., 2010).

The unique structure of self-assembled carbon scaffolds is attributed to the non-covalent interactions between the carbon-based building blocks (Li et al., 2020b). These interactions, including π–π stacking, hydrogen bonding, and van der Waals forces, allow for the formation of complex networks and hierarchically porous structures (You et al., 2014). The resulting materials exhibit high surface areas and porosity, providing a large number of active sites for photocatalytic reactions (Tie et al., 2019; Huang et al., 2018). These active sites enable efficient absorption of light and interaction with reactants, facilitating the transfer of charge carriers and reaction intermediates (Mo et al., 2018). The unique structure of self-assembled carbon scaffolds also allows for the tunability of their optical and electronic properties, making them suitable for a wide range of photocatalytic applications (Ghamsari et al., 2021).

The synthesis of self-assembled carbon scaffolds typically involves the assembly of carbon-based building blocks, such as graphene oxide, carbon nanotubes, or organic molecules (Calbo et al., 2018). The tunable properties of self-assembled carbon scaffolds offer a significant advantage for photocatalytic applications (Robertson et al., 2022). The electrical conductivity of self-assembled carbon scaffolds is another critical property that can be tuned for photocatalytic applications. Scaffolds with high electrical conductivity can facilitate the transfer of charge carriers and reaction intermediates. For example, scaffolds with integrated conductive materials, such as carbon nanotubes or graphene, can enhance their photocatalytic activity. The high electrical conductivity of self-assembled carbon scaffolds is a critical property that enhances their photocatalytic performance (Lin et al., 2016). The scaffolds' conductivity enables efficient charge transport and separation, reducing recombination losses and improving photocatalytic efficiency. This property is particularly important for photocatalytic reactions, such as water splitting and carbon dioxide reduction, where efficient charge separation is crucial for high reaction rates and yields. The electrical conductivity of self-assembled carbon scaffolds can be further enhanced by introducing dopants or functional groups. For example, the incorporation of nitrogen or sulfur dopants can improve the scaffolds' electrical conductivity and photocatalytic activity. Similarly, the introduction of functional groups, such as hydroxyl or carboxyl groups, can enhance the scaffolds' surface reactivity and photocatalytic efficiency. These modifications enable the optimization of self-assembled carbon scaffolds for specific photocatalytic reactions, making them a promising material for industrial applications (Nie et al., 2017). Self-assembled carbon scaffolds have been successfully applied in various photocatalytic reactions, including water splitting, carbon dioxide reduction, and organic pollutant degradation. These scaffolds have shown high photocatalytic efficiency, stability, and recyclability, making them a promising material for large-scale industrial applications. For example, self-assembled carbon scaffolds have been used for the photocatalytic reduction of carbon dioxide to methanol, with high reaction rates and selectivity. The high surface area of self-assembled carbon scaffolds provides many active sites for photocatalytic reactions, enabling efficient absorption of light and interaction with reactants. The scaffolds' porosity also facilitates mass transport and diffusion of reactants and products, further enhancing photocatalytic efficiency. Moreover, the electrical conductivity of these scaffolds enables efficient charge transport and separation, reducing recombination losses and improving photocatalytic performance (Sun et al., 2020; Zang, 2015).

2.4 An Environmental and Economic Perspective

The development of self-assembled carbon scaffolds has revolutionized the field of photocatalysis, offering a sustainable and efficient solution for environmental remediation and energy production (Akram et al., 2024; Kommula et al., 2022). These scaffolds have shown enhanced photoactivity compared to traditional photocatalysts, enabling efficient degradation of pollutants and conversion of solar energy into chemical energy (Tian et al., 2022). From an environmental perspective, self-assembled carbon scaffolds offer several advantages, including high surface area, tunable porosity, and excellent stability (La et al., 2021). These properties enable efficient adsorption and degradation of pollutants, reducing the environmental impact of industrial activities (Vidyasagar et al., 2020).

Moreover, self-assembled carbon scaffolds have been shown to be economically viable, offering a cost-effective solution for large-scale industrial applications (Gong et al., 2019). The use of carbon-based materials reduces the dependence on precious metals, lowering production costs and enabling widespread adoption (Wong et al., 2023). Additionally, the tunability of self-assembled carbon scaffolds enables optimization for specific applications, reducing material waste and improving overall efficiency (Iqbal et al., 2018). The environmental benefits of self-assembled carbon scaffolds are further enhanced by their ability to facilitate carbon dioxide reduction and water splitting (Dasgupta et al., 2017). These reactions enable the conversion of solar energy into chemical energy, offering a sustainable solution for energy production (Zhou et al., 2019). Furthermore, the use of self-assembled carbon scaffolds in photocatalytic applications reduces the generation of hazardous byproducts, minimizing environmental harm (Xiang et al., 2019; Hao & Wu, 2022; Kundu et al., 2020; Liu et al., 2022; Kandy, 2020; Tian et al., 2020; Wei et al., 2019; Ma et al., 2022; Weingarten et al., 2014).

2.5 Conclusions

The self-assembled carbon scaffolds have emerged as a groundbreaking material for photocatalytic applications, offering a unique combination of high surface area, porosity, and electrical conductivity. These scaffolds, composed of carbon-based building blocks assembled through non-covalent interactions, exhibit exceptional photocatalytic efficiency, stability, and recyclability. Their tunable properties enable optimization for specific reactions, such as water splitting, carbon dioxide reduction, and organic pollutant degradation. The advantages of self-assembled carbon scaffolds over traditional materials, including enhanced efficiency, stability, and recyclability, make them an attractive alternative for large-scale industrial applications. Further research and development will be crucial in unlocking their full potential, paving the way for widespread adoption in photocatalytic applications. As we continue to explore the capabilities of self-assembled carbon scaffolds, we may uncover even more innovative solutions to global energy and environmental challenges. The future of photocatalysis looks bright with these scaffolds leading the way. By harnessing the power of self-assembled carbon scaffolds, we can create a more sustainable future, one photocatalytic reaction at a time. The self-assembled carbon scaffolds offer a sustainable and economically viable solution for environmental remediation and energy production. Their enhanced photoactivity, tunability, and stability make them an attractive alternative to traditional photocatalysis. Computational estimation and optimization of photocatalytic activity have emerged as powerful tools for designing and developing efficient photocatalysts. Degradation processes, including photocatalytic degradation of organic pollutants, water splitting, and CO_2 reduction, have significant potential for environmental remediation, renewable energy, and chemical synthesis.

References

Adhikari, R. (2022). *A Study on Thermally Stable Transition Metal Complexes Loaded Hierarchically Porous Carbon for Catalysis and Mechanism*. The University of Alabama.

Akimov, A. V., Asahi, R., Jinnouchi, R., & Prezhdo, O. V. (2015). What makes the photocatalytic CO_2 reduction on N-Doped Ta_2O_5 efficient: Insights from nonadiabatic molecular dynamics. *Journal of the American Chemical Society*, 137(35), 11517–11525. https://doi.org/10.1021/jacs.5b07454.

Akram, M. Y., Ashraf, T., Tong, L., Yin, X., Dong, H., & Lu, H. (2024). Architecting high-performance photocatalysts: A review of modified 2D/2D graphene/g-C_3N_4 heterostructures. *Journal of Environmental Chemical Engineering*, 12, 113415.

Ameta, R., Solanki, M. S., Benjamin, S., & Ameta, S. C. (2018). Chapter 6 – Photocatalysis. In S. C. Ameta & R. Ameta (Eds.), *Advanced Oxidation Processes for Waste Water Treatment* (pp. 135–175). Academic Press. https://doi.org/10.1016/B978-0-12-810499-6.00006-1.

Bhattacharyya, A. & Dash, R. (2022). Effect of temperature in ZnO and related computational studies. *Preprints*. https://doi.org/10.20944/preprints202209.0382.v1.

Calbo, J., Sancho-García, J. C., Orti, E., & Arago, J. (2018). Quantum-chemical insights into the self-assembly of carbon-based supramolecular complexes. *Molecules*, 23(1), 118.

Cerrato, E., Gionco, C., Paganini, M. C., Giamello, E., Albanese, E., & Pacchioni, G. (2018). Origin of visible light photoactivity of the CeO_2/ZnO heterojunction. *ACS Applied Energy Materials*, 1(8), 4247–4260.

Chen, S., Wang, L., Liu, Y., Zhu, Y., Zhang, Y., & Hu, C. (2020). Surface modification and functionalization of photocatalysts for enhanced performance. *Journal of Physical Chemistry C*, 124, 14141–14151.

Chen, X., Liu, L., Yu, P. Y., & Mao, S. S. (2018a). Understanding photocatalysis through computational methods. *Chemical Society Review*, 47, 2891–2911.

Chen, X., Shen, S., Guo, L., & Mao, S. S. (2018b). Semiconductor-based photocatalytic hydrogen generation: Challenges and opportunities. *Chemical Reviews*, 118, 6501–654.

Choi, J., Reddy, D. A., Islam, M. J., Ma, R., & Kim, T. K. (2016). Self-assembly of CeO_2 nanostructures/reduced graphene oxide composite aerogels for efficient photocatalytic degradation of organic pollutants in water. *Journal of Alloys and Compounds*, 688, 527–536.

Dasgupta, A., Rajukumar, L. P., Rotella, C., Lei, Y., & Terrones, M. (2017). Covalent three-dimensional networks of graphene and carbon nanotubes: Synthesis and environmental applications. *Nano Today*, 12, 116–135.

Dash, R., Gurjar, M., Kumari, N., Harish, T., & Bhattacharyya, A. S. (2023). Preferential growth and effect of temperature in (Ni, Zn) co-doped CuO. *MRS Advances*, 8, 947–952. https://doi.org/10.1557/s43580-023-00643-w.

Davis, K. P. (2021). *Sol-gel Synthesis and Band Gap Engineering of Zinc Oxide Nanostructures*. The University of North Carolina at Greensboro.

de P. R. Moreira, I., Illas, F., & Martin, R. L. (2002). Effect of Fock exchange on the electronic structure and magnetic coupling in NiO. *Physical Review B*, 65, 1551.

Dharmarajan, N. P., Vidyasagar, D., Yang, J. H. et al. (2024). Bio-inspired supramolecular self-assembled carbon nitride nanostructures for photocatalytic water splitting. *Advanced Materials*, 36(2), 2306895.

Fagiolari, L. & Bella, F. (2019). Carbon-based materials for stable, cheaper and large-scale processable perovskite solar cells. *Energy & Environmental Science*, 12(12), 3437–3472.

Faisal, M., Ismail, A. A., Ibrahim, A., Bouzid, A. H., & Al-sayari, S. A. (2013). Highly efficient photocatalyst based on Ce doped ZnO nanorods: Controllable synthesis and enhanced photocatalytic activity. *Chemical Engineering Journal*, 229, 225–233.

Fang, C., Hu, X., Du, X. et al. (2023). Self-assembled 3D hollow carbon nitride with electron delocalization for enhanced photocatalytic hydrogen evolution. *Applied Catalysis A: General*, 652, 119032.

Gaurav, K., Kumari, N., Bhattacharyya, A. S., & Samdarshi, S. K. (2019). Photoactivity of tungsten and niobium composite of titania under UV irradiation. *Carbon- Science and Technology*, 11, 101–110.

Gavilán, H., Serrano, M. B., & Cabanelas, J. C. (2023). Nanomaterials and their synthesis for a sustainable future. *New Materials for a Circular Economy*, 149, 233–310.

Ghamsari, M., Madrakian, T., Afkhami, A., & Ahmadi, M. (2021). Self-assembled graphene-based microfibers with eclectic optical properties. *Scientific Reports*, 11(1), 5451.

Gong, C., Sun, S., Zhang, Y. et al. (2019). Hierarchical nanomaterials via biomolecular self-assembly and bioinspiration for energy and environmental applications. *Nanoscale*, 11(10), 4147–41482.

Guo, X., Zhang, J., Yuan, L. et al. (2023). Biologically assisted construction of advanced electrode materials for electrochemical energy storage and conversion. *Advanced Energy Materials*, 13(24), 2204376.

Guo, Y., Li, H., Ma, W., Shi, W., Zhu, Y., & Choi, W. (2020). Photocatalytic activity enhanced via surface hybridization. *Carbon Energy*, 2(3), 308–349.

Gurjar, M. K., Harsh, T., Dash, R., & Bhattacharyya, A. S. (2023). Abnormalities and preferences in divalent (Cu, Ni) co-doping in ZnO at lower proportions. *MRS Advances*, 8, 925–930. https://doi.org/10.1557/s43580-023-00588-0.

Hao, H., Zhang, L., Wang, W., & Zeng, S. (2018). Modification of heterogeneous photocatalysts for selective organic synthesis. *Catalysis Science & Technology*, 8(5), 1229–1250.

Hao, Z. & Wu, J. (2022). Self-assembled zeolitic imidazolate framework/polyimide nanofibers for efficient air pollution control. *ACS Applied Nano Materials*, 5(2), 2343–2349.

Harsh, T. K., Samdarshi, S. K., Deshpande, U. et al. (2023). Performance of mixed phase TiO_2 film samples immobilized on a 2-D flat plate surface for photo catalytic reusability application. In *ISES Solar World Congress*.

Herrmann, J. M. (2019). Heterogeneous photocatalysis: Fundamentals and applications for water purification. *Catalysis Today*, 335, 2–14.

Huang, R. L., Huang, W. Q., Li, D. F. et al. (2018). Self-assembled hierarchical carbon/g-C_3N_4 composite with high photocatalytic activity. *Journal of Physics D: Applied Physics*, 51(13), 135501.

Iqbal, W., Yang, B., Zhao, X. et al. (2018). Controllable synthesis of graphitic carbon nitride nanomaterials for solar energy conversion and environmental remediation: The road travelled and the way forward. *Catalysis Science & Technology*, 8(18), 4576–4599.

Islam, M., Lantada, A. D., Mager, D., & Korvink, J. G. (2022). Carbon-based materials for articular tissue engineering: From innovative scaffolding materials toward engineered living carbon. *Advanced Healthcare Materials*, 11(1), 2101834.

Jian, X., Liu, X., Yang, H. M. et al. (2016). Construction of carbon quantum dots/proton-functionalized graphitic carbon nitride nanocomposite via electrostatic self-assembly strategy and its application. *Applied Surface Science*, 370, 514–521.

Kandy, M. M. (2020). Carbon-based photocatalysts for enhanced photocatalytic reduction of CO_2 to solar fuels. *Sustainable Energy & Fuels*, 4(2), 469–484.

Khan, M.A., Cantù, E., Tonello, S., Serpelloni, M., Lopomo, N. F., & Sardini, E. (2019). A review on biomaterials for 3D conductive scaffolds for stimulating and monitoring cellular activities. *Applied Sciences*, 9(5), 961.

Ko, J. W., Choi, W. S., Kim, J., Kuk, S. K., Lee, S. H., & Park, C. B. (2017). Self-assembled peptide-carbon nitride hydrogel as a light-responsive scaffold material. *Biomacromolecules*, 18(11), 3551–3556.

Kommula, B., Durairaj, P., Mishra, S. et al. (2022). Self-assembled oligothiophenes for photocatalytic hydrogen production and simultaneous organic transformation. *ACS Applied Nano Materials*, 5(10), 14746–14758.

Kong, L., Zhang, C., Wang, J., Qiao, W., Ling, L., & Long, D. (2016). Nanoarchitectured Nb_2O_5 hollow, Nb_2O_5@carbon and NbO_2@carbon core-shell microspheres for ultrahigh-rate intercalation pseudocapacitors. *Scientific Reports*, 6, 21177.

Kumar, G., Neha, K., Bhattacharyya, A. S., Samrat, P., Uday, D., & Samdarshi, S. K. (2022). Distinctive role of hexagonal tungsten (VI) oxide in tungsten (VI) oxide/ ceria hetero-junction for efficient photocatalysis under visible photon flux. *Solar Energy Materials & Solar Cells*, 234, 111405. https://doi.org/10.1016/j.solmat.2021.111405.

Kumari, N., Gaurav, K., Dash, R., & Bhattacharyya, A. S. (2024). Doped mixed phase transition metal oxides for photocatalysis. In N. V. Kulkarni & B. I. Kharissov (Eds.), *Handbook of Emerging Materials for Sustainable Energy* (pp. 333–346). Elsevier.

Kumari, N., Harsh, T. K., Bhattacharyya, A. S., Gaurav, K., Verma, R., & Samdarshi, S. K. (2023a). Enhanced luminescence and photocatalytic activity of ceria doped ZnO under UV illumination prepared via chemical precipitation technique. *Luminescence*, 38(7), 1282–1286. https://doi.org/10.1002/bio.4396.

Kumari, N., Samdarshi, S. K., Verma, R. et al. (2023b). Superior functionality of niobium pentoxide nano-rod/tripod photocatalyst synthesized using polyethyleneimine as a soft template for the abatement of methylene blue under UV and visible irradiation. *Environmental Science and Pollution Research*, 30, 122458–122469. https://doi.org/10.1007/s11356-023-31001-w.

Kundu, S., Bramhaiah, K., & Bhattacharyya, S. (2020). Carbon-based nanomaterials: In the quest of alternative metal-free photocatalysts for solar water splitting. *Nanoscale Advances*, 2(11), 5130–5151.

La, D. D., Nguyen, T. A., Nguyen, X. S. et al. (2021). Self-assembly of porphyrin on the surface of a novel composite high performance photocatalyst for the degradation of organic dye from water: Characterization and performance evaluation. *Journal of Environmental Chemical Engineering*, 9(5), 106034.

Li, C., Li, Q., Kaneti, Y. V., Hou, D., Yamauchi, Y., & Mai, Y. (2020a). Self-assembly of block copolymers towards mesoporous materials for energy storage and conversion systems. *Chemical Society Reviews*, 49(14), 4681–4736.

Li, R., Zhang, Y., Zhu, Y., Chen, H., Liu, Y., & Hu, C. (2019a). Design and construction of nanostructured photocatalysts for environmental applications. *ACS Catalysis*, 9, 848–869.

Li, S., Peng, Y., Hu, C., & Chen, Z. (2020b). Self-assembled synthesis of benzene-ring-grafted g-C_3N_4 nanotubes for enhanced photocatalytic H_2 evolution. *Applied Catalysis B: Environmental*, 279, 119401.

Li, W., Wang, Y., Zhang, X., & Chen, J. (2020c). Machine learning for photocatalytic materials design. *ACS Catalysis*, 10, 14133–14141.

Li, X., Yu, J., & Jaroniec, M. (2019b). Hierarchical photocatalysts for water splitting and CO_2 reduction. *Chemical Society Reviews*, 48, 255–275.

Li, Y., Zhao, Y., Cheng, H. et al. (2016). Nitrogen-doped graphene quantum dots with oxygen-rich functional groups. *Journal of the American Chemical Society*, 138(30), 9326–9334. https://doi.org/10.1021/jacs.6b05153.

Li, Z., Wang, L., Li, Y., Feng, Y., & Feng, W. (2019c). Carbon-based functional nanomaterials: Preparation, properties and applications. *Composites Science and Technology*, 179, 10–40.

Lin, Z. & Huang, J. (2021). A hierarchical $H_3PW_{12}O_{40}/TiO_2$ nanocomposite with cellulose as scaffold for photocatalytic degradation of organic pollutants. *Separation and Purification Technology*, 264, 118427.

Lin, Z., Zeng, Z., Gui, X., Tang, Z., Zou, M., & Cao, A. (2016). Carbon nanotube sponges, aerogels, and hierarchical composites: Synthesis, properties, and energy applications. *Advanced Energy Materials*, 6(17), 1600554.

Liu, X., Ji, Y., Du, Y. et al. (2022). A "green" all-organic heterostructure functionalized by self-assembled fullerene small molecule with enhanced photocatalytic activity. *Applied Surface Science*, 585, 152738.

Luo, L. M., Zhan, W., Zhang, R. H. et al. (2019). Ternary CoAuPd and binary AuPd electrocatalysts for methanol oxidation and oxygen reduction reaction: Enhanced catalytic performance by surface reconstruction. *Journal of Power Sources*, 412, 142–152.

Ma, B., Blanco, M., Calvillo, L. et al. (2021). Hybridization of molecular and graphene materials for CO_2 photocatalytic reduction with selectivity control. *Journal of the American Chemical Society*, 143(22), 8414–8425.

Ma, R., Xue, Y., Ma, Q., Chen, Y., Yuan, S., & Fan J. (2022). Recent advances in carbon-based materials for adsorptive and photocatalytic antibiotic removal. *Nanomaterials*, 12(22), 4045.

Ma, X. & Cheng, H. (2023). Self-introduction of carbon nitride quantum dots into carbon nitride planar structure for enhanced photocatalytic hydrogen production. *Applied Catalysis B: Environmental*, 339, 123101.

Mo, Z., Xu, H., Chen, Z. et al. (2018). Self-assembled synthesis of defect-engineered graphitic carbon nitride nanotubes for efficient conversion of solar energy. *Applied Catalysis B: Environmental*, 225, 154–161.

Morales-García, Á., Viñes, F., Sousa, C., & Illas, F. (2023). Toward a rigorous theoretical description of photocatalysis using realistic models. *Journal of Physical Chemistry Letters*, 14(15), 3712–3720. https://doi.org/10.1021/acs.jpclett.3c00359.

Nair, R. G., Bharadwaj, P. J., & Samdarshi, S. K. (2016). Design improvement and performance evaluation of solar photocatalytic reactor for industrial effluent treatment. *Ecotoxicology and Environmental Safety*, 134, 301–307.

Nam, Y. S., Magyar, A. P., Lee, D. et al. (2010). Biologically templated photocatalytic nanostructures for sustained light-driven water oxidation. *Nature Nanotechnology*, 5(5), 340–344.

Neha, K., Kumar, G., Samrat, P., Bhattacharyya, A. S., & Samdarshi, S. K. (2020). Dependence of photoactivity of niobium pentoxide (Nb_2O_5) on crystalline phase and electrokinetic potential of the hydrocolloid. *Solar Energy Materials and Solar Cell*, 208, 110408. https://doi.org/10.1016/j.solmat.2020.110408.

Nie, W., Peng, C., Zhou, X. et al. (2017). Three-dimensional porous scaffold by self-assembly of reduced graphene oxide and nano-hydroxyapatite composites for bone tissue engineering. *Carbon*, 116, 325–337.

Nithiyanantham, U., Ede, S. R., Anantharaj, S., & Kundu, S. (2015). Self-assembled $NiWO_4$ nanoparticles into chain-like aggregates on DNA scaffold with pronounced catalytic and supercapacitor activities. *Crystal Growth & Design*, 15(2), 673–686.

Pavlenko, V., Żółtowska, S., Haruna, A. B. et al. (2022). A comprehensive review of template-assisted porous carbons: Modern preparation methods and advanced applications. *Materials Science and Engineering: R: Reports*, 149, 100682.

Prakash, J., Kumar, P., Saxena, N. et al. (2023). CdS based 3D nano/micro-architectures: Formation mechanism, tailoring of visible light activities and emerging applications in photocatalytic H_2 production, CO_2 reduction and organic pollutant degradation. *Journal of Materials Chemistry A*, 11(19), 10015–10064.

Raisina, D., Driessen, R. T., Ogieglo, W., & Lammertink, R. G. H. (2015). Intrinsic photocatalytic assessment of reactively sputtered TiO_2 films. *ACS Applied Materials & Interfaces*, 7, 8727.

Ran, F., Yang, X., & Shao, L. (2018). Recent progress in carbon-based nanoarchitectures for advanced supercapacitors. *Advanced Composites and Hybrid Materials*, 1, 32–55.

Ripoll, J. (2011). Derivation of the scalar radiative transfer equation from energy conservation of Maxwell's equations in the far field. *Journal of the Optical Society of America*, 28, 1765.

Robertson, M., Zagho, M. M., Nazarenko, S., & Qiang, Z. (2022). Mesoporous carbons from self-assembled polymers. *Journal of Polymer Science*, 60(14), 2015–2042.

Rosencher, E. & Vinter, B. (2002). *Optoelectronics*. Cambridge: Cambridge University Press.

Sahani, S., Tripathi, K. M., Lee, T. I. et al. (2022). Recent advances in photocatalytic carbon-based materials for enhanced water splitting under visible-light irradiation. *Energy Conversion and Management*, 252, 115133.

Samanta, B., Morales-García, Á., Illas, F. et al. (2022). Challenges of modeling nanostructured materials for photocatalytic water splitting. *Chemical Society Reviews*, 51(9), 3794–3818. https://doi.org/10.1039/d1cs00648g.

Serrano, M. C., Gutiérrez, M. C., & del Monte, F. (2014). Role of polymers in the design of 3D carbon nanotube-based scaffolds for biomedical applications. *Progress in Polymer Science*, 39(7), 1448–1471.

Singh, A. K., Mathew, K., Zhuang, H. L., & Hennig, R. G. (2015). Computational screening of 2D materials for photocatalysis. *The Journal of Physical Chemistry Letters*, 6(6), 1087–1098.

Soufi, G. J. & Iravani, S. (2020). Eco-friendly and sustainable synthesis of biocompatible nanomaterials for diagnostic imaging: Current challenges and future perspectives. *Green Chemistry*, 22(9), 2662–2687.

Sousa, C., Tosoni, S., & Illas, F. (2013). Theoretical approaches to excited-state-related phenomena in oxide surfaces. *Chemical Reviews*, 113, 4456—4495.

Spadavecchia, F., Cappelletti, G., Ardizzone, S., Ceotto, M., & Falciola, L. (2011). Electronic structure of pure and N-doped TiO_2 nanocrystals by electrochemical experiments and first principles calculations. *The Journal of Physical Chemistry C*, 115(14), 6381–6391.

Sun, H., Dong, B., Su, G. et al. (2015). Modification of TiO_2 nanotubes by WO_3 species for improving their photocatalytic activity. *Applied Surface Science*, 343, 181–187.

Sun, J., Yao, F., Dai, L. et al. (2020). Task-specific synthesis of 3D porous carbon nitrides from the cycloaddition reaction and sequential self-assembly strategy toward photocatalytic hydrogen evolution. *ACS Applied Materials & Interfaces*, 12(36), 40433–40442.

Sun, M. H., Huang, S. Z., Chen, L. H. et al. (2016). Applications of hierarchically structured porous materials from energy storage and conversion, catalysis, photocatalysis, adsorption, separation, and sensing to biomedicine. *Chemical Society Reviews*, 45(12), 3479–3563.

Tan, C., Cao, X., Wu, X. J. et al. (2017). Recent advances in ultrathin two-dimensional nanomaterials. *Chemical Reviews*, 117(9), 6225–6331. https://doi.org/10.1021/acs.chemrev.6b00558.

Tian, J., Yu, J., Tang, Q. et al. (2022). Self-assembled supramolecular materials for photocatalytic H_2 production and CO_2 reduction. *Materials Futures*, 1(4), 042104.

Tian, J., Zhang, Y., Du, L. et al. (2020). Tailored self-assembled photocatalytic nanofibres for visible-light-driven hydrogen production. *Nature Chemistry*, 12(12), 1150–1156.

Tie, W., Jin, J., Bhattacharyya, S. S. et al. (2019). Covalent bonding of ZnO nanostructures with dispersible carbon nanotubes for self-assembly photocatalytic heterostructures. *Applied Surface Science*, 492, 219–227.

Titirici, M. M. (2013). *Sustainable Carbon Materials from Hydrothermal Processes*. John Wiley & Sons.

Varshni, Y. P. (1967). Temperature dependence of the energy gap in semiconductors. *Physica*, 34, 149–154.

Venkateswaran, M. R., Khosravi, A., Zarepour, A., Iravani, S., & Zarrabi, A. (2024). Self-healing materials in biomedicine and the circular economy. *Environmental Science: Nano*, 11(7), 2771–2802.

Verma, R. & Samdarshi, S. K. (2015). Correlating oxygen vacancies and phase ratio/interface with efficient photocatalytic activity in mixed phase TiO_2. *Journal of Alloys and Compounds*, 629, 105–112.

Vidyasagar, D., Kumari, V., Bhoyar, T., & Umare, S. S. (2020). Unveiling morphology altered photoactivity of microspherical carbon nitride scaffolds. *Applied Surface Science*, 526, 146661.

Visan, A., Rafieian, D., Ogieglo, W., & Lammertink, R. G. H. (2014). Modeling intrinsic kinetics in immobilized photocatalytic micro-reactors. *Applied Catalysis B: Environmental*, 150—151, 93—100.

Visan, A., van Ommen, J. R., Kreutzer, M. T., & Lammertink, R. G. H. (2019). Photocatalytic reactor design: Guidelines for kinetic investigation. *Industrial & Engineering Chemistry Research*, 58(14), 5349–5357. https://doi.org/10.1021/acs.iecr.9b00381.

Wagh, N. K., Shinde, S. S., Lee, C. H. et al. (2020). Densely colonized isolated Cu-N single sites for efficient bifunctional electrocatalysts and rechargeable advanced Zn-air batteries. *Applied Catalysis B: Environmental*, 268, 118746.

Wang, D., Mueses, M. A., Márquez, J. A. C. et al. (2021). Engineering and modeling perspectives on photocatalytic reactors for water treatment. *Water Research*, 202(117421), 117421. https://doi.org/10.1016/j.watres.2021.117421.

Wang, H., Zhang, L., Chen, Z. et al. (2014). Semiconductor heterojunction photocatalysts: Design, construction, and applications. *Chemical Society Review*, 43, 5234–5244.

Wei, Q., Xiong, F., Tan, S. et al. (2017). Porous one-dimensional nanomaterials: Design, fabrication and applications in electrochemical energy storage. *Advanced Materials*, 29(20), 1602300.

Wei, W., Ouyang, S., & Zhang, T. (2020). Perylene diimide self-assembly: From electronic structural modulation to photocatalytic applications. *Journal of Semiconductors*, 41(9), 091708.

Wei, Y., Zhu, Y., & Jiang, Y. (2019). Photocatalytic self-cleaning carbon nitride nanotube intercalated reduced graphene oxide membranes for enhanced water purification. *Chemical Engineering Journal*, 356, 915–925.

Weingarten, A. S., Kazantsev, R. V., Palmer, L. C. et al. (2014). Self-assembling hydrogel scaffolds for photocatalytic hydrogen production. *Nature Chemistry*, 6(11), 964–970.

Whitby, R. L. (2014). Chemical control of graphene architecture: Tailoring shape and properties. *ACS Nano*, 8(10), 9733–9754.

Wong, K. J., Foo, J. J., Siang, T. J., & Ong, W. J. (2023). Shining light on carbon aerogel photocatalysts: Unlocking the potentials in the quest for revolutionizing solar-to-chemical conversion and environmental remediation. *Advanced Functional Materials*, 33(50), 2306014.

Xiang, Y., Xu, Z., Wei, Y. et al. (2019). Carbon-based materials as adsorbent for antibiotics removal: Mechanisms and influencing factors. *Journal of Environmental Management*, 237, 128–138.

Xie, S., Zhang, Q., Liu, G., & Wang, Y. (2020). Photocatalytic CO_2 reduction and utilization: Challenges and opportunities. *Energy & Environmental Science*, 13, 1316–1332.

Yang, K., Xu, X., Lin, X., & Yang, W. (2018). Computational design of photocatalytic materials. *Journal of Physical Chemistry C*, 122, 13434–13441.

You, B., Jiang, J., & Fan, S. (2014). Three-dimensional hierarchically porous all-carbon foams for supercapacitor. *ACS Applied Materials & Interfaces*, 6(17), 15302–15308.

Yuan, Z., Xiao, X., Li, J., Zhao, Z., Yu, D., & Li, Q. (2018). Self-assembled graphene-based architectures and their applications. *Advanced Science*, 5(2), 1700626.

Zaleska-Medynska, A. (2018). *Metal Oxide-based Photocatalysis: Fundamentals and Prospects for Application*. Elsevier.

Zanatta, A. R. (2019). Revisiting the optical bandgap of semiconductors and the proposal of a unified methodology to its determination. *Scientific Reports*, 9, 11225.

Zang, L. (2015). Interfacial donor–acceptor engineering of nanofiber materials to achieve photoconductivity and applications. *Accounts of Chemical Research*, 48(10), 2705–2714.

Zhang, Y., Poon, K., Masonsong, G. S., Ramaswamy, Y., & Singh, G. (2023). Sustainable nanomaterials for biomedical applications. *Pharmaceutics*, 15(3), 922.

Zhou, S., Zhou, L., Zhang, Y., Sun, J., Wen, J., & Yuan, Y. (2019). Upgrading earth-abundant biomass into three-dimensional carbon materials for energy and environmental applications. *Journal of Materials Chemistry A*, 7(9), 4217–4229.

Zhou, X., Liu, B., Chen, Y., Guo, L., & Wei, G. (2020). Carbon nanofiber-based three-dimensional nanomaterials for energy and environmental applications. *Materials Advances*, 1(7), 2163–2181.

Zou, Y., Gong, Y., Lin, B., & Mellott, N. P. (2016). Photodegradation of methylene blue in the visible spectrum: An efficient W^{6+} ion doped anatase titania photocatalyst via a solvothermal method. *Vacuum*, 126, 63–69.

3

Carbon Scaffold-based Novel Nanophotocatalysts: Industrial Synthesis, Degradation Mechanism, and Sustainability

*Bhaskar Sharma[1], Subhasish Roy[1], Swagata Baruah[2], Sagar Sharma[3], Samiyara Begum[4], Manash J. Deka[5], Monjumoni Das[1,6], and Pallabi Borah[1]**

[1] *Department of Chemistry, Assam Don Bosco University, Tapesia Gardens, Kamarkuchi, Assam, India*
[2] *Department of Chemistry, Morigaon College, Morigaon, Assam, India*
[3] *Department of Chemistry, S. B. Deorah College, Guwahati, Assam, India*
[4] *Department of Chemistry, University of Science and Technology, Baridua, Meghalaya, India*
[5] *Department of Chemistry, Guwahati Collge, Guwahati, Assam, India*
[6] *Department of Chemistry, Sibsagar College, Sivasagar, Assam, India*

* Corresponding author: pallabi.borah@dbuniversity.ac.in

3.1 Introduction

Nanoparticles are ultrafine particles sized between 1 and 100 nm. When particles are fabricated within this nano range, they always offer magical physical and chemical properties. Due to the very small particle size and greater surface area associated with nanoparticles, they constantly become superior candidates over their bulk counterparts in terms of their applications in the area of catalysis (Astruc, 2020), material science, agricultural and food science (Duncan et al., 2011), environmental cleanup and remediation (Patil et al., 2016), electronic sector (Hossain et al., 2023), biological science (Zaman & Manghwar et al., 2024; Salata, 2004), innovative energy technology (Al Mahmud, 2023), drug delivery for the treatment of life-threatening diseases (Pala et al., 2020; Yusuf et al., 2023), etc. Among the various nanocomposites used till date, carbon-based nanocomposites find special interest due to their low cost, high abundance in the earth's crust, and versatile applications. Important example of carbon nanomaterials includes zero dimensional (0D) fullerenes and quantum dots, one dimensional (1D) carbon nanotube (CNTs), two-dimensional (2D) graphene, and their derivatives, three-dimensional (3D) nano diamonds, etc. (Mahor et al., 2021). These carbon-based nanomaterials

Photocatalysts based on Carbon Scaffolds: Environmental and Industrial Applications, First Edition. Edited by Ramesh Oraon et al.
Published 2026 by Wiley-VCH GmbH.

provide outstanding thermal, mechanical, electronic-electrical, optical, photoluminescence, photo-chemical, electro-chemical, and biological properties due to which they are becoming eye-catching candidates in diverse ranges of applications, which are highlighted in Table 3.1.

Photocatalysis is the speeding up of a chemical reaction in the presence of light and catalyst. When reactants are exposed to light of an appropriate wavelength, photocatalysts actively initiate chemical reactions without undergoing any alteration. Photocatalysts can efficiently be involved in important redox reactions with neighboring molecules due to their aptitude to generate electron-hole pairs and active intermediates after absorbing photons or light. Interestingly, photocatalytic reactions can be performed under ambient reaction conditions, such as at room temperature or under atmospheric pressure, which greatly reduces the operating cost and hence simplifies operation procedures. The chlorophyll part of a plant leaf is a good example of a photocatalyst, which is involved in the complete conversion of water and carbon dioxide into oxygen and glucose in the presence of sunlight during photosynthesis (Simon et al., 2019). Therefore, photocatalysis can mimic our popularly known natural process of photosynthesis. However, photosynthetic procedures differ from photocatalytic procedures in terms of their thermodynamic point of view. For example, photosynthetic processes are generally thermodynamically

Table 3.1 Applications of various carbon-based nanomaterials.

S. no.	Type of carbon-based nanoparticles	Applications	Reference
1	0D fullerene (C_{60})	Extensively employed in drug delivery process, photodynamic therapy, and imaging agents, etc.	Bosi et al. (2003)
2	1D CNTs	Used in transparent displays, quantum wires, field emission transistors, tissue engineering, aerospace materials, energy storage, water filters, thin-film electronics, etc.	Jagadeesan et al. (2020) and Ali et al. (2023)
3	2D graphene and their derivatives	Applied in smooth drug delivery systems, biosensors for detection of virus, anti-bacterial agent, imaging systems, cancer therapy and diagnosis, as a potent theranostic, removal of harmful ions in environmental remediation, etc.	Sharma et al. (2022a); Banerjee (2018) and Magne et al. (2022)
4	3D nano diamonds	Nanodiamonds are nontoxic and hence are promising materials in the area of tribology, drug delivery, bioimaging, tissue engineering and for protein mimics, etc.	Mochalin et al. (2012)

Source: Adapted from Mahor et al. (2021).

unfavorable and uphill ($\Delta G > 0$) and need photochemical energy input to happen, whereas photocatalytic processes are thermodynamically favorable and downhill ($\Delta G < 0$) and can be greatly accelerated by the use of catalysts (Osterloh, 2017). An important requirement for any species to behave as photocatalyst is its dual ability to act as an oxidizing and reducing agent simultaneously in the presence of light. The concept of photocatalysis originated from the ground-breaking experiment of two renowned Japanese scientists, Akira Fujishima and Kenichi Honda, in the early 1970s, when they successfully performed photolysis of water in the presence of TiO_2 and ultraviolet light (Fujishima & Honda, 1972). In their experiment, it was observed that semiconductor-based TiO_2 electrodes can act as photocatalyst in the presence of UV light for catalyzing water splitting to produce hydrogen and oxygen in an efficient process (Fujishima & Honda, 1972). After this discovery, the phenomenon of photocatalysis gained significant importance across the globe and led to enormous and innovative advancements in the areas of green technology, environmental science, organic synthesis, material science, energy storage, and also in the production of clean energy (Osterloh, 2008; Xiang et al., 2012a; Li et al., 2019; Romero & Nicewicz, 2016). In the recent past, many photocatalysts such as TiO_2-RuO_2 composite powders (Morais et al., 2021), mercuric thiocyanate, ruthenium (III) complexes (Tennakone et al., 1985), ZnO (Sun et al., 2023), Fe_2O_3 (Li et al., 2020), Ni and Al oxides (Anbarasu et al., 2020), CdS (Cheng et al., 2018), ZnS (Li et al., 2004), Rh- and Ir-doped metal oxides, etc. have been introduced for various photocatalytic transformations (Bickley et al., 1973; Matsumura et al., 1984; Bandara et al., 1999; Tian et al., 2012; Koike & Akita, 2014; Kudo et al., 2019). However, many of the process either require high catalyst loading or are expensive, which limits their synthetic utility in modern science. These problems can be solved by incorporating small amounts of the active catalyst species into some appropriate matrix. In present science, green and sustainable approaches are mostly emphasized to meet socio-economic requirements. Literature review revealed that the catalytic activity of a particular species can be further tuned when they are brought into the nano-range. Recent advancements in nanotechnology have led to the development of new types of photocatalysts, known as nanophotocatalysts. Due to the high surface area-to-volume ratio and excellent quantum confinement properties of nanophotocatalysts, they always offer additional advantages as compared to their bulk counterparts (Radhika et al., 2019). The band energy gap between the valence band and conduction band precisely depends on the photocatalytic activity of the material, which is again interconnected with particle size. Therefore, by controlling the size of the nanophotocatalyst, appropriate tuning of wavelength is possible in different regions of light, which is actually required to target a specific goal in photocatalysis. Due to its wonderful optical and conducting properties, nanophotocatalysis provides various applications in modern science, which are presented in Figure 3.1.

Among the various supporting materials reported for binding nanophotocatalyst, carbon-nano-based scaffolds are particularly gaining attraction and hence are considered to be superior in several aspects over the other nanophotocatalysts. The reasons may be due to the facts: (1) Carbon-based nanocomposite can be easily obtained from either renewable biomass or waste materials and hence is

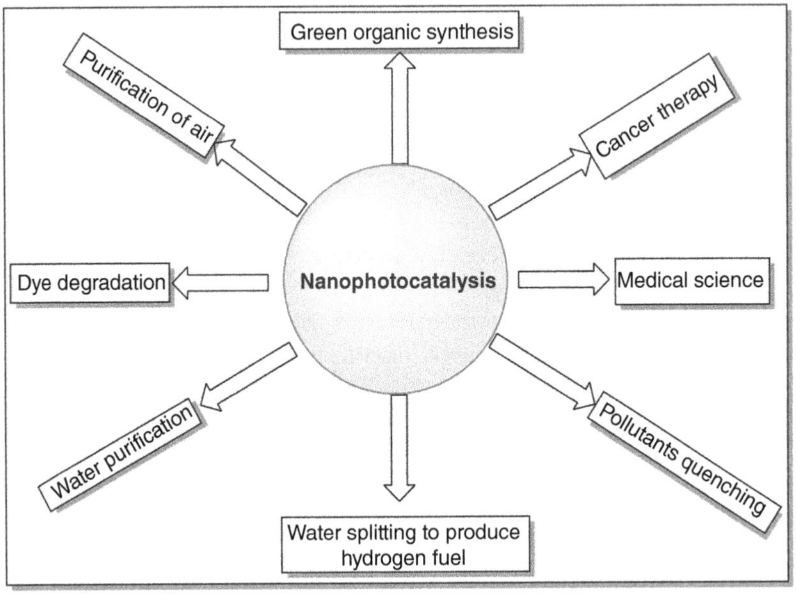

Figure 3.1 Multidimensional applications of nanophotocatalysis.

more sustainable as compared to other materials. (2) Carbon-based nanoparticles like carbon dots (CDs), graphene, and its derivatives, as well as single-wall carbon nanotubes (SWCNTs) and multiwall carbon nanotubes (MWCNTs), offer unique electronic structures due to which they readily generate electron-hole pairs and hence can efficiently promote photo-redox catalysis. (3) A simple synthetic procedure, low cost, high reproducibility, and scalability of carbon-based nanophotocatalysts meet the criteria of a green and sustainable approach that is directed according to the "Environmental, Social, and Governance (ESG) Principles." (4) Carbon-based nanoparticles provide tunable properties *via* structural modification, which in turn can greatly enhance their catalytic activity and selectivity as required for a specific purpose. (5) Generally, carbon-based nanocomposites are less toxic, more stable, and show superb biocompatibility over classical the photocatalysts (Wen et al., 2023). In addition to these, carbon-based nanocomposites comprising of g-C_3N_4, graphene, CNT, and carbon quantum dots may serve as heterogeneous photocatalyst. Some of the important applications of carbon scaffolds based nanophotocatalyst include: reduction of CO_2 to reduce carbon footprint, fixation of nitrogen for enhancing fertility in agricultural sectors, conversion of toxic compounds to important harmless materials, etc. (Xu et al., 2022); visible light mediated photocatalytic degradation (Park et al., 2019); hazardous pollutants degradation and wastewater treatment (Sheoran et al., 2022), etc.

Due to greater demand of carbon-based novel nanophotocatalyst in present technology, development of their green and sustainable synthetic routes is always challenging. Again, during their synthesis, selection of starting materials as well as type of synthetic approach play key role for achieving specific applications and goals.

Some commonly used techniques used for the synthesis of carbon scaffold-based novel nanophotocatalysts include: chemical vapor deposition (CVD), physical vapor deposition, sol–gel method, hydrothermal or solvothermal synthesis, chemical precipitation, electrochemical deposition, atomic layer deposition, microwave-assisted synthesis, ultrasonic assisted synthesis, bio-inspired synthesis, etc. (Cui et al., 2021; Dutta et al., 2022).

3.2 Carbon Scaffold-based Nanophotocatalysts

Scarcity of energy and environmental pollution emerges as a prodigious threat for mankind since late 90s. It highlighted the necessity of sustainable development in the near future. The development of pollution-controlling technologies to clear out the environment and alternative green energy sources has become a topic of interest for many researchers. Semiconductor photocatalysis has the capacity to tie up the light energy (obtain from either natural or man-made sources) and making it accessible to anyone and thus, it appeared as a prime candidate in this array of sustainable development. The major domains covered by photocatalysis include splitting of hydrogen and oxygen from water and providing a green source of fuel, disruption of harmful pollutants, by imitating photosynthesis process of nature they can convert CO_2 into useful organic products, surface modification, improvement of water attracting properties, generation of electricity from light, etc. From the above utility of photocatalytic materials, it is observed that semiconductor photocatalysis can be emerged as a promising and adaptable contender with their multifolded applications in the area of environmental sustainability (Ahmad et al., 2023).

Again, from the last century, the carbon scaffold nanomaterials have emerged as an interesting area in research for their versatile advantages. Carbon scaffolds have the characteristics like electron carrying capacity, adaptive permeability, chemical and thermal stability, hydrophobic behavior, easy recovery from reaction mixture, and cost effectiveness. Such versatility helps them to become a frontrunner in the advancement of photolytic catalysis mechanism. The most commonly used nanomaterials are CNTs, graphene, CDs, porous carbon materials, etc. These nanomaterials are highly advantageous in a variety of catalytic applications (Khan, 2021; Kumar et al., 2017).

By observing the distinct potential of carbon-based nanostructures in the area of heterogeneous catalysis, the researchers try to extend their application in photocatalysis. Their studies lead to the conclusion that carbon scaffold-based nanophotocatalysts are highly important in multidimensional aspects. The integration of carbon scaffold with semiconductor photocatalyst produces some active catalytic components that accelerate the electron transfer capacity, induce charge separation, upgradation in photocatalytic activities and light absorption, etc. Apart from this, these photocatalysts have the prospect in innumerable environmental remediation technologies (Zhai et al., 2015; Zhao et al., 2015; Ng et al., 2012).

The integration of carbon-based scaffolds with semiconductor materials can develop a photocatalyst with various applications. All semiconductor materials are not suitable in this process. From various literature, it is found that the following are the most reported semiconductor materials that are used to combine with carbon scaffold photocatalysts:

i. **Titanium Dioxide (TiO_2):** From various studies titanium dioxide appeared as the most promising semiconductor material in the process of photocatalysis. It gives some tremendous properties like chemical stability, photo activity, and low cost. Apart from this, when titanium dioxide is added to carbon scaffold, it provides better absorption of light and separation of charge (Xu et al., 2019).
ii. **Zinc Oxide (ZnO):** Apart from titanium dioxide, another good candidate of semiconductor material that can provide excellent photocatalytic property is zinc oxide. The prime reason of such behavior is the wide bandgap of zinc oxide that can create electron-hole pairs on light irradiation. The carbon scaffolds integrated with zinc oxide can provide better stability and can induce the photocatalytic performance (Xu et al., 2019).
iii. **Metal Sulfides:** Metal sulfides like CdS, ZnS, etc. have the potential to be applied as a semiconductor material. Their band structures favor photocatalysis and allow the maximum utilization of the visible light. Metal sulfide-incorporated carbon bases have exclusive properties like prevention of aggregation, induce chemical stability, and thus provide excellent photocatalytic behavior (Zhang et al., 2022).
iv. **Metal-free Graphitic Carbon Nitride (G-C_3N_4):** These are the semiconductor materials that can enhance the photocatalytic efficiency when integrated with carbon scaffolds. They are metal free and have an appropriate bandgap. They can improve the utilization of light and also can improve the charge separation (Zhai et al., 2015; Zhao et al., 2015).
v. **Perovskite Materials:** Hybrid organic and inorganic perovskites are also useful semiconductor materials. They have unique optoelectronic properties. The combination of such perovskite material with carbon scaffolds can generate heterojunctions and lead to improved charge separation and hence photocatalytic activities (Deschler et al., 2019).
vi. **Other Metal Oxides:** Apart from zinc and titanium, some other metal oxides are also tested to be used as semiconductor material by many researchers. In this context, iron oxide (Fe_2O_3) and tungsten oxides (WO_3) are most widely used materials. Integrated metal oxide carbon scaffold has synergistic effects, resulting in excellent photocatalytic behavior (Xu et al., 2019).
vii. **Quantum Dot:** Semiconductor quantum dots (CdSe, PbS) are also considered favorable candidates for photocatalytic mechanism. Their excellent optoelectronic properties, adjustable bandgap, and high quantum yield make them suitable for photocatalytic processes. When integrated with carbon scaffolds, they can provide better light absorption, charge separation resulting in an excellent photocatalytic movement (Yan et al., 2011).

Here, we have highlighted only a few semiconductor materials commonly integrated with carbon scaffolds in photocatalysis. In the quest of definite applications, requirements and specific properties, the researchers will explore other semiconductor materials and their integration with carbon scaffolds to provide better photocatalytic performance.

Carbon scaffold nanophotocatalysts incorporate a variety of materials, each of these materials have some distinctive properties and utilization in photocatalysis. In the following section, we overview some prominent types:

i. **CNT-based Photocatalysts:** CNTs are the most common nanoparticles used in photocatalysis. Their broad surface area and high electrical conductivity make them ideal scaffolds for the incorporation of semiconductor materials.

   The semiconductor materials are integrated into the carbon scaffolds by deposition technique or coating techniques. In the deposition method, the semiconductor particles such as TiO_2 and ZnO are placed on the CNT's surface through chemical vaporization method or solution-based techniques. This incorporation provides some improved catalytic activity by enlarging the surface area for the catalytic mechanism. Furthermore, it enhances the charge separation capacity and excellent electrical transportation of CNTs (Georgakilas et al., 2007).

   In the coating process, the CNT surface is coated with the thin film of semiconductor materials to develop some heterojunctions. These junctions promote charge transfer and decrease the recombination losses. It improves the light absorption, allowing the utilization of broader range of solar radiation in the photocatalytic process. Apart from this, CNT photocatalysts have some other applications like evolution of hydrogen, degradation of pollutant, and reduction of CO_2 (Ren et al., 2021).

ii. **Photocatalysts Comprising Graphene-based Materials:** The 2D structure of graphene induces electrical conductivity and generates a high surface, which makes graphene a promising candidate in the photocatalytic platform. Here, the semiconductor nanomaterials are deposited onto the graphene sheets and accelerate the separation of charged particles and thus increase their photolytic action. Such photocatalysts are also useful in splitting of water to hydrogen and oxygen, pollution reduction, and dye-sensitized solar units (Bie et al., 2021).

   The semiconductor materials are incorporated by the following methods:

   a. In this method, chemical functionalization or adsorption techniques are used to anchor metal oxides (TiO_2 or ZnO) or sulfides (CdS) on to graphene sheets. Graphene can strongly interact with the semiconductor materials because of their large surface area and outstanding conductivity. This strong binding helps to promote charge transfer as well as photocatalytic activity.
   b. In this method, graphene is integrated into semiconductor nanocomposites. Here, it serves as a support for semiconductor particles. It improves the activity of semiconductor material in terms of stability and dispersibility. The high electron transport capacity of graphene enables excellent charge transfer and accelerates the photocatalytic behavior (Xiang et al., 2012b).

iii. **Carbon Dot-based Photocatalysts:** Carbon dot-based nanomaterials can be synthesized from various sources of carbon. They have unique optical as well as electronic properties; they can efficiently promote light absorption and charge transfer. CDs are widely used as a sensitizer in the semiconductor surface that provide improved light absorption as well as photoinduced electron transfer mechanism. Apart from this, it can also be used as a co-catalyst for semiconductor photocatalyst that facilitates the charge separation and can enhance the utilization of photogenerated electron transporter. In addition to this, carbon dot photocatalysts have some dynamic applications like reduce CO_2, reduce pollutant, water splitting, energy transfer to semiconductor, high photoluminescence quantum yield, optical properties and broad absorption spectra, etc. that induce their excellent photolytic activity (Asadzadeh-Khaneghah & Habibi-Yangjeh et al., 2020).

iv. **Photocatalysts Comprising Porous Carbon Material:** Porous carbon-based materials can be used efficiently in many photocatalytic reactions. This class of carbon materials have high surface area and controllable pore structures, which provide many active sites in the photolytic reactions. Aerogels and activated carbon are examples of porous carbon material, they have large surface area and can serve as a support for semiconductor nanocatalyst. The tunable porous structure increases mass transfer and increases the accessibility of reactants to active sites. Such integrations help to improve their properties to promote light absorption, charge transfer, and separation in the photocatalytic processes (Lopes et al., 2021; Chakraborty et al., 2022).

The porous carbon material can also be combined with hierarchical nanocomposites that provide better light harvesting, improved electron transport, and stability. Again, these types of nano semiconductor materials offer multiple pathways for charge transfer and enhance the utilization of photogenerated carriers. In addition to this, porous carbon-based nanophotocatalysts are also useful in water purification, air pollution control, and an energy storage (Lopes et al., 2021).

From the above study, it is observed that the nanophotocatalysts generated by the integration of semiconductor materials with carbon scaffolds represent a promising class of materials with advanced properties. This advance class of materials offer a variety of opportunities that can resolve many environmental and energy challenges to promote sustainable technological developments.

3.3 Industrial Synthesis Methods

The need for bulk-scale and industrial-grade photocatalysts has prompted researchers to consider carbon scaffold-based materials owing to their cost-effectiveness, environmental suitability, and sustainability. Carbon, in its various forms, has been widely utilized for a variety of applications, such as optical purposes, solar devices, etc. due to its exceptional electrochemical and physicochemical properties (Wang et al., 2022).

The flexibility and fascinating properties of the carbon scaffold-based nanostructures have garnered significant attention in the area of catalytic processes involving visible-light. The compositions and structural features of these structures can be altered to get enhanced properties. A number of fabrication methods have been designed to create various photocatalysts based on carbon materials with different structures and arrangements. These advancements have the potential to be used in many catalytic processes where visible light is used as a source of energy (Khan, 2021).

The remarkable physiochemical properties of carbon scaffold-based nanomaterials (CNMs) make them suitable candidates in photocatalysis. Examples of such CNMs include CDs, CNTs, graphene, carbon nanofibers (CNFs), graphitic carbon nitride (g-C_3N_4), fullerene (C_{60}), and others. These CNMs exhibit excellent structural integrity against high temperature and reactive chemicals. The high surface area, optical properties, excellent electrical conductivity, and superior mechanical strength of CNMs play a crucial role in photocatalytic applications. Different fabrication techniques like ball-milling, arc discharge, laser ablation, and CVD are employed to create CNMs. CNMs increase the efficiency of the active sites of photocatalysts by serving as carriers, improving adsorption sites, and acting as electron acceptors or transmission channels.

There are different fabrication techniques that have been extensively explored to synthesize several carbon scaffold-based materials as potential photocatalysts:

3.3.1 Synthesis of Low-dimensional Carbon Nanostructures (Carbon Dots)

Depending on the synthesis techniques, structures, and properties, CDs can be classified into different types:

Carbon Quantum Dots (CQDs): The size of the small carbon nanoparticles, CQDs are in general less than 10 nm, often synthesized from carbon-rich precursors.

Graphene Quantum Dots (GQDs): These are CDs derived from graphene or graphene oxide sheets, possessing unique electronic and optical properties.

Carbonized Polymer Dots (CPDs): These CDs are formed from the carbonization of polymer precursors, offering tunable properties and biocompatibility.

The synthesis of CDs involves optimizing various parameters such as time, temperature, and voltage. After synthesis, further purification steps are often required to obtain the specific properties of CDs. Purification methods can range from simple treatments like centrifugation, dialysis, and filtration to more complex techniques such as high-performance liquid chromatography (HPLC), silica gel column chromatography, electrophoresis, etc. Here's an overview of some common synthesis methods:

3.3.1.1 Hydrothermal/Solvothermal Synthesis

The most common synthetic method for CDs is hydrothermal. Here, the raw materials are dissolved in water or other suitable solvent and then heated in a Teflon-lined stainless autoclave at a definite temperature over a period. This method offers several advantages, including simplicity, uniform synthesis with high yields, and the ability to easily introduce other elements or adjust the diversity of CDs.

3.3.1.2 Microwave Synthesis

The microwave synthesis method of CDs provides several advantages such as shorter reaction times and higher yields compared to conventional heating methods. Microwave irradiation technique is used to rapidly heat the reaction mixture to form the CDs.

3.3.1.3 Electrochemical Synthesis

This method facilitates exact controlling of the size and properties of CDs. Here, synthesis of CDs is done by electrochemically oxidizing carbon precursors, such as graphite electrodes, in a suitable electrolyte solution. Although less common, CQDs can be prepared through electrochemical carbonization. Zhang and colleagues reported the electrochemical carbonization of low molecular weight alcohols using platinum electrodes. They observed the increase in graphitization and size CQDs against the applied potential. These CQDs exhibited high sorption capacity for pollutants due to their abundant functionality, high surface area, and stability making them valuable for environmental pollution management.

3.3.1.4 Microplasma Method

This method allows to control the size and morphology of CDs. This synthetic method can be carried out at relatively low temperatures. Microplasma discharge is used to generate highly energetic species, which react with carbon precursors to form CDs.

3.3.1.5 Chemical Oxidation

Chemical oxidation methods involve the oxidation of carbonaceous materials, such as carbon black or graphite, using strong oxidizing agents. This approach allows for the preparation of CDs of well-defined sizes and properties.

In summary, the synthesis and purification of CDs involve optimizing various parameters and employing different methods depending on the desired characteristics and the specific applications requirements. Hydrothermal/solvothermal synthesis is widely used due to its simplicity and effectiveness, while electrochemical methods offer alternative routes for specific types of CDs with unique properties.

3.3.2 Synthesis of High-dimensional Carbon Nanomaterials

The synthesis of high-dimensional carbon nanomaterials involves various methods tailored to produce structures such as graphene, CNTs, graphitic carbon nitride ($g\text{-}C_3N_4$), CNFs, and other carbon-based structures with extended dimensions (Wang et al., 2009). Here are some common synthesis techniques for different high-dimensional carbon nanomaterials:

3.3.2.1 Graphene

The carbon atoms in grapheme are arranged is a single layer 2D honeycomb lattice. Graphene exhibits high surface area, exceptional mechanical as well as electrical conductivity, making them promising candidates for photocatalytic applications. Consequently, metal nanoparticles, especially Au and Ag nanoparticles, are well suited for enhancing the photo-based properties of graphene and can be effectively utilized for photocatalytic applications. By combining graphene with metal nanoparticles, particularly Au and Ag, synergistic effects can be achieved, leading to improved photocatalytic efficiency and expanded applications in various catalytic processes. Some synthesis methods for graphene are:

3.3.2.1.1 Mechanical Exfoliation

Also known as the "Scotch tape method," this technique involves repeatedly peeling off graphene layers from bulk graphite using adhesive tape, resulting in monolayer or few-layer graphene.

3.3.2.1.2 Chemical Vapor Deposition

To grow large-area graphene films on metal substrates using hydrocarbons as precursors and metal catalysts, CVD can be used.

3.3.2.1.3 Chemical Reduction of Graphene Oxide

Graphene oxide, derived from graphite oxide, can be chemically reduced to produce reduced graphene oxide (rGO) sheets. This method offers a scalable approach to obtain graphene-like materials.

3.3.2.2 Carbon Nanotubes

CNTs are cylindrical nanostructures composed of rolled-up graphene sheets. CNTs have high aspect ratios, exceptional electrical conductivity, in addition to large surface areas, making them suitable for various photocatalytic processes. To enhance photocatalytic efficiency, several procedures have been investigated, including:

3.3.2.2.1 Increasing the Surface Area for Metal on CNT Surfaces

This can be achieved through methods such as deposition techniques or surface functionalization, which aim to maximize the contact area between the metal nanostructures and the CNTs, thereby promoting efficient charge transfer and utilization of light energy.

3.3.2.2.2 Introducing Defects in Carbon-based Materials to Generate Space-charge Separation

By creating defects in carbon-based materials, such as vacancies or heteroatom doping, it is possible to generate additional active sites for photocatalytic reactions and facilitate the separation of photogenerated charge carriers. This helps reduce the recombination rate of charges, leading to enhanced photocatalytic performance.

3.3.2.2.3 Incorporating Metal Nanostructures onto CNTs to Prevent Agglomeration

Metal nanostructures can be dispersed on CNT surfaces through various methods, such as deposition or chemical functionalization. Not only does it help to prevent the agglomeration of metal particles, which could inhibit catalytic activity but also provides synergistic effects between the metal nanostructures and CNTs, leading to improved photocatalytic properties.

3.3.2.2.4 Chemical Vapor Deposition

CVD is one of the most widely used methods for CNT synthesis. It involves the catalytic decomposition of hydrocarbons at elevated temperatures using metal catalysts such as iron, nickel, or cobalt.

3.3.2.2.5 Arc Discharge

This method is carried out in an inert gas atmosphere of high temperature and high pressure to form CNTs. A high-current electric arc is produced between two parallel graphite electrodes that produce the extreme high temperature and pressure.

3.3.2.2.6 Laser Ablation

Laser ablation involves irradiating a carbon target (e.g., graphite) with a high-power laser beam in an inert gas atmosphere. The vaporized carbon condenses to form CNTs.

3.3.2.2.7 Template-assisted Growth

CNTs can be synthesized using templates such as porous alumina or zeolites, where carbon is deposited inside the nanopores to form nanotubes.

3.3.2.3 Graphitic Carbon Nitride

The structure of 2D carbon-based material, g-C_3N_4 is uniquely resembling graphite. It exhibits a suitable bandgap for photocatalysis, high chemical stability, and good visible light absorption and thus become a promising photocatalyst for numerous applications.

Zhang et al. proposed a simple and cost-effective approach to enhance the performance of photocatalytic degradation of g-C_3N_4. They employed sacrificial templates to introduce pores within the g-C_3N_4 structure, thereby increasing its porosity as well as surface area. This modification enhances photocatalytic efficiency and also mass transfer capability of engineered g-C_3N_4, leading to improved performance in photocatalytic degradation reactions.

To enhance the photon harvesting capability of g-C_3N_4, researchers are engaged in various bandgap engineering approaches, and also atom-level as well as molecular-level doping techniques. These strategies aim to modify the electronic structure of g-C_3N_4, thereby improving its photocatalytic performance. Some of the commonly employed methods are:

3.3.2.3.1 Bandgap Engineering
This approach involves tailoring the bandgap of g-C_3N_4 by modifying its structure or composition. For instance, heteroatom doping (e.g., nitrogen or sulfur doping) can alter the electronic band structure, leading to improved absorption of a broader range of photons, including visible light. Additionally, strategies such as surface modification or nanoscale engineering can be employed to tune the bandgap and optimize the light absorption properties of g-C_3N_4.

3.3.2.3.2 Atom-level Doping
Here, foreign atoms are introduced into g-C_3N_4 lattice at atomic scale to significantly influence the electronic properties. For example, nitrogen doping has been widely investigated as it can introduce additional energy levels within the bandgap, facilitating the separation of photogenerated charge carriers and enhancing photocatalytic activity. Other dopants such as boron, sulfur, and phosphorus have also been explored for their potential to improve performance.

3.3.2.3.3 Molecular-level Doping
This involves incorporating molecular species, such as organic molecules or metal complexes, onto the surface or within the pores of g-C_3N_4. These molecules can act as sensitizers or co-catalysts, enhancing light absorption, charge separation, and catalytic activity. Molecular doping strategies offer exact tuning the functionality and composition of the photocatalyst, enabling tailored design for specific applications.

3.3.2.4 Other Carbon-based Structures

3.3.2.4.1 Carbon Nanofibers
CNFs can be synthesized via CVD using metal catalysts on substrates such as silicon wafers or alumina fibers. Hydrocarbon gases are decomposed at high temperatures, leading to the growth of CNFs on the catalyst particles. CNFs can also be synthesized using electrospinning. Here, high voltage is applied to a polymer melt or solution leading to the formation of continuous nanofibers. Carbonization of these polymer nanofibers at high temperatures yields CNFs.

3.3.2.4.2 Carbon Nanowires
These can be synthesized using methods similar to CNT growth but with different catalysts and growth conditions to achieve wire-like structures.

3.3.2.4.3 Carbon Nanobelts, Nanoribbons
These structures can be synthesized by CVD or template-assisted methods, where the shape and dimensions of the template dictate the final structure.

Overall, the synthesis of high-dimensional carbon nanomaterials involves a combination of techniques such as CVD, chemical reduction, template-assisted growth, and electrospinning, tailored to the specific requirements of the desired carbon nanostructure. Each method offers unique advantages in terms of scalability, control over dimensions, and properties of the synthesized materials.

3.4 Degradation Mechanism

In general, semiconductor materials (e.g., TiO_2, ZnO, etc.) can act as photocatalysts as they have filled valence bands and vacant conduction bands, which facilitate the transfer of electrons from the valence band to the conduction band when exposed to solar light. Exposure to light causes charge excitation and separation, migration of electrons to the conduction band, where surface redox reactions occur, followed by charge recombination. The performance of inorganic photocatalysts is affected by low generation rate of electrons and holes, and their recombination, thus leading to low quantum efficiency. Further, these semiconductors also have relatively wide bandgaps due to which they respond to a UV-light, which is a small fraction of solar radiation. Also, the aggregation of the photocatalyst particles reduces their active surface area and hinders oxidation reactions, possibly yielding harmful byproducts. The photocatalytic activity can be enhanced if photocatalysts are used in conjunction with carbon scaffolds such as graphene. These composites of graphene with semiconductors facilitates improve charge generation by effectively separating photogenerated electron-hole pairs thus reducing charge recombination. The degradation of organic pollutants using semiconductor photocatalysts (graphene-based) can be illustrated with the following schematic diagram (Figure 3.2).

A brief description of degradation of pollutants by carbon scaffold-based photocatalyst is given below:

3.4.1 Graphene/MO_2 (M=Ti, Zn, etc.) Photocatalysts

TiO_2 is extensively studied for water purification because of its strong oxidizing capabilities and chemical stability. However, its rapid charge recombination and specific activation under UV light due to its wide bandgap restrict its cost-effective application. Graphene–TiO_2 nanocomposites have demonstrated improved photodegradation performance, as graphene enhances surface area and charge separation.

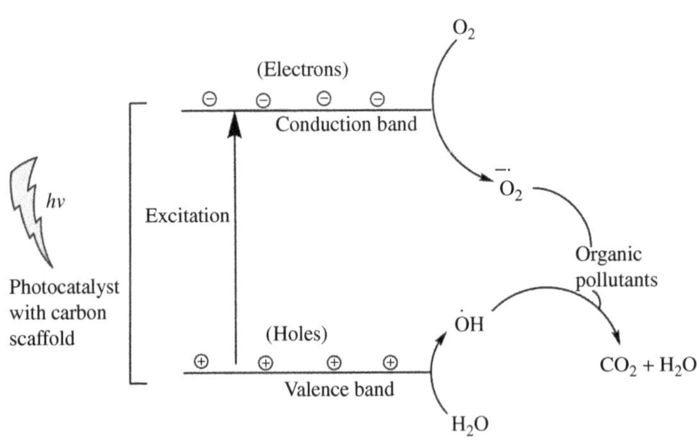

Figure 3.2 Photocatalytic degradation of pollutants.

This graphene–TiO_2 composite has been effectively used in the removal of pollutants like benzene and dyes, including Rhodamine B and methylene blue.

For instance, the increased photodegradation of methylene blue (MB) is due to multitude of factors including the monolayer graphene structure, well-distributed TiO_2 nanoparticles on the graphene matrix, and the role of graphene oxide as a charge collector separator. This facilitates the transfer of charge carriers and promotes the generation of superoxide and hydroxyl radicals, ultimately leading to its degradation of methylene blue (Zhou et al., 2011; Garrafa-Gálvez et al., 2019). Similarly, graphene–ZnO composites have shown enhanced photocatalytic performance, which can be attributed to graphene's large surface area and superior conductivity. Different synthesis methods such as hydrothermal, ultrasonic, spin-coating have been employed to prepare these composites resulting in improved degradation of pollutants under UV and visible light. For example, a ZnO/RGO composite, synthesized through a one-step photochemical method, achieved a much higher photodegradation efficiency – 12 times greater – for methylene blue under both visible and UV light compared to pristine (Xue & Zou, 2018). Additionally, metal oxides photocatalysts such as NiO, ZrO_2, and SnO_2 exhibit wide bandgaps, while metal oxides such as In_2O_3, CeO_2 possess slightly narrower bandgaps allowing for visible light absorption. In both cases, integration with graphene further reduces the bandgap, thereby enhancing photocatalytic degradation performance.

3.4.2 Graphitic Carbon Nitride

Graphitic carbon nitride (g-C_3N_4) and graphene oxide composites have been explored for environmental remediation. It has a bandgap of 2.7 eV, and its use in photocatalytic degradation is limited by the rapid recombination of electron-hole pairs. Wan et al. developed a recyclable C_3N_4/graphene oxide aerogel composite, which showed excellent adsorption properties and visible-light photocatalytic activity for the removal of organic pollutants. Recently, (g-C_3N_4)/GO nanosheets have been applied to treat organic wastewaters containing hexavalent uranium (U(VI)) under simulated sunlight, showing improved U(VI) removal and tannic acid (TA) degradation compared to standalone g-C_3N_4 and GO (Chenet al., 2020).

3.4.3 Carbon Dots

CDs have been used as photocatalysts in various applications, including the degradation of organic compounds and hydrogen production through water splitting. CDs are used both independently and with other materials (like TiO_2) to degrade various organic pollutants, such as dyes, antibiotics, etc. They offer various advantages such as photochemical stability, fine-tuned structure, surface functionalization, low-cost along with strong light absorption. It is worth mentioning that heteroatom doping can be effectively used in improving photocatalytic performance of CDs by enhancing their optical properties and photoluminescence. For instance, doping with nitrogen and oxygen atoms in CDs reduce recombination rates and promote the generation of reactive oxygen species (ROS) at the surface by inducing an electric field, which causes the bands near the surface of CDs to bend upward. Additionally, surface charge tuning through N-doping can enhance the adsorption of pollutants via electrostatic interactions. For instance, N-doped CDs are reported to be efficient photodegradation of various dyes and small organic compounds (Sabet & Mahdavi, 2019; Zhanga et al., 2020). CD-driven photocatalytic degradation mechanisms closely resemble those of semiconductors like TiO_2. Photons excite electrons across the CDs' bandgap, generating electron-hole pairs (e^-h^+), which then migrate to active sites on the CDs' surface. These charge carriers react with H_2O and O_2, producing ROS such as •OH and $\bullet O_2^-$, known for their ability to degrade organic pollutants like dyes. Some studies suggest direct degradation of pollutants by holes (h^+), supported by investigations using photoluminescence quenching with rhodamine B. Apart from above two pathways, singlet $^1O_2$ production via energy transfer from photoexcited CDs may also contribute toward degradation of organic pollutants.

Thus, the mechanistic advantages of carbon scaffold-based photocatalysts, including improved charge separation, enhanced light absorption, and efficient ROS generation, highlight their potential for advancing environmental remediation and sustainable energy applications. These insights drive further exploration and optimization of carbon-based photocatalytic systems for real-world pollutant degradation.

3.5 Sustainability and Applications

The irreversible degradation of the environment in the name of development has compelled the world to adopt a new approach to development without destroying the ecosystem, i.e., sustainable development. Sustainability means the application of the resources for the current generation and making sure that they are available for the coming generations (United Nations, 1987). The excessive exploitation of natural resources leads to deforestation, climate change, soil erosion, and global warming as a result sustainable development or sustainability is becoming the need of modern time (Lampert, 2019). As the current generation is becoming more aware of their surroundings, we are observing a revolutionary shift not only in the research and development sectors but also in the business sectors. In this respect for the business sectors, environmental, social, and corporate governance (ESG) acts as a guiding framework to account for the sustainability parameter of any business model. The economic growth of any country is largely determined by industrial development, which in turn depends on the advancement of science and technology (de Souza Barbosa et al., 2023). However, it should be kept in mind that to become the winner in the long race, the industries have to adhere to the principles of sustainability. Innovation plays a critical role to achieve sustainable development of science and technology (Sachs et al., 2024). In chemical industries, catalysts have pivotal roles as they not only increase the rates of reactions but in many cases, they make the reactions more specific thereby reducing the byproducts and making the process more environmentally friendly. In this regard, purification of water and air via photocatalytic decomposition of the toxic chemicals present in water, air to less toxic chemicals ca also be considered (Horváth et al., 2022; He et al., 2021). In the past few decades, with the advent of nanoscience, groundbreaking developments are being observed in almost all the branches of science and technology (Sharma et al., 2022b). Catalysis has also felt the tremor of this paradigm shift. Nanocatalysts are collections of nanoparticles or nanostructures, and they are more active than conventional heterogeneous catalysts. It is well known that the properties of nanoparticles are distinct from their bulk counterparts (Saikia & Ramakrishnan, 2023).

Photocatalysts are semiconducting materials that change the rate of reactions when exposed to light. Though photocatalysts particularly nanophotocatalysts an emerging field, their roles in green synthesis have been extensively studied. The applications of photocatalysts encompass a wide range of photochemical reactions. They are used in the production of hydrogen from water, in wastewater treatment, in air purification, in photovoltaics, and in solar fuel generation from the reduction of CO_2 (Hassaan et al., 2023). This section will give a brief account of the applications of nanophotocatalysts particularly carbon scaffold-based nanophotocatalysts such as graphene-based photocatalysts, porous organic polymers, and their applications in clean energy production, carbon sequestration, and environmental remediation.

Clean energy has become a necessity of the world, and photocatalysts play a pivotal role in achieving this goal, i.e., to harvest clean energy from renewable sources using sustainable methods. At present the major source of energy is fossil fuels, a nonrenewable source, and excessive consumption of them has disturbed the ecological balance (Takanabe et al., 2017). Production of hydrogen seems to be the most promising technology to meet global energy demands as it has a higher energy density than conventional fuels such as fossil fuels. Presently, at the industrial scale, hydrogen is produced from fossil fuels via the syn gas route, which is not a sustainable approach considering its impact on the environment (Megía, 2021). Thus, considering the abundance of water on the earth's surface it can be proposed that the photolysis of water may be the holy grail approach for hydrogen production toward sustainable energy. In 1971 Fujishima and Honda reported the first-ever successful photoelectrochemical splitting of water to hydrogen and oxygen using light and TiO_2 as a photoelectrode after that there has been an exhaustive effort to pave the way toward clean energy either by designing novel photocatalysts for water splitting (Fujishima et al., 1972). These include inorganic solids, polyoxometalates, metal–organic frameworks, graphene, fullerene, and graphitic nitride-based photocatalysts. There has been very limited success in the photolysis of water for several reasons some of which are unfavorable thermodynamics, sluggish kinetics, and recombination of electrons and holes (Wang & Domen, 2020). Thus, photocatalytic production of hydrogen from water is still an active research area.

The first-ever use of organic materials for solar light-driven water splitting was reported by Shozo Yanagida and coworkers in 1990. They used ploy(p-phenylene)-based photocatalysts and UV radiation to split water after that several organic polymers were designed for efficient water splitting (Yanagida et al., 1985). In one of the studies, Liu et al. designed acetylene-linked conjugated polymers to study the sacrificial photocatalytic hydrogen evolution from water (L. Lunjie et al., 2021). Conjugated organic polymers or organic semiconductors despite their tunable properties received less attention as photocatalysts for water splitting as compared to their inorganic semiconductors. It was only after the discovery of graphitic nitride (g-C_3N_4) that organic-based semiconductors received significant attention and breakthroughs in water photolysis using visible light from sunlight. The discovery of g-C_3N_4 due to its various unique properties such as small bandgap (2.7 eV), ease of synthesis, and environmentally benign make it an ideal candidate for further exploration in the area of sustainable and clean energy production (Dong et al., 2022). However, the pristine g-C_3N_4 has some bottlenecks such as limited active sites, high charge carrier recombination, and low specific surface areas, which have to be overcome before it is used as a catalyst for water photolysis. The physicochemical properties of the g-C_3N_4 can be altered via elemental doping, which alters the optical bandgap of the material (Dong et al., 2014). For instance, incorporation of Co not only results in the narrowing of the bandgap but also increases the specific surface areas thereby making the separation of photoinduced carriers more efficient (Chen et al., 2017). In another study, it was observed that elemental doping by Fe not only changes the absorption spectrum but also makes the materials more efficient for hydrogen evolution (Gao et al., 2017). Similarly, results were obtained when g-C_3N_4 was doped with nonmetal, for instance, Jiang et al. observed that the photocatalytic performance was significantly increased after g-C_3N_4 was doped with oxygen (Jiang et al., 2019). Similarly, porous organic polymers have shown promising applications in the photocatalytic water splitting (Xu et al., 2018a). They are covalently connected polymer materials with high porosity and specific surface area. Porous organic polymers are tunable, which makes them better than zeolites, silicates, metal–organic frameworks, and metal–organic cages (Mohamed et al., 2022). Conjugated microporous polymers (Jiang et al., 2008), covalent triazine frameworks (Jiang et al., 2018), and covalent organic frameworks (Cote, 2005) are the three classes of porous organic polymers. To enhance the photocatalytic activities of conjugated microporous polymers, several approaches were designed (Li et al., 2016). It was observed by Jiang's group that by adjusting the ratio of donor and acceptor units, donor-acceptor conjugated microporous polymers exhibit excellent tunable photocatalytic activity. By adjusting the ratios of donor and acceptors, various conjugated microporous polymers were further designed (Xu et al., 2018b). In one such study, it was found that incorporation nitrogen-doped acceptor ligand (Li et al., 2016) and the introduction of the alkynyl group (Zhang et al., 2017) into the conjugated polymers effectively enhanced the activity of photocatalysts for water splitting. Covalent triazine frameworks due to their electron-withdrawing character have shown interesting photocatalytic activity for water splitting under visible light irradiation (Qu et al., 2016; Mier et al., 2017). For some of the synthesized covalent triazine frameworks, the photocatalytic activity for the hydrogen evolution reaction was competitive to the g-C_3N_4 (Wang et al., 2009) and titanium dioxide nanocrystals (Chen et al., 2011). However, there was not much success in the oxygen evolution reaction for these materials. The covalent organic framework has uniform porosity and high crystallinity as compared to other porous organic polymers (Ding & Wang, 2013).

They are a new class of crystalline and porous photocatalysts for water splitting and have shown a promising visible-light-driven hydrogen evolution reaction in the presence of a sacrificial electron donor (Banerjee et al., 2018). In a recent study, Yang et al. showed that by systematically engineering the N-sites, morphology, and architecture of the β-ketoamine covalent organic framework, they achieved the overall water-splitting activity (Yang et al., 2023). Persistent efforts have been made to facilitate the hydrogen and oxygen evolution reactions using porous organic polymers (Xu et al., 2018). Though in their early stages of development, porous organic polymers are up against cutting-edge materials like metal oxides and graphitic carbon nitrides in the water-splitting market. Without a doubt, the inclusion of porous organic polymers in the family of photocatalysts will expand our possibilities for investigating the diverse range of photocatalysts and aid in the development of novel, environmentally friendly photocatalysts by the scientific community.

Carbon dioxide sequestration is another area where nanophotocatalysts have potential applications. Various studies have established the correlation between global warming and greenhouse gas emissions (Humphrey et al., 2018). Among the greenhouse gases H_2O, CH_4, and CO_2, CO_2 alone is responsible for more than 50% of the greenhouse effect. The accumulation of CO_2 in the atmosphere has catastrophic effects on the climate such as global warming, ocean acidification, desertification, etc. One way to mitigate the negative effects of rising CO_2 levels is to cut the emission of CO_2 from nonrenewable energy sources such as fossil fuels. This can be achieved by switching our energy sources from nonrenewable to renewable such as solar and wind energy. The second possible way is the carbon neutral energy scheme, and, in this scheme, the carbon dioxide will provide raw materials for chemicals. This will not only reduce the CO_2 levels in the environment but also help to mitigate the energy crisis and balance the CO_2 levels in the atmosphere. The photochemical reduction of CO_2 to lower carbon fuels is possible and there has been an exhaustive study for the photocatalyst design to carry out the reduction reaction. One such material is the metal–organic frameworks (MOFs), which are crystalline and porous nanocatalysts. They are composed of inorganic metal centers connected with the organic centers or ligands (Zhang et al., 2023). In the photocatalytic reduction of CO_2, MOFs can act as photocatalysts, co-catalysts, and hosts or scaffolds. It was found that functionalization can improve the photocatalytic activity of the MOFs (Hussain et al., 2021) and the first case of visible-light-driven photocatalytic reduction of CO_2 was in 2011 reported by Lin and co-workers, where they functionalized the Ti-MOF by amine group (Wang et al., 2011). In another study, an Ir-based ligand was designed and incorporated in the MOFs and observed that the incorporation not only enhanced the photocatalytic activity of the MOFs for CO_2 reduction but also increased the stability and recyclability (Pope et al., 2004). Several studies were carried out on Fe, Zr, Co, Co-based MOFs and they show a rise in the photocatalytic activity of the MOFs (Hayat et al., 2024). Thus, MOFs are promising photocatalysts for the capturing and reduction of CO_2. Their porosity, stability, and tunable character have gained significant attention over the past few decades.

Wastewater treatment is another potential application of photocatalysts (Ren et al., 2021). An efficient and sustainable oxidation technology for wastewater treatment is photocatalysis (Ren et al., 2021). Conventional wastewater treatment technologies consume large amounts of energy. Photocatalysts have the potential to oxidize the organic and inorganic pollutants in wastewater to harmless substances. The organic pollutants in water are categorized into dyes, surfactants, plasticizers, organohalides, etc. Most of the organic pollutants are toxic and thermodynamically stable. In a recent study, composite material of TiO_2/ZnO/rGO showed excellent activity in the degradation of methylene blue, methyl orange, and Rhodamine B under UV irradiation (Nguyen et al., 2020). The biochar-supported K-doped g-C_3N_4 composites were found to have excellent photocatalytic activity toward naphthalene degradation under visible light irradiation (Li & Lin, 2020). In another study, nanocomposites such as Au@Ni/rGO have shown excellent photocatalytic activity under natural sunlight for the degradation of phenols and their derivatives (Darabdhara & Das, 2019). In a recent study, it was found that α-Fe_2O_3/g-C_3N_4 (Kadi et al., 2020a) and CuO/g-C_3N_4 (Kadi et al., 2020b) nanocomposites were able to photo-reduced Hg (II). In the medical field also carbon-based nanocatalyst has potential applications. For instance, g-C_3N_4@Co-TiO_2 nanofiber was found to be effective in the inactivation of *E. coli* under visible light irradiation (Song et al., 2020).

Thus, the applications of carbon-based nanophotocatalysts are not limited to a particular area. Their applications cover almost all aspects of contemporary science. Though the subject is relatively new, further rational development of photocatalysts is required to attain sustainability.

3.6 Challenges and Future Scope

The gradual rise in the use of carbon-based nanomaterials (CNMs) for their application in environmental remediation, energy conversion, sustainable development, and industrial processes has become imperative in recent years. These materials have their unique identity and offer advantages in terms of their high surface area, chemical stability, and modifiable attributes, which make them attractive materials for easy customization and usage according to the requirements of various photocatalytic processes (Oliveira et al., 2017). Over the years, numerous carbon scaffold-based nanostructures have been developed, such as fullerene, CNTs, graphene, mesoporous carbon, and others, which act as excellent support materials for photocatalysis. Their unique properties make them a material in high demand (Fu et al., 2023). Some of the major requirements for CNMs to be used as a photocatalyst are such that their chemical interaction should render strong chemical and photo stability and must provide economic viability. These carbon-based nanostructured materials have been used as photocatalysts in many important applications such as the degradation of organic pollutants, wastewater treatment, especially for recovering heavy metals, air purification, energy storage, biomedical applications, electronics, and others. CNMs effectively break down the pollutants into harmless byproducts through photocatalytic reactions. They are used for efficient selective chemical synthetic methodologies with reduced energy consumption and environmental impact. For instance, many new methods have been developed recently for photocatalytic hydrogen production using CNMs. However, there still lies a lot of challenges in achieving maximum photocatalytic hydrogen production. Factors such as charge carrier recombination, degradation efficiency, charge separation, and interfacial charge transfer inhibition significantly reduces the effect of the photocatalytic process on exposure to visible light. Further, low hydrogen storage is also an important challenge as their storage requires low volumetric hydrogen to be compressed to either liquid or gas. The energy required to store hydrogen escalates the cost of production of photocatalytic H_2 (Mohapatra et al., 2023). Thus, some of the common challenges that hamper the effectiveness of CNMs in various fields include efficiency enhancement, visible light utilization, stability, durability, selective catalysis, and scalable synthesis.

CNMs have to be structurally more stable as they undergo harsh reaction conditions. It is generally seen that carbon materials are stable but the integration of semiconductor nanoparticles onto the carbon scaffolds leads to the leaching of active sites, which degrades the performance of the photocatalyst. Metals like Ag, Au, Pd, and Pt and their various possible combinations with carbon scaffolds have shown to be very useful in various photocatalytic applications. The metal nanomaterial carbon framework combination reduces the charge recombination rate of electron-hole pairs and also makes the surface of the photocatalyst hydrophobic (Somasundaram et al., 2019). For example, it was reported that loading of AgNPs on graphene sheets significantly improved the photocatalytic property of the graphene sheets (Ehtisham Khan et al., 2015). Metal doping onto the g-C_3N_4 sheets can lower the bandgap to 2.67–1.58 eV by tuning the electronic structure of g-C_3N_4. It was also found that graphene when doped with nitrogen shows a very high oxygen reduction reaction activity and enhances the durability of the material with better electrocatalytic activity than commercial Pt/C (Khan, 2021). The future scope lies in developing cost-effective novel doping strategies to enhance the properties of CNM photocatalysts for specific applications.

Although carbon materials have a high surface area for photocatalytic reactions, the efficiency of charge separation and transfer at the semiconductor-carbon interface remains a critical issue. There is a need to increase the interaction between the semiconductor nanoparticles and the carbon scaffold to facilitate efficient charge transfer for improving photocatalytic performance. A wide variety of photocatalysts such as titanium dioxide and zinc oxide show their optimal activity only under UV light irradiation. The development of a more efficient CNM photocatalyst that utilizes visible light is a challenge that needs attention for achieving efficient solar-driven photocatalysts. For instance, g-C_3N_4 materials show high stability at 600 °C in air and have a tunable bandgap of 1.8–2.7 eV, which enables it to harness the visible light at 460–698 nm wavelength (Khan, 2021).

In many photocatalytic reactions, especially those in organic synthesis and pollutant degradation, selectivity plays a crucial role. But CNM photocatalysts show deficiencies in selectivity, resulting in low yields, and formation of unwanted byproducts or incomplete conversion of reactants. It is a challenge to devise strategies aimed at augmenting the selectivity of these catalysts toward specific target reactions while minimizing side reactions, thereby ensuring their effective utilization in practical applications. Currently, many of the synthetic routes involve complex and multiple steps and require expensive precursors, which puts a limit on their scalability and economic viability. The focus should be on developing newer efficient synthetic routes that promote bulk production of CNMs while maintaining the desired chemical and physical properties of the photocatalyst, for their universal application.

Carbon scaffold restructuring can further increase its surface area and provide additional active sites for catalytic reactions. Synthesis of porous carbon structures with specific morphologies can improve mass transport and facilitate efficient reaction diffusion, leading to enhanced photocatalytic performance. For example, activated carbon that is a highly porous material with having vast surface area acts as an excellent photocatalyst that can capture organic molecules from air as well as from water. Combining activated carbon with metal nanoparticles diminishes the charge recombination of photo-generated electron-hole pairs and boosts the photocatalytic activity of the resulting nanocomposite (Khan, 2021).

Another aspect of improving the performance of CNM photocatalysts is by considerable modifications of their surface chemistry. Functional groups such as hydroxyl, carboxyl, or amine and others can be incorporated into the carbon scaffold surface to improve the dispersion and stability of semiconductor nanoparticles and promote charge transfer processes. Future studies should investigate advanced nano structuring techniques to design CNM photocatalysts with new structural morphology showing optimized properties. It has been demonstrated, for instance, that surface modification of CNTs through particular chemical functionalization greatly enhances their dispersibility, selectivity, sorption characteristics, and surface-to-volume ratios (Alamry et al., 2022).

The most challenging aspect of CNM photocatalyst is the high production cost and mass production of these photocatalysts. Most of the synthetic methods are still confined to the laboratory stage. The transformation of the laboratory techniques to an industrial mass production stage can be achieved by a better understanding of the reaction dynamics and mechanistic pathways. With the advancement of characterization and spectroscopic techniques, it is desirable to develop in-situ methods that enable real-time monitoring of structural changes, surface reactions, and charge carrier dynamics during photocatalysis, which can help understand the mechanistic pathways, reaction kinetics and also elucidate the underlying principles governing their performance. Spectroscopic techniques such as time-resolved laser fluorescence spectroscopy, XAFS spectroscopy, etc. provide insights into the structure and composition of metal ions on nanomaterials, including bond distances and coordination numbers. These details are crucial for distinguishing sorption mechanisms such as inner- or outer-sphere surface complexation, co-precipitation, and oxidation/reduction processes, etc. (Tan et al., 2010; Wang et al., 2022).

To conclude, the scientific community anticipates significant advancements in the performance of CNM photocatalysts, which pose a substantial challenge. Clarifying catalysis mechanisms is essential to enhance their performance. It is desirable to develop practical approaches for fabrication, scalability in manufacturing, and commercial viability. Establishing novel methods utilizing environmentally friendly and sustainable carbon-based photocatalysts with broader applicability is an absolute necessity.

3.7 Conclusion

In conclusion, carbon scaffold-based novel nanophotocatalysts have enhanced photocatalytic activity, outstanding thermal and mechanical stability, and most importantly possess some unique properties, due to which they are becoming potential and superior candidates over the existing traditional materials. Starting from the discovery of photocatalysis, the concept and importance of CNMs have been covered and explained with suitable examples. The book chapter also demonstrates the important photocatalytic activity of carbon nanophotocatalysts for some important transformations such as dye degradation, toxic material decomposition, reduction of carbon footprint, water splitting, wastewater treatment, etc., which are presently considered in the domain of clean technology in the real world. In addition, mechanistic investigation details for some nano-photocatalytic processes have been described, which in turn can help readers to have a clear insight into this topic. However, vast amount of research is going on across the globe for the efficient designing of advanced carbon-based nanophotocatalyst, which can solve some actual problems encountered in our daily lives and hence can address some global challenges.

References

Ahmad, I., Zou, Y., Yan, J. et al. (2023). Semiconductor photocatalysts: A critical review highlighting the various strategies to boost the photocatalytic performances for diverse applications. *Advances in Colloid and Interface Science*, 311, 102830.

Al Mahmud, M. Z. (2023). A concise review of nanoparticles utilized energy storage and conservation. *Journal of Nanomaterials*, Article ID 5432099 https://doi.org/10.1155/2023/5432099. Academic Editor : Vidya Nand Singh.

Ali, A., Koloor, S. S. R., Alshehri, A. H., & Arockiarajan, A. (2023). Carbon nanotube characteristics and enhancement effects on the mechanical features of polymer-based materials and structures – A review. *Journal of Materials Research and Technology*, 24, 6495–6521.

Anbarasu, S., Ilangovan, S., Usharani, K. et al. (2020). Visible light mediated photocatalytic activity of Ni-doped Al_2O_3 nanoparticles. *Surfaces and Interfaces*, 18, 100416.

Asadzadeh-Khaneghah, S. & Habibi-Yangjeh, A. (2020). g-C_3N_4/carbon dot-based nanocomposites serve as efficacious photocatalysts for environmental purification and energy generation: A review. *Journal of Cleaner Production*, 276, 124319.

Astruc, D. (2020). Introduction: Nanoparticles in catalysis. *Chemical Reviews*, 120(2), 461–463. http://doi.org/10.1021/acs.chemrev.8b00696

Bandara, J., Mielczarski, J. A., & Kiwi, J. (1999). Photosensitized degradation of azo dyes on Fe, Ti, and Al oxides. Mechanism of charge transfer during the degradation. *Langmuir*, 15(22), 7680–7687, http://doi.org/10.1021/la990030j

Banerjee, A. N. (2018). Graphene and its derivatives as biomedical materials: Future prospects and challenges. *Interface Focus*, 8, 20170056. http://doi.org/10.1098/rsfs.2017.0056

Banerjee, T., Gottschling, K., Savasci, G., Ochsenfeld, C., & Lotsch, B. V. (2018). H_2 evolution with covalent organic framework photocatalysts. *ACS Energy Letters*, 3, 400–409

Bickley, R. I., Munuera, G., & Stone, F. S. (1973). Photoadsorption and photocatalysis at rutile surfaces: II. Photocatalytic oxidation of isopropanol. *Journal of Catalysis*, 31, 398–407.

Bie, C., Yu, H., Cheng, B., Ho, W., Fan, J., & Yu, J. (2021). Design, fabrication, and mechanism of nitrogen-doped graphene-based photocatalyst. *Advanced Materials*, 33(9), 2003521.

Bosi, S., Da Ros, T., Spalluto, G., & Prato, M. (2003). Fullerene derivatives: An attractive tool for biological applications. *European Journal of Medicinal Chemistry*, 38(11-12), 913–923.

Chakraborty, R., Vilya, K., Pradhan, M., & Nayak, A. K. (2022). Recent advancement of biomass-derived porous carbon-based materials for energy and environmental remediation applications. *Journal of Materials Chemistry A*, 10(13), 6965–7005.

Chen, P.-W., Li, K., Yu, Y.-X., & Zhang, W.-D. (2017). Cobalt-doped graphitic carbon nitride photocatalysts with high activity for hydrogen evolution. *Applied Surface Science*, 392, 608–615.

Chen, T., Zhang, J., Ge, H. L. et al. (2020). Efficient extraction of uranium in organics-containing wastewater over g-C_3N_4/GO hybrid nanosheets with type-II band structure. *Journal of Hazardous Materials*, 384, 121383.

Chen, X., Liu, L., Yu, P. Y., & Mao, S. S. (2011). Increasing solar absorption for photocatalysis with black hydrogenated titanium dioxide nanocrystals. *Science*, 331, 746–750.

Cheng, L., Xiang, Q., Liao, Y., & Zhang, H. (2018). CdS-based photocatalysts. *Energy Environmental Sciences*, 11, 1362–1391.

Cote, A. P. (2005). Porous, crystalline, covalent organic frameworks. *Science*, 310, 1166–1170. https://doi.org/10.1126/science.1120411

Cui, L., Ren, X., Sun, M., Liu, H., & Xia, L. (2021). Carbon dots: Synthesis, properties and applications. *Nanomaterials*, 11(12), 3419. https://doi.org/10.3390/nano11123419

Darabdhara, G. & Das, M. R. (2019). Dual responsive magnetic Au@Ni nanostructures loaded reduced graphene oxide sheets for colorimetric detection and photocatalytic degradation of toxic phenolic compounds. *Journal of Hazardous Materials*, 368, 365–377.

de Souza Barbosa, A., da Silva, M. C. B. C., & da Silva, L. B. et al. (2023). Integration of environmental, social, and governance (ESG) criteria: Their impacts on corporate sustainability performance. *Humanities & Social Sciences Communications*, 10, 410.

Deschler, F., Neher, D., & Schmidt-Mende, L. (2019). Perovskite semiconductors for next generation optoelectronic applications. *APL Materials*, 7(8).

Ding, S.-Y. & Wang, W. (2013). Covalent organic frameworks (COFs): From design to applications. *Chemical Society Reviews*, 42, 548–568.

Dong, G., Zhang, Y., Pan, Q., & Qiu, J. (2014). A fantastic graphitic carbon nitride ($g-C_3N_4$) material: Electronic structure, photocatalytic and photoelectronic properties. *Journal of Photochemistry and Photobiology C: Photochemistry Reviews*, 20, 33–50

Dong, J., Zhang, Y., Hussain, M. I., Zhou, W., Chen, Y., & Wang, L.-N. (2022). $g-C_3N_4$: Properties, pore modifications, and photocatalytic applications. *Nanomaterials*, 12, 121.

Duncan, T. V. (2011). Applications of nanotechnology in food packaging and food safety: Barrier materials, antimicrobials and sensors. *Journal of Colloid and Interface Science*, 363(1), 1–24.

Dutta, V., Verma, R., Gopalkrishnan, C. et al. (2022). Bio-inspired synthesis of carbon-based nanomaterials and their potential environmental applications: A state-of-the-art review. *Inorganics*, 10, 169. https://doi.org/10.3390/inorganics10100169

Ehtisham Khan, M. E., Khan, M. M., & Moo Hwan Cho, M. H. (2015). Biogenic synthesis of a Ag–graphene nanocomposite with efficient photocatalytic degradation, electrical conductivity and photoelectrochemical performance. *New Journal of Chemistry*, 39, 8121–8129.

Fu, T., Zhang, B., Gao, X. et al. (2023). Recent progresses, challenges, and opportunities of carbon-based materials applied in heavy metal polluted soil remediation. *Science of the Total Environment*, 856, 158810.

Fujishima, A. & Honda, K. (1972). Electrochemical photolysis of water at a semiconductor electrode. *Nature*, 238, 37–38, https://doi.org/doi.org/10.1038/238037a0

Gao, J., Wang, Y., Zhou, S., Lin, W., & Kong, Y. (2017). A facile one-step synthesis of Fe-doped $g-C_3N_4$ nanosheets and their improved visible light photocatalytic performance. *ChemCatChem*, 9(9), 1708–1715.

Garrafa-Gálvez, H. E., Alvarado-Beltrán, C. G., Almaral-Sánchez, J. L. et al. (2019). Graphene role in improved solar photocatalytic performance of TiO_2-RGO nanocomposite. *Chemical Physics*, 521, 35–43.

Georgakilas, V., Gournis, D., Tzitzios, V., Pasquato, L., Guldi, D. M., & Prato, M. (2007). Decorating carbon nanotubes with metal or semiconductor nanoparticles. *Journal of Materials Chemistry*, 17(26), 2679–2694.

Hassaan, M. A., El-Nemr, M. A., Elkatory, M. R. et al. (2023). Principles of photocatalysts and their different applications: A review. *Topics in Current Chemistry (Z)*, 381, 31.

Hayat, A., Rauf, S., Alwan, B. Al. et al. (2024), Recent advance in MOFs and MOF-based composites: synthesis, properties, and applications. *Materials Today Energy*, 41, 101542.

He, F., Jeon, W., & Choi, W. (2021). Photocatalytic air purification mimicking the self-cleaning process of the atmosphere. *Nature Communications*, 12, 2528.

Horváth, E., Gabathuler, J., Bourdiec, G. et al. (2022). Solar water purification with photocatalytic nanocomposite filter based on TiO_2 nanowires and carbon nanotubes. *npj Clean Water*, 5, 10.

Hossain, N., Mobarak, M. H., Mimona, M. A. et al. (2023). Advances and significances of nanoparticles in semiconductor applications – A review. *Results in Engineering*, 19, 101347.

Humphrey, V., Zscheischler, J., Ciais, P., Gudmundsson, L., Sitch, S., & Seneviratne, S. I. (2018). Sensitivity of atmospheric CO_2 growth rate to observed changes in terrestrial water storage. *Nature*, 560, 628–631.

Hussain, M. Z., Yang, Z., Huang, Z., Jia, Q., Zhu, Y., & Xia, Y. (2021). Recent advances in metal–organic frameworks derived nanocomposites for photocatalytic applications in energy and environment. *Advanced Science*, 8, 2100625.

Jagadeesan, A. K., Thangavelu, K., & Dhananjeyan, V. (2020). Carbon nanotubes: Synthesis, properties and applications. In P. V. Pham, P. Goel, S. Kumar & K. Yadav (Eds.), *From the Edited Volume- "21st Century Surface Science – A Handbook."* https://doi.org/10.5772/intechopen.92995

Jiang, J.-X., Su, F., Trewin, A. et al. (2008). Synthetic control of the pore dimension and surface area in conjugated microporous polymer and copolymer networks synthetic control of the pore dimension and surface area in conjugated microporous polymer and copolymer networks. *Journal of the American Chemical Society*, 130(24), 7710–7720.

Jiang, Q., Sun, L., Bi, J. et al. (2018). MoS_2 Quantum dots-modified covalent triazine-based frameworks for enhanced photocatalytic hydrogen evolution. *ChemSusChem*, 11, 1108–1113.

Jiang, Y., Sun, Z., Tang, C., Zhou, Y., Zeng, L., & Huang, L. (2019). Enhancement of photocatalytic hydrogen evolution activity of porous oxygen doped g-C_3N_4 with nitrogen defects induced by changing electron transition. *Applied Catalysis B: Environmental*, 240, 30—38.

Kadi, M. W., Mohamed, R. M., Ismail, A. A., & Bahnemann, D. W. (2020a). Soft and hard templates assisted synthesis mesoporous CuO/g-C_3N_4 heterostructures for highly enhanced and accelerated Hg(II) photoreduction under visible light. *Journal of Colloid and Interface Science*, 580, 223–233.

Kadi, M. W., Mohamed, R. M., Ismail, A. A., & Bahnemann, D. W. (2020b). Performance of mesoporous Fe_2O_3/g-C_3N_4 heterojunction for photoreduction of Hg(II) under visible light illumination. *Ceramics International*, 46, 23098–23106.

Khan, E. M. (2021). State-of-the-art developments in carbon-based metal nanocomposites as a catalyst: Photocatalysis. *Nanoscale Advances*, 3, 1887–1900.

Kudo, A., Yoshino, S., Tsuchiya, T. et al. (2019). Z-scheme photocatalyst systems employing Rh- and Ir-doped metal oxide materials for water splitting under visible light irradiation. *Faraday Discussions*, 215, 313–328.

Kumar, S., Kumar, A., Bahuguna, A., Sharma, V., & Krishnan, V. (2017). Two-dimensional carbon-based nanocomposites for photocatalytic energy generation and environmental remediation applications. *Beilstein Journal of Nanotechnology*, 8(1), 1571–1600.

Lampert, A. (2019). Over-exploitation of natural resources is followed by inevitable declines in economic growth and discount rate. *Nature Communications*, 10, 1419.

Li, F. & Lin, M. (2020). Synthesis of biochar-supported K-doped g-C_3N_4 photocatalyst for enhancing the polycyclic aromatic hydrocarbon degradation activity. *International Journal of Environmental Research and Public Health*, 17, 2065.

Li, J., Xu, Y., Liu, Y., Wu, D., & Sun, Y. (2004). Synthesis of hydrophilic ZnS nanocrystals and their application in photocatalytic degradation of dye pollutants. *China Particuology*, 2, 266–269.

Li, L., Cai, Z., Wu, Q. et al. (2016). Rational design of porous conjugated polymers and roles of residual palladium for photocatalytic hydrogen production. *Journal of the American Chemical Society*, 138, 7681–7686.

Li, X., Yu, J., Jaroniec, M., & Chen, X. (2019). Cocatalysts for selective photoreduction of CO_2 into solar fuels. *Chemical Reviews*, 119(6), 3962–4179, https://doi.org/10.1021/acs.chemrev.8b00400

Li, Y., Zhu, S., Liang, Y. et al. (2020). Synthesis of α-Fe_2O_3/g-C_3N_4 photocatalyst for high-efficiency water splitting under full light. *Materials & Design*, 196, 109191.

Magne, T. M., Oliveira Vieira, T. d., Alencar, L. M. R. et al. (2022). Graphene and its derivatives: Understanding the main chemical and medicinal chemistry roles for biomedical applications. *Journal of Nanostructure in Chemistry*, 12, 693–727.

Mahor, A., Singh, P. P., Bharadwaj, P. et al. (2021). Carbon-based nanomaterials for delivery of biologicals and therapeutics: A cutting-edge technology. *C - Journal of Carbon Research*, 7, 19. https://doi.org/10.3390/c7010019

Matsumura, M., Hiramoto, M., Iehara, T., & Tsubomura, H. (1984). Photocatalytic and photoelectrochemical reactions of aqueous solutions of formic acid, formaldehyde, and methanol on platinized cadmium sulfide powder and at a cadmium sulfide electrode. *Journal of Physical Chemistry*, 88, 248–250.

Mochalin, V. N., Shenderova, O., Ho, D., Gogotsi, Y., & Gogotsi, Y. (2012). The properties and applications of nanodiamonds. *Nature Nanotechnology*, 7(1), 11–23.

Mohamed, M. G., El-Mahdy, A. F. M., Kotp, M. G., & Kuo, S. W. (2022). Advances in porous organic polymers: Syntheses, structures, and diverse applications. *Materials Advances*, 3, 707–733.

Mohapatra, L., Cheon, D., & Yoo, H. S. (2023). Carbon-based nanomaterials for catalytic wastewater treatment: A review. *Molecules*, 28, 1805.

Morais, E., Modhrain, C., Ravindranathan Thampi, K., & Sullivan, J. A. (2021). RuO_2/TiO_2 photocatalysts prepared via a hydrothermal route: Influence of the presence of TiO_2 on the reactivity of RuO_2 in the artificial photosynthesis reaction. *Journal of Catalysis*, 401, 288–296

Ng, Y. H., Ikeda, S., Matsumura, M., & Amal, R. (2012). A perspective on fabricating carbon-based nanomaterials by photocatalysis and their applications. *Energy & Environmental Science*, 5(11), 9307–9318.

Nguyen, C. H., Tran, M. L., Tran, T. T. V., & Juang, R.-S. (2020). Enhanced removal of various dyes from aqueous solutions by UV and simulated solar photocatalysis over TiO_2/ZnO/rGO composites. *Separation and Purification Technology*, 232, 115962.

Osterloh, F. E. (2008). Inorganic materials as catalysts for photochemical splitting of water. *Chemistry of Materials*, 20(1), 35–54. https://doi.org/10.1021/cm7024203

Osterloh, F. E. (2017). Photocatalysis versus Photosynthesis: A sensitivity analysis of devices for solar energy conversion and chemical transformations. *ACS Energy Letters*, 2(2), 445–453.

Pala, R., Anju V. T., Dyavaiah, M., Busi, S., & Nauli, S, M. (2020). Nanoparticle-mediated drug delivery for the treatment of cardiovascular diseases. *International Journal of Nanomedicine*, 15, 3741–3769, https://doi.org/10.2147/IJN.S250872. PMID: 32547026; PMCID: PMC7266400

Park, S. J., Das, G. S., Schütt, F. et al. (2019). Visible-light photocatalysis by carbon-nano-onion-functionalized ZnO tetrapods: Degradation of 2,4-dinitrophenol and a plant-model-based ecological assessment. *NPG Asia Materials*, 11, 8. https://doi.org/10.1038/s41427-019-0107-0

Patil, S. S., Shedbalkar, U. U., Truskewycz, A., Chopade, B. A., & Ball, A. S. (2016). Nanoparticles for environmental clean-up: A review of potential risks and emerging solutions. *Environmental Technology & Innovation*, 5, 10–21.

Pope, S. J. A., Coe, B. J., Faulkner, S., Bichenkova, E. V., Yu, X., & Douglas, K. T. (2004). Self-assembly of heterobimetallic d–f hybrid complexes: Sensitization of lanthanide luminescence by d-block metal-to ligand charge-transfer excited states. *Journal of the American Chemical Society*, 126, 9490–9491.

Qu, A., Xu, X., Zhang, Y. et al. (2016). A nitrogen-rich mesoporous polymer for photocatalytic hydrogen evolution from water. *Reactive and Functional Polymers*, 102, 93–100.

Radhika, N. P., Selvin, R., Kakkar, R., & Umar, A. (2019). Recent advances in nano-photocatalysts for organic synthesis. *Arabian Journal of Chemistry*, 12, 4550–4578.

Ren, G., Han, H., Wang, Y. et al. (2021). Recent advances of photocatalytic application in water treatment: A review. *Nanomaterials*, 11, 1804.

Romero, N. A. & Nicewicz, D. A. (2016). Organic photoredox catalysis. *Chemical Reviews*, 116, 17, 10075–10166. https://doi.org/10.1021/acs.chemrev.6b00057

Sabet, M. & Mahdavi, K. (2019). Green synthesis of high photoluminescence nitrogen-doped carbon quantum dots from grass via a simple hydrothermal method for removing organic and inorganic water pollutions. *Applied Surface Science*, 463, 283–291.

Sachs, J. D., Lafortune, G., Fuller, G., & Drumm, E. (2024). Roles of innovation in achieving the sustainable development goals: A bibliometric analysis. *Journal of Innovation & Knowledge*, 9(2), 100472.

Saikia, J. & Ramakrishnan, V. (2023). *Chapter 7, Peptide Nanocatalysts*. Academic Press.

Salata, O. V. (2004). Applications of nanoparticles in biology and medicine. *Journal of Nanobiotechnology*, 2, 3, https://doi.org/10.1186/1477-3155-2-3

Shao, L., Gao, Y., & Yan, F. (2011). Semiconductor quantum dots for biomedicial applications. *Sensors*, 11(12), 11736–11751.

Sharma, S., Bashir, S., Kais, R., & Subramaniam, R. T. (2022a). A review on graphene and its derivatives as the forerunner of the two-dimensional material family for the future. *Journal of Materials Science*, 57(11), 1–43.

Sharma, S., Kumar, R., Raizada, P. et al. (2022b). An overview on recent progress in photocatalytic air purification: Metal-based and metal-free photocatalysis. *Environmental Research*, 214, 113995.

Sheoran, K., Kaur, H., Siwal, S. S., Saini, A. K., Vo, D.-V. N., & Thakur, V. K. (2022). Chemosphere, recent advances of carbon-based nanomaterials (CBNMs) for wastewater treatment: Synthesis and application. *Chemosphere*, 299, 134364.

Si, Q. S., Guo, W. Q., Wang, H. Z., Liu, B. H., & Ren, N. Q. (2020). Carbon quantum dots-based semiconductor preparation methods, applications and mechanisms in environmental contamination. *Chinese Chemical Letters*, 31(10), 2556–2566.

Simon, E. J., Dickey, J. L., Reece, J. B., & Campbell, N. A. (2019). *Campbell Essential Biology*. (7th ed.). Pearson.

Song, J., Wu, X., Zhang, M. et al. (2020). Highly flexible, core–shell heterostructured, and visible-light driven titania-based nanofibrous membranes for antibiotic removal and E. coli inactivation. *Chemical Engineering Journal*, 379, 122269.

Sun, Y., Zhang, W., Li, Q., Liu, H., & Wang, X. (2023). Preparations and applications of zinc oxide based photocatalytic materials. *Advanced Sensor and Energy Materials*, 2, 100069.

Takanabe, K. (2017). Photocatalytic water splitting: Quantitative approaches toward photocatalyst by design. *ACS Catalysis*, 7(11), 8006.

Tan, X., Fang, M., & Wang, X. (2010). Sorption speciation of lanthanides/actinides on minerals by TRLFS, EXAFS and DFT study: A review. *Molecules*, 15, 8431–8468.

Tennakone, K. (1985). Semiconducting and photocatalytic properties of mercuric thiocyanate. *Solar Energy Materials*, 12, 69–75.

Tian, C., Zhang, Q., Wu, A. et al. (2012). Cost-effective large-scale synthesis of ZnO photocatalyst with excellent performance for dye photodegradation. *Chemical Communications*, 48, 2858–2860.

United Nations (1987). *Report of the World Commission on Environment and Development: Our Common Future*. https://sustainabledevelopment.un.org/content/documents/5987our-common-future.pdf.

Wang, C., Xie, Z. G., deKrafft, K. E., & Lin, W. B. (2011). Doping metal–organic frameworks for water oxidation, carbon dioxide reduction, and organic photocatalysis. *Journal of the American Chemical Society*, 133, 13445–13454.

Wang, Q. & Domen, K. (2020). Particulate photocatalysts for light-driven water splitting: mechanisms, challenges, and design strategies. *Chemical Reviews*, 120, 919–985.

Wang, S. Q., Shi, L., Yu, S. J. et al. (2022). Effect of Shewanella oneidensis MR-1 on U(VI) sequestration by montmorillonite. *Journal of Environmental Radioactivity*, 242, 106798.

Wang, X., Maeda, K., Thomas, A. et al. (2009). A metal-free polymeric photocatalyst for hydrogen production from water under visible light. *Nature Materials*, 8, 76–80.

Wen, J., Zhou, L., Tang, Q., Xiao, X., & Sun, S. (2023). Photocatalytic degradation of organic pollutants by carbon quantum dots functionalized g-C_3N_4: A review. *Ecotoxicology and Environmental Safety*, 262, 115133.

Xiang, Q., Yu, J., & Jaroniec, M. (2012a). Synergetic effect of MoS_2 and graphene as cocatalysts for enhanced photocatalytic H_2 production activity of TiO_2 nanoparticles. *Journal of the American Chemical Society*, 134(15), 6575–6578, https://doi.org/10.1021/ja302846n

Xiang, Q., Yu, J., & Jaroniec, M. (2012b). Graphene-based semiconductor photocatalysts. *Chemical Society Reviews*, 41(2), 782–796.

Xu, C., Anusuyadevi, P. R., Aymonier, C., Luque, R., & Marre, S. (2019). Nanostructured materials for photocatalysis. *Chemical Society Reviews*, 48(14), 3868–3902.

Xu, C., Zhang, W., Tang, J., Pan, C., & Yu, G. (2018a). Porous organic polymers: An emerged platform for photocatalytic water splitting. *Frontiers in Chemistry*, 6, 1–12.

Xu, Y., Li, S., Chen, M., Zhang, J., & Rosei, F. (2022). Carbon-based nanostructures for emerging photocatalysis: CO_2 reduction, N_2 fixation, and organic conversion. *Trends in Chemistry*, 4(11), 984–1004.

Xu, Y., Mao, N., Zhang, C. et al. (2018b). Rational design of donor-5-acceptor conjugated microporous polymers for photocatalytic hydrogen production. *Applied Catalysis B: Environmental*, 228, 1–9.

Xue, B. & Zou, Y. (2018). High photocatalytic activity of ZnO–graphene composite. *Journal of Colloid and Interface Science*, 529, 306–313.

Yanagida, S., Kabumoto, A., Mizumoto, K., Pac, C., & Yoshino, K. (1985). Poly (p-phenylene)-catalysed photoreduction of water to hydrogen. *Journal of the Chemical Society, Chemical Communications*, 474–475.

Yang, Y., Chu, X., Zhang, H. Y. et al. (2023). Engineering β-ketoamine covalent organic frameworks for photocatalytic overall water splitting. *Nature Communications*, 14, 593.

Yusuf, A., Almotairy, A. R. Z., Henidi, H., Alshehri, O. Y., & Aldughaim, M. S. (2023). Nanoparticles as drug delivery systems: A review of the implication of nanoparticles' physicochemical properties on responses in biological systems. *Polymers (Basel)*, 15(7), 1596, https://doi.org/10.3390/polym15071596. PMID: 37050210; PMCID: PMC10096782 (Academic Editor: Hassan, S. and Miri, A.K.)

Zaman, W. & Manghwar, H. (2024). Nanoparticles and their biological applications: Recent advances in 2022–2023. *Microorganisms*, 12(3), 489.

Zhai, Y., Zhu, Z., & Dong, S. (2015). Carbon-based nanostructures for advanced catalysis. *ChemCatChem*, 7(18), 2806–2815.

Zhang, Q., Yan, S., Yan, X., & Lv, Y. (2023). Recent advances in metal–organic frameworks: Synthesis, application and toxicity. *Science of The Total Environment*, 102, 165944.

Zhang, S., Ou, X., Xiang, Q., Carabineiro, S. A., Fan, J., & Lv, K. (2022). Research progress in metal sulfides for photocatalysis: From activity to stability. *Chemosphere*, 303, 135085.

Zhang, X. H., Wang, X. P., Xiao, J. et al. (2017). Synthesis of 1,4-diethynylbenzene-based conjugated polymer photocatalysts and their enhanced visible/near-infrared-light-driven hydrogen production activity. *Journal of Catalysis*, 350, 64–71.

Zhang, Z., Yi, G., Li, P. et al. (2020). A minireview on doped carbon dots for photocatalytic and electrocatalytic applications. *Nanoscale*, 12, 13899–13906.

Zhao, Z., Sun, Y., & Dong, F. (2015). Graphitic carbon nitride-based nanocomposites: A review. *Nanoscale*, 7(1), 15–37.

Zhou, K., Zhu, Y., Yang, X., Jiang, X., & Li, C. (2011). Preparation of graphene-TiO_2 composites with enhanced photocatalytic activity. *New Journal of Chemistry*, 35, 353–359.

Section 2

Remediations and Ecological Aspects

4

Persistent Pollutants in the Biosphere and Their Remediation Through Carbon Scaffold-based Nanophotocatalysts: Development and Challenges

Shashikant Shivaji Vhatkar[1] *and Ramesh Oraon*[2]*

[1] *Center for Materials for Electronics Technology (C-MET), Panchwati, Pune, India*
[2] *Department of Chemistry, Central University of Jharkhand, Ranchi, India*

* Corresponding author: ramesh.oraon@cuj.ac.in

4.1 Introduction

Persistent pollutants, referred to as persistent organic pollutants (POPs) or persistent environmental contaminants, are hazardous substances that endure in the environment without decomposition (Alharbi et al., 2018; Muir & Howard, 2006). These pollutants persist in ecosystems for prolonged durations, accumulating in soil, water, and creatures, thereby posing substantial threats to human health and biodiversity (Ashraf, 2017). Their tenacity enables them to traverse extensive distances by air and water, resulting in pervasive contamination on a worldwide scale (Beyer et al., 2000). Persistent pollutants (Figure 4.1) encompass industrial compounds such as polychlorinated biphenyls (PCBs), pesticides like DDT, and leftovers from combustion and industrial activities, including dioxins and furans (Pariatamby & Kee, 2016).

These compounds frequently demonstrate bioaccumulative characteristics, resulting in their concentration within the tissues of species and amplification via the food chain, potentially leading to detrimental effects such as endocrine disruption, cancer, and reproductive abnormalities (Jones & de Voogt, 1999). Moreover, POPs are directly associated with several health complications, such as cancer, neurological diseases, reproductive troubles, and immune system suppression (El-Shahawi et al., 2010). Mercury pollution in fish can cause neurological damage, especially in developing fetuses and young children (Ahmed et al., 2021). Likewise, exposure to PCBs has been linked to endocrine disruption and developmental complications. These contaminants accumulate in human tissues over time, presenting long-term

Photocatalysts based on Carbon Scaffolds: Environmental and Industrial Applications, First Edition. Edited by Ramesh Oraon et al.
Published 2026 by Wiley-VCH GmbH.

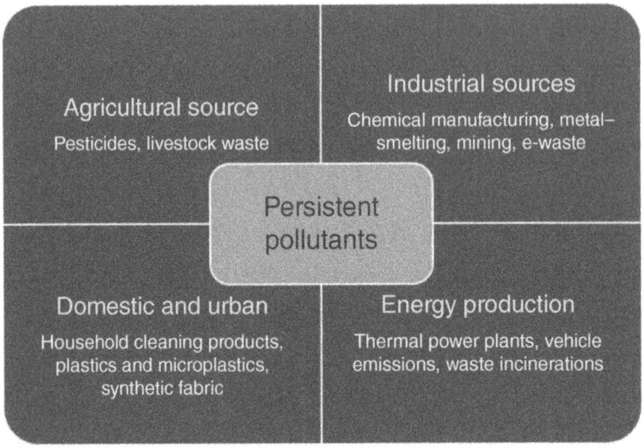

Figure 4.1 Several persistent pollutants with their sources.

risks even at minimal exposure levels. By eliminating these pollutants from soil, water, and air, we can mitigate their effects on human populations and avert chronic health issues. Persistent contaminants also adversely impact wildlife (Hao et al., 2021). Numerous hazardous agents disrupt reproductive processes, resulting in population declines across multiple species. DDT caused the weakening of eggshells in avian species including the bald eagle and peregrine falcon, nearly leading to their extinction until remediation measures facilitated population recovery (Jagannath et al., 2008). Furthermore, the bioaccumulation of heavy metals such as lead and mercury within aquatic food chains can result in neurological damage in fish, avians, and mammals (Corsolini et al., 2005). Pollution-induced species loss disturbs ecosystems, impacting predator-prey dynamics and essential ecosystem services, including pollination and water purification, upon which humans depend. Persistent pollutants have been documented to lead to prolonged ecosystem degradation through the contamination of soil, aquatic systems, and the atmosphere (Gong et al., 2021). Upon entering an environment, these chemicals might persist for decades, complicating their degradation by natural processes. Industrial waste containing PCBs can infiltrate groundwater, compromising its safety for consumption and agricultural use. Soil contamination by heavy metals can impede plant growth, diminish agricultural production, and contribute to desertification (Masih et al., 2014). If neglected, such contamination can render large regions uninhabitable for both humans and wildlife. The existence of enduring contaminants places substantial economic strains on society. Healthcare expenditures linked to pollution-induced ailments, diminished agricultural yields from soil degradation, and decreased fishery outputs owing to bioaccumulation in aquatic organisms all result in economic losses (Milic et al., 2019). Furthermore, remediation costs escalate with the prolonged presence of pollutants in the environment. Investing in initial remediation initiatives is more economically advantageous than confronting the prolonged repercussions of environmental deterioration. Thus, remediation techniques include bioremediation (utilizing microorganisms to decompose pollutants), phytoremediation (employing plants to absorb toxins), and chemical treatments, which can effectively eliminate persistent pollutants from the environment. Implementing more stringent environmental legislation, advocating for cleaner industrial practices, and investing in green technologies are essential measures to avert future contamination (Matthies et al., 2016). Historically, preindustrial societies predominantly utilized natural materials and engaged in small-scale production, resulting in significantly lower pollution emissions compared to the periods before and following industrialization. Moreover, natural ecosystems possessed adequate capacity to decompose or disseminate these contaminants throughout time. Rivers, woodlands, and wetlands functioned as natural filters, diminishing the buildup of pollutants in the environment. Notwithstanding worldwide initiatives, enduring pollutants continue to pose a substantial environmental concern. Failure to promptly regulate, remediate, and substitute toxic chemicals with sustainable alternatives would perpetuate their detrimental long-term effects on biodiversity, ecosystems, and human health. This chapter will examine the principal advancements, foundations, problems, and upcoming trends in the remediation of POPs by photocatalytic degradation of nanomaterials, with a specific focus on carbon scaffolds. The economic aspects of implementation have been briefly examined.

4.2 Remediation Mechanisms

Various remediation strategies have been developed to address the hazards of persistent pollutants. These methods seek to eliminate, neutralize, or confine pollutants in soil, water, and air. Remediation methods can be classified into three primary categories: physical, chemical, and biological. An illustrative featuring the remediation mechanism is shown in Figure 4.2.

4.2.1 Physical

These procedures entail the extraction or segregation of contaminants from the environment without modifying their chemical structure (Fei et al., 2022). Contaminated soil is physically excavated and transported to secure landfills, typically for the disposal of heavy metals, including PCBs and dioxins. They have been executed at Love Canal (USA) (Gill & Mix, 2020), during the 2011 nuclear disaster in Fukushima (Japan) (Bin Feng et al., 2022), and in response to lead pollution in Kabwe (Zambia). The elevated expense and possible danger of secondary contamination have been significant drawbacks.

4.2.1.1 Soil Excavation

One of the principal advantages of this strategy is its capacity to promptly mitigate risk (Jang & Townsend, 2001). This technique rapidly eliminates contaminants from the location, mitigating environmental and health hazards. It is also efficacious for places with significant contamination. Additionally, it inhibits the migration of contaminants into groundwater, rivers, and adjacent ecosystems (El-Shahawi et al., 2010).

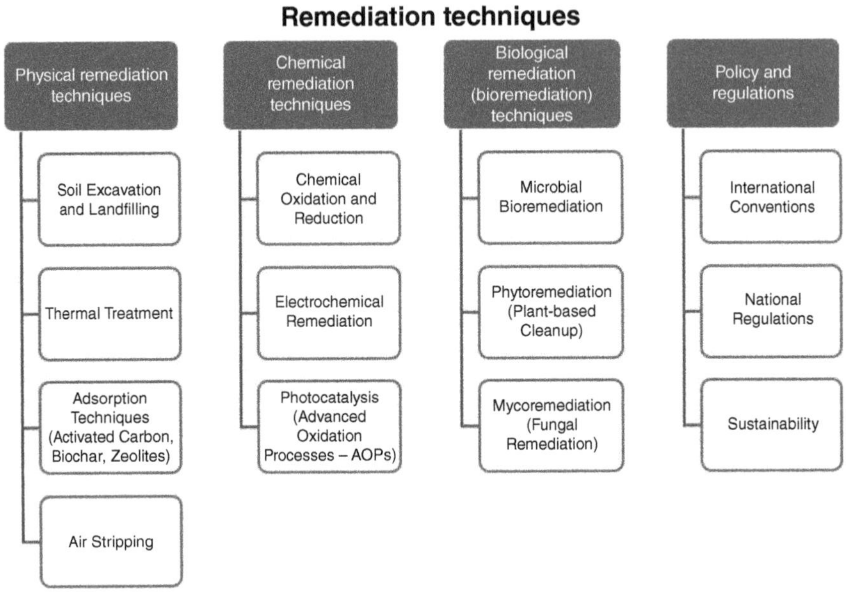

Figure 4.2 Different routes of remediating persistent pollutants.

Additionally, this technique can be integrated with other procedures; for example, soil may be pretreated prior to landfilling, or alternative treatment methods such as thermal desorption or bioremediation might be employed before disposal.

4.2.1.1.1 Process of Soil Excavation and Landfilling

The process of soil excavation and landfilling involves the following key steps:

A. **Site Assessment and Contamination Evaluation**
   The degree of contamination is assessed via soil sampling and laboratory analysis, focusing on the types of contaminants, their concentrations, and their depth within the soil. The determination of whether excavation is the optimal cleanup method is based on environmental and financial considerations.
B. **Excavation of Contaminated Soil**
   Heavy machinery, including excavators, bulldozers, and backhoes, is employed to excavate the contaminated soil. The excavation depth is dictated by the extent of contaminant penetration. During excavation, dust suppression techniques, such as water sprays, are employed to mitigate airborne contaminants.
C. **Transportation and Temporary Storage**
   The excavated soil is put onto trucks and conveyed to a hazardous waste landfill or treatment facility. Occasionally, the dirt may be temporarily retained in containment units prior to final disposal or treatment.
D. **Disposal in Secure Landfills**
   Contaminated soil is deposited in engineered landfills constructed to avert leakage into groundwater. These landfills are equipped with liner systems, leachate

collecting systems, and gas collection systems to capture hazardous emissions. The landfill undergoes continuous monitoring to avert environmental concerns.

E. **Site Restoration**

Subsequent to excavation, the site may be backfilled with uncontaminated soil to rehabilitate it for secure utilization. Vegetation can be rebuilt to stabilize the region and mitigate erosion. Notwithstanding the aforementioned benefits, excavation and transportation incur significant costs, particularly for extensive pollution. It affects ecosystems, as soil removal obliterates natural habitats, thereby impacting biodiversity. The remedy is not permanent, as it merely relocates the pollution rather than eradicating it. Should landfill barriers fail, contaminants may infiltrate groundwater. Furthermore, these technologies are significantly energy- and carbon-intensive because of the substantial consumption of fossil fuel energy, which contributes to carbon emissions during excavation, transportation, and landfill activities (*Organic Pollutants in Leachates from Landfill Sites - Sawhney - 1984 - Journal of Environmental Quality - Wiley Online Library*, n.d.). Complementary methods such as soil washing—cleaning polluted soil on-site using water and chemical agents prior to reuse—and bioremediation, which employs bacteria or plants (phytoremediation) to decompose toxins before excavation, are frequently utilized. In addition to these methods, in situ stabilization techniques (involving the injection of chemicals into the soil that bind with pollutants, thereby affecting their mobility) and thermal desorption (heating the soil to eliminate volatile contaminants prior to disposal, thereby reducing waste volume) are employed to mitigate environmental damage.

4.2.1.2 Thermal

This method uses elevated temperatures to eliminate or neutralize contaminants by incinerating hazardous waste at high temperatures to decompose dangerous substances (Trivedi & Majumdar, 2013). Another method, termed thermal desorption, is the heating of contaminated soil to evaporate contaminants, which are subsequently collected and treated. This method is efficacious for POPs, hydrocarbons, and industrial chemicals.

4.2.1.3 Adsorption Techniques (Activated Carbon, Biochar, Zeolites)

Adsorbents such as activated carbon or biochar capture toxins from water or air and are employed for the treatment of wastewater containing heavy metals, PFAS, and organic pollutants. Adsorption refers to the accumulation of chemicals (adsorbates) from a liquid or gaseous phase onto the surface of a solid object (adsorbent) (Titchou et al., 2021). It transpires through physical forces (physisorption), such as van der Waals interactions, or through chemical bonds (chemisorption). The efficacy of adsorption is contingent upon elements such as surface area, pore dimensions, surface chemistry, pH, and temperature (Wagner et al., 2021). Consequently, they may be customized for a diverse array of persistent pollutants, rendering them highly effective, particularly for addressing low-concentration contaminants in air, water, and soil. Moreover, their incorporation into established treatment systems renders this procedure straightforward to implement. Moreover, in comparison to thermal or chemical treatments, adsorption techniques necessitate minimal energy, rendering them a more favorable option than chemical or thermal methods.

A. **Activated Carbon (Activated Charcoal)**
The predominant adsorbent employed for the elimination of organic chemicals, heavy metals, and PFAS due to its extensive surface area and porosity (Jeirani et al., 2017). It is utilized in both air and water filtration systems and is particularly effective in the remediation of dioxins, PCBs, VOCs, PFAS, and dye-affected systems. Activated carbon and ion exchange resins were employed in drinking water treatment to eliminate PFAS in regions such as Michigan and North Carolina (Helmer et al., 2022).

B. **Biochar**
Biochar is a carbon-dense substance derived from biomass (such as wood and crop leftovers) during pyrolysis (Ni et al., 2020). It serves as a sustainable and economical substitute for activated carbon, demonstrating efficacy in the remediation of pesticides, medicines, heavy metals, and hydrocarbons. Recently, biochar has been utilized in India and Africa to mitigate pesticide and metal contamination in agricultural fields (Yadav et al., 2015).

C. **Zeolites**
Zeolites are naturally occurring or synthesized aluminosilicate minerals characterized by a porous crystalline structure (Wang et al., 2022). They demonstrate exceptional capabilities for cation exchange and the removal of heavy metals. They are proficient in the remediation of lead, cadmium, ammonium, and radioactive metals. Currently, zeolites are employed throughout Eastern Europe to remediate soil and water damaged by mining activities, such as lead and arsenic (Doula et al., 2012).

D. **Clay Minerals (e.g., Bentonite, Kaolinite)**
Finer-grained, stratified minerals exhibit elevated cation exchange capability. Consequently, they are employed to immobilize contaminants in soil and water (Ewis et al., 2022). These are very efficacious in the remediation of heavy metals, dyes, and ammonium compounds (Novikau & Lujaniene, 2022).

E. **Metal-Organic Frameworks (MOFs)**
These advanced materials have ultra-high porosity and adjustable chemistry; they remain under investigation but demonstrate significant potential for the selective adsorption of POPs and gases (Naghdi et al., 2023).

F. **Synthetic Polymers and Resins**
These are designed for the targeted capture of certain pollutants, particularly in industrial wastewater treatment, and are applicable to organic solvents, insecticides, and medicines.

4.2.1.4 Air Stripping

Air stripping is employed for the removal of volatile contaminants in water, such as benzene and trichloroethylene. The contaminated water is aerated, leading to the evaporation of volatile contaminants, which are subsequently collected for treatment.

4.2.2 Chemical

These approaches employ chemical reactions to decompose or neutralize pollutants and utilize subsequent techniques to remediate POPs.

4.2.2.1 Chemical Oxidation and Reduction

This technique enables the utilization of potent oxidizing agents (e.g., hydrogen peroxide, ozone, or permanganate) or reducing agents (e.g., zero-valent iron) to convert hazardous metals into less mobile or less poisonous forms (Venny et al., 2012). This method is proven efficient for the remediation of petroleum hydrocarbons, insecticides, and industrial solvents. In situ chemical oxidation—"in situ" denotes on-site or in place. This process is destructive, in contrast to treatments that just immobilize or transfer pollutants. The procedure employs reactive oxidizing agents that chemically modify contaminants via redox processes, making them innocuous or more amenable to biodegradation. Practical applications encompass the remediation of soil and groundwater. Hydrogen peroxide and persulfate have been employed to remediate diesel and gasoline spills adjacent to refineries and fuel storage tanks (Liu et al., 2023), tainted with chlorinated chemicals from previous dry-cleaning activities. Table 4.1 presents a compilation of prevalent oxidants, including their characteristics and efficacy.

4.2.2.1.1 Mechanism

The oxidant is injected into the contaminated area by injection wells. Upon contact with pollutants, the oxidants generate reactive radicals (such as hydroxyl and sulfate radicals), oxidize the contaminants, cleaving molecular bonds and transforming them into innocuous molecules.

$$\text{Pollutant} + \text{Oxidant} \rightarrow CO_2 + H_2O + \text{Inorganic salts} + \text{Heat}$$

These reactions are frequently exothermic and require meticulous regulation to avert gas production or subsurface pressure accumulation. The following distribution systems have been instituted to efficiently address pollutants:

Table 4.1 List of common oxidants and their effectiveness against various persistent pollutants.

Oxidants	Reaction	Effective against	Reference
Hydrogen peroxide (H_2O_2)	Frequently associated with iron (Fe^{2+}) in the Fenton reaction to generate hydroxyl radicals (•OH), which are extremely reactive and nonselective.	BTEX, TCE, phenols, and PAHs	Liu et al. (2023)

(Continued)

Table 4.1 (Continued)

Oxidants	Reaction	Effective against	Reference
Permanganate ($KMnO_4$ or $NaMnO_4$)	A reliable oxidizing agent employed for chlorinated solvents such as trichloroethylene (TCE), perchloroethylene (PCE), and vinyl chloride. Responds gradually yet consistently under subsurface circumstances. Produces a discernible purple plume in the treated areas	Trichloroethylene (TCE), PCE, and vinyl chloride	Pierri (2021)
Persulfate ($S_2O_8^{2}$)	Initiated by thermal energy, pH modification, or catalysts (e.g., Fe^{2+}) to generate sulfate radicals ($SO_4 \bullet^-$)	Chlorinated solvents, PAHs, pesticides, and MTBE	Naghdi et al. (2023)
Ozone (O_3)	Gaseous oxidant introduced directly or solubilized in water	Useful for volatile organics, though it degrades quickly in subsurface environments	O'Connor et al. (2018)

- **Direct-push injection:** used for small-scale sites or hotspots.
- **Injection wells:** installed to reach deeper contamination zones.
- **Recirculation systems:** extract contaminated groundwater, mix with oxidant, and re-inject.
- **Soil mixing:** for shallow sites; oxidant is mechanically mixed into the soil.

This technique offers numerous advantages over physical methods, including in situ treatment (eliminating the need for excavation or pumping and reducing surface disruption), rapid degradation (contaminants are eradicated within hours or days), versatility (effective against various pollutant types in soil and groundwater), scalability (applicable to both minor spills and extensive industrial sites), and compatibility with other methods, allowing integration with bioremediation, soil vapor extraction, or surfactant flushing for improved outcomes. Despite these advantages, this approach presents specific obstacles. This encompasses oxidant demand, in which natural organic matter and soil minerals deplete oxidants, hence diminishing their efficacy. In heterogeneous soils, oxidants may fail to reach all contaminants due to nonuniform distribution, resulting in restricted contact. Additionally, certain processes emit gases, posing a risk of soil heave or surface eruptions if not properly handled, resulting in gas accumulation. Moreover, incomplete oxidation may yield hazardous intermediates (e.g., chloroform). Ultimately, potent oxidants pose handling and storage risks; they may necessitate tailored safety protocols that incorporate regular groundwater monitoring to prevent the migration of byproducts or unreacted oxidants off-site. Furthermore, pH fluctuations and thermal production during oxidation may influence local ecology and soil chemistry.

4.2.2.2 Electrochemical Remediation

Electrochemical remediation is a sophisticated environmental decontamination method that uses electric current or potential to remediate polluted soil, groundwater, and wastewater (Liu et al., 2023). It is especially efficacious in eliminating persistent pollutants, including heavy metals, organic contaminants, halogenated chemicals, and radionuclides such as uranium and cesium. This technology has garnered global recognition because of its exceptional precision, minimal chemical consumption, and versatility across diverse site circumstances, particularly in locations where conventional techniques may be impractical. For instance, it is employed for the cleanup of heavy metals and organics from manufacturing zones and refineries, remediation of tailings ponds contaminated with metals like arsenic and cadmium, treatment of mixed waste leachates, including PAHs, PCBs, and solvents. This technique uses electric current to extract metals and organic pollutants from soil and water. These approaches, including electrocoagulation and electrokinetics, are employed to eliminate heavy metals from wastewater by applying an electric field that directs pollutants toward extraction wells in the soil. It operates through multiple fundamental processes:

- **Electromigration:** movement of charged contaminants (ions) toward electrodes.
- **Electroosmosis:** movement of pore water through soil due to electric current.
- **Electrophoresis:** movement of charged colloids or particles.
- **Electrochemical reactions:** oxidation/reduction (redox) reactions occurring at the electrodes.

4.2.2.2.1 Core Mechanisms of Electrochemical Remediation

A. Electrokinetic Transport

It applies mainly to soils:

- **Contaminants like heavy metal cations (e.g., Pb^{2+}, Zn^{2+})** move toward the cathode.
- **Anions (e.g., arsenate, chromate)** move toward the anode.
- Water also moves, helping to mobilize nonionic contaminants through electroosmosis.

B. Electrode Reactions (Redox Chemistry)

Contaminants can be oxidized or reduced at or near the electrodes:

- **Anodic oxidation:** organic pollutants like PAHs, phenols, or chlorinated solvents are broken down into harmless compounds.
- **Cathodic reduction:** metal ions like Cr^{6+} can be reduced to Cr^{3+}, which is less toxic and more easily precipitated.

C. pH and Redox Front Manipulation

Electrochemical gradients modify pH and redox conditions, causing certain contaminants to precipitate or become mobilized for removal. These are exceptionally efficient in low-permeability soils (e.g., clay) where traditional flushing methods are inadequate. Exact regulation of pollutant transport and degradation is achievable. Additionally, minimal chemical input diminishes the necessity for supplementary reagents. Moreover, minimal excavation and its compatibility with other technologies such as phytoremediation or chemical oxidation facilitate the simultaneous removal of several toxins, rendering this technology preferable to traditional methods. Nonetheless, issues such as elevated energy usage, electrode corrosion, and clogging complicate this technology and escalate operational costs. The approach is constrained for nonpolar organics and may require augmentation with surfactants or oxidants. Moreover, scaling up is intricate, particularly for extensive or profound pollution plumes.

4.2.3 Biological Remediation

These methods use microorganisms, plants, or enzymes to degrade or absorb persistent pollutants.

4.2.3.1 Microbial Bioremediation

Microbial bioremediation employs bacteria and fungi to decompose or sequester contaminants (Chakraborty & Das, 2016). Depending on the oxygen levels in the environment, aerobic bacteria are utilized in oxygen-rich settings to decompose hydrocarbons and organic contaminants, whereas anaerobic bacteria are employed to remediate chlorinated solvents, PCBs, and heavy metals in oxygen-depleted environments. Pseudomonas bacteria are utilized for the degradation of oil spills.

4.2.3.2 Phytoremediation (Plant-based Cleanup)

It is further classified into three systems

1. Phytoextraction: plants absorb heavy metals (e.g., arsenic, lead) from soil.
2. Phytostabilization: plants prevent the spread of pollutants by stabilizing contaminated soil.
3. Phytodegradation: plants break down organic pollutants like pesticides.

For instance, sunflowers were used to extract radioactive contaminants at Chernobyl (Andriyashina et al., 2015).

4.2.3.3 Mycoremediation (Fungal Remediation)

Certain fungi (e.g., Pleurotus species) break down hydrocarbons, pesticides, and even plastics. They are particularly used in oil spill cleanup and soil remediation (Chandra & Enespa, 2019).

4.2.4 Policy and Regulatory Mechanisms

In addition to technical solutions, governmental regulations, and international treaties are essential for cleanup.

4.2.4.1 International Conventions and Treaties

The Basel Convention (1989) governs the international transport of hazardous waste among nations. The Stockholm Convention on POPs (2001) seeks to eradicate or diminish POPs globally (Lallas, 2001). The Minamata Convention on Mercury (2013) mitigates mercury pollution originating from industrial activity (UN Treaty Collection, 2013).

4.2.4.2 Sustainable Alternatives and Pollution Prevention

Creation of biodegradable materials to substitute enduring contaminants. Implementation of circular economy ideas to minimize waste.

4.2.4.3 National Regulations

Environmental Protection Agencies, such as the EPA in the United States and the CPCB in India, implement cleanup standards and pollution control regulations. The Superfund Program (U.S.) identifies and remediates hazardous waste sites.

4.3 Persistent Pollutants: Pre- and Postindustrial Era

Table 4.2 features several sources in pre- and postindustrial era of POPs.

This chapter will examine the fundamentals of photocatalysts, particularly nanophotocatalysts, including their development, mechanisms, problems, performance enhancement tactics, recovery and reusability, and future possibilities concerning carbon scaffolds. Furthermore, several mechanisms involved, photocatalysts utilizing nanotechnology, and their examination, culminating in carbon scaffold-based

Table 4.2 Criteria-based scenario featuring pre- and postindustrial era of POPs.

Criteria	Preindustrial era	Postindustrial era
Sources	Natural or from localized human activities (e.g., wood burning, agricultural runoff). Combustion of wood and coal produced some heavy metals	Chemicals produced on an industrial scale include synthetic pesticides, plasticizers, flame retardants, and PCBs. Extensive introduction of synthetic chemicals in the twentieth century, including DDT, PCBs, PFAS, dioxins, and BPA

(*Continued*)

Table 4.2 (Continued)

Criteria	Preindustrial era	Postindustrial era
Nature of Pollutants	Biodegradable or nonpersistent (e.g., organic waste, animal manure, soot)	High concentrations or mobilization of heavy metals such as lead and mercury. POPs are identified in human tissue, breast milk, and food worldwide
Environmental and Health Awareness	Little understanding of toxicology, no monitoring systems	Imperceptible poisons, Pollutants last for decades notwithstanding prohibitions (e.g., DDT remains detectable today)
Environmental impact	Local pollution (e.g., smog in cities) existed. Food chains were relatively clean	Global environmental challenges such as bioaccumulation and long-range transport
Regulatory and Technological Response	None	Advancement of remediation technology (chemical oxidation, bioremediation, adsorption) and heightened public awareness and environmental advocacy

systems, have been succinctly addressed. Ultimately, prospective avenues concerning practical applications, like bio-inspired or green manufactured nanophotocatalysts, have also been delineated.

4.4 Economic Aspects of Nanophotocatalysts

A comprehensive analysis of the economic factors associated with the utilization of nanophotocatalysts for the remediation of persistent pollutants, encompassing both advantages and obstacles, is presented in Table 4.3:

4.4.1 Life Cycle Cost Analysis

Table 4.4 features various nanophotocatalysts and their economic implications.

4.4.2 Environmental Cost Savings

Despite greater initial costs, nanophotocatalysts provide enduring economic benefits. No secondary waste is produced, in contrast to adsorption or chemical treatment procedures that generate sludge or byproducts. They require reduced chemical

Table 4.3 Economic aspects featuring different nanophotocatalysts.

Raw material costs	Common basic materials such as TiO_2 and ZnO are comparatively inexpensive. Nonetheless, the incorporation of noble metals (e.g., Au, Ag, Pt) or rare elements escalates expenses. Carbon-based nanomaterials, such as graphene and carbon nanotubes, may be more expensive due to intricate manufacturing processes.
Synthetic techniques	Sol–gel, hydrothermal, and chemical vapor deposition techniques may be energy-demanding and protracted. Green synthesis methods, such as utilizing plant extracts, are emerging as economical and environmentally sustainable alternatives.
Scale-up costs	The transition from laboratory-scale to industrial-scale manufacturing of nanophotocatalysts is costly and necessitates specialized infrastructure and experienced manpower.
Cost of light source	Visible-light responsive nanophotocatalysts (e.g., doped TiO_2) alleviate this burden by harnessing solar energy, hence reducing operational expenses. Utilization of UV light using artificial UV lamps elevates electricity expenses.
Catalysts reusability	Magnetic or immobilized photocatalysts facilitate straightforward recovery and reutilization, hence diminishing recurring expenses. The durability and reliability of the catalyst are essential economic considerations.
System maintenance costs	Systems utilizing nano-based materials require oversight to avert blockage, deterioration, or nanoparticle leaching.

Table 4.4 Nanophotocatalysts and their economic implications.

Stage	Economic implications
Synthesis	High initial cost (materials + energy)
Deployment	Medium: depends on the site, system type, and light source
Operation	Low to moderate (solar energy = low cost)
Maintenance	Medium (monitoring for catalyst loss, equipment upkeep)
Disposal/Recovery	Can be low if catalysts are recoverable; otherwise, it adds to long-term cost

Table 4.5 Comparative cost analysis for different techniques in the remediation of POPs.

Technique	Initial cost	Operational cost	Waste by-products	reusability
Nanophotocatalysis	High	Low (solar-driven)	Minimal	High
Activated carbon adsorption	Medium	Medium	Spent carbon/ sludge	Limited
Chemical oxidation (e.g., Fenton)	Low	High (chemical and disposal)	High	No
Membrane filtration	High	Medium-high	Concentrated brine	Medium

input relative to conventional oxidation or precipitation techniques. It also encourages minimal energy consumption during solar activation. Reduced treatment duration and diminished pollutant residuals yield enhanced efficiency.

4.4.3 Comparative Cost Analysis of Different Techniques

Table 4.5 features comparative cost analysis of several techniques.

High capital investment deters adoption in low-income or rural regions. The absence of standardization in production and application impedes market scalability. The potential discharge of nanoparticles into ecosystems may result in future regulations, hence escalating compliance costs. Finally, there is a necessity for techno-economic evaluations to substantiate investment and return on investment (ROI).

Cost reduction can be attained through public-private partnerships to co-develop nanotechnology for water treatment, with government subsidies and grants for sustainable remediation technologies. The promotion of green synthesis and the utilization of plentiful resources, such as clay-TiO_2 composites, should extend beyond laboratory settings. The development of hybrid systems (e.g., adsorption combined with nanophotocatalysis) may serve as a mechanism to enhance cost-effectiveness.

4.5 Photocatalysis, Prior to Nanophotocatalysts

Prior to the emergence of nanophotocatalysts, the remediation of POPs depended on traditional physical, chemical, and biological treatment approaches, which have been previously addressed. Although these approaches demonstrated different degrees of efficacy, they frequently presented limitations, including incomplete degradation, elevated costs, secondary contamination, and restricted efficiency against stubborn chemicals. Until the mid-to-late 20th century, the understanding of persistent pollutants was inadequate. The notions of bioaccumulation, long-range atmospheric transport, and endocrine disruption originated between the 1960s and 1990s. Environmental science was deficient in the methodologies required to identify trace-level contaminants in soil, water, and biological systems. Although conventional methods were essential for initial pollution management, they frequently proved reactive, inefficient, and unsustainable in addressing contemporary persistent pollutants, including per- and polyfluoroalkyl substances (PFAS), chlorinated solvents, pesticides such as DDT, PCBs, and pharmaceutical residues. Researchers recognized the capability of photocatalytic degradation to entirely mineralize contaminants. The nanoscale provided enhanced surface area, reactivity, and solar energy absorption. This resulted in the creation of nanophotocatalysts that can overcome the shortcomings of previous techniques.

4.5.1 Fundamental of Photocatalysis and Mechanism

4.5.1.1 Photocatalysis (Advanced Oxidation Processes - AOPs)

Photocatalysis is a cutting-edge environmental remediation technique that uses light energy (usually from sunlight or UV light) and a photocatalyst to degrade persistent pollutants into nontoxic byproducts such as carbon dioxide and water (Noor et al., 2022). It's particularly effective against a broad range of organic contaminants in water and air, and is gaining ground as a sustainable, green technology. Apart from this, they are eco-friendly, recyclable, and work with a wide range of organic compounds. As a result, they have been deployed in wastewater treatment plants, industrial effluent treatment, air purification, and solar-driven photocatalytic systems. A list of a few widely used photocatalyst materials has been tabulated below (Table 4.6).

Table 4.6 Few states of art/widely established nanophotocatalysts with challenges.

Sr. No	Material	Notes
1	Titanium dioxide (TiO_2), Doped TiO_2 (Dong et al., 2015)	Extensively reported; stable, inexpensive, but only UV-active
2	Zinc oxide (ZnO) (Khan et al., 2019)	UV-active, but prone to photocorrosion
3	g-C_3N_4 (Jitae et al., 2025)	Visible light-active; metal-free and eco-friendly
4	Composite materials (Zhang et al., 2020)	Combined with graphene, carbon nanotubes, or magnetic supports for better performance and recyclability

Table 4.7 Different pollutants with type and examples using photocatalysts.

Type	Examples	Reference
Organic dyes	Methylene blue, Rhodamine B	Yu et al. (2017)
Pharmaceuticals	Antibiotics, hormones, ibuprofen	Nassri et al. (2023)
Pesticides	Atrazine, paraquat, chlorpyrifos	Maldani et al. (2022)
Phenolic compounds	Bisphenol A (BPA), phenol, cresols	Meylemans et al. (2012)
Volatile organic compounds (VOCs)	Toluene, benzene, and formaldehyde	Yang et al. (2019)
Emerging contaminants	PFAS, microplastics (with hybrid systems)	Giroux et al. (2024)

4.5.1.1.1 Mechanism

Its mechanism involves a three-step degradation process, which is as follows:

Step 1: Light Absorption

When ultraviolet or visible light interacts with the photocatalyst, it energizes electrons, promoting an electron (e^-) from the valence band to the conduction band, so creating a vacancy known as a hole (h^+).

Step 2: Formation of Reactive Species

Electrons and holes produce reactive oxygen species (ROS): electrons convert O_2 to superoxide ($O_2\bullet^-$), whereas holes oxidize water or OH^- to hydroxyl radicals ($\bullet OH$).

Step 3: Pollutant Degradation

The ROS decompose organic pollutants, such as dyes, insecticides, and PAHs, into CO_2, H_2O, and benign mineral residues.

Additionally, photocatalysts are categorized into two types: homogeneous and heterogeneous. The former is dissolved in a solution (such as Fenton's reagent), whereas the latter is a solid photocatalyst that directly interacts with pollutants in either liquid or gas phases, due to its reusability and simplicity of separation. A variety of photocatalysts has been extensively developed to combat different pollutants, with a few examples listed in Table 4.7:

4.6 Nanophotocatalysts

Nanophotocatalysts have markedly enhanced the cleanup of persistent pollutants by overcoming the constraints of traditional approaches and offering efficient, sustainable, and novel alternatives (Boulkhessaim et al., 2022). Conventional photocatalysts such as bulk TiO_2 necessitate UV light, which constitutes around 5% of sunshine. Nanophotocatalysts, particularly doped or composite materials, are currently engineered to function in visible light, facilitating solar-driven degradation of pollutants (Boulkhessaim et al., 2022). This significantly lowers energy expenses and renders the process environmentally benign and sustainable. Moreover, owing to their elevated surface-area-to-volume ratio, nanophotocatalysts offer an increased number of active sites for the interaction between pollutants and reactive molecules. They provide improved charge separation, diminishing electron-hole recombination, hence enhancing photocatalytic efficacy. This leads to accelerated breakdown and more thorough mineralization of persistent pollutants into innocuous end-products such as CO_2 and H_2O. Moreover, in contrast to adsorption or chemical precipitation techniques, nanophotocatalysis does not generate harmful sludge or concentrated waste (Gaur et al., 2022). Pollutants are decomposed instead of being relocated to another medium, resulting in cleaner and more sustainable cleanup. Nanophotocatalysts have demonstrated significant efficacy in the degradation of POPs, pharmaceutical residues, endocrine-disrupting compounds, perfluorinated substances (PFAS), dyes, and pesticides. Their capacity to produce potent oxidizing radicals enables them to cleave even resilient chemical bonds, which most traditional approaches cannot address. Advanced designs, such as magnetic nanophotocatalysts (e.g., Fe_3O_4–TiO_2), provide straightforward separation and recovery posttreatment through the application of magnetic fields, enhancing reusability and recovery. Numerous nanocatalysts sustain efficacy across several cycles, hence decreasing long-term operational expenses. Moreover, nanophotocatalysts can be customized by doping (to diminish bandgap), composite creation (to improve light absorption), and surface functionalization (to target certain contaminants). This adaptability facilitates application across diverse water and soil matrices as well as various environmental situations. Furthermore, nanophotocatalysts can be integrated with other technologies such as adsorption (nano-adsorbents combined with photocatalysis), membrane filtration (photocatalytic membranes), and electrochemical systems (photoelectrocatalysis), thereby enhancing the overall performance and adaptability of the system through a synergistic approach. From a financial perspective, although the initial synthesis costs are substantial, the high efficiency, low chemical input, reusability, and minimal waste formation render nanophotocatalysis more economically feasible in the long term, particularly for large-scale or remote applications (Tahir et al., 2022). Continued research and corporate interest are enhancing the scalability of nanophotocatalysts, photocatalytic reactors for water purification, solar-powered purification devices, and functional coatings for self-cleaning and pollutant degradation on surfaces. Urban areas, sectors, and governmental bodies have commenced investments in nano-enhanced environmental remediation technology. A compilation of nanophotocatalysts is presented in Table 4.8.

Despite these compelling advantages, photocatalysts remain inefficient due to their restricted light accessibility. Subsequent electron-hole recombination diminishes efficiency unless mitigated by dopants or co-catalysts. The UV-dependence of predominant photocatalysts further diminishes their accessibility due to the limited fraction of solar energy they utilize. Moreover, diminished solar intensity may further impede

Table 4.8 List of nanophotocatalysts along with their target pollutants with application.

Nanophotocatalyst	Target pollutants	Application	Reference
Ag/TiO_2, Fe-TiO_2	Arsenic, dyes, drugs	Rural drinking water	Upadhyay et al. (2024)
N-TiO_2, g-C_3N_4	Antibiotics, phenols	Industrial effluent	Islam (2025)
ZnO/TiO_2	Pesticides, pharmaceuticals	Agricultural runoff	Yassin et al. (2024)
Magnetic TiO_2	PCBs	PCB remediation	Dirinck et al. (2011)
TiO_2 concrete	Organics, NOx	Urban runoff	Lin et al. (2020)
CdS/TiO_2 composites	PFAS, antibiotics	Portable systems	Christensen et al. (2022)

kinetics, particularly in the presence of elevated pollution levels. Finally, scalability difficulties have posed significant challenges when deployment is necessitated in large-scale or real-world situations (Liu et al., 2023; O'Connor et al., 2018).

As a result, carbon scaffold-based nanophotocatalysts have garnered extensive international interest in recent years owing to their remarkable efficacy, adaptability, and capacity to address the shortcomings of conventional photocatalysts (Busquets et al., 2016). Consequently, they are undergoing active research and pilot programs worldwide. The materials—graphene, carbon nanotubes (CNTs), carbon quantum dots (CQDs), biochar, and graphitic carbon nitride (g-C_3N_4)—function as supporting frameworks or co-catalysts that significantly improve the photocatalytic efficacy of semiconductors such as TiO_2 and ZnO. A significant limitation of conventional photocatalysts (e.g., pure TiO_2) is the swift recombination of photogenerated electron-hole pairs, which diminishes catalytic effectiveness. Carbon-based scaffolds such as graphene and carbon nanotubes (CNTs) offer an exceptional conductive framework, facilitating rapid electron transport and effective charge separation (Baptista-Pires et al., 2021). This markedly enhances the generation of reactive oxygen species (ROS) that decompose enduring contaminants. Moreover, the majority of traditional photocatalysts are exclusively active under UV light, but carbon-based materials, particularly graphitic carbon nitride (g-C_3N_4) and carbon quantum dots, facilitate the reduction of the bandgap in composites, hence permitting photocatalysis driven by visible light. This enhances the utilization of available solar energy, rendering the process more sustainable. The transition to visible-light activation is essential for practical, cost-effective solar cleanup systems. Numerous carbon scaffolds (e.g., biochar, graphene oxide) can be produced from renewable, biomass-derived, or waste sources. Their synthesis may adhere to green chemistry principles, rendering them appealing for ecologically sensitive applications. This also diminishes production costs in comparison to costly noble metal co-catalysts (such as Pt or Au). Furthermore, due to their extensive surface area and porous architecture, the adsorption of contaminants is significantly enhanced, along with the dispersion of metal oxide nanoparticles and mass transfer within catalytic systems (Boulkhessaim et al., 2022; Trivedi & Majumdar, 2013). They may also undergo chemical modifications (e.g., oxygenation, doping with nitrogen or sulfur) to improve photocatalytic performance or to target certain pollutants. Furthermore, carbon scaffolds can be readily combined with TiO_2, ZnO, WO_3, SnO_2, $BiVO_4$, CdS, g-C_3N_4, and several other nanophotocatalysts (Ding et al., 2016). They create heterojunctions or Z-scheme systems, resulting in enhanced photocatalytic efficacy. Studies have shown that carbon-based nanocomposites can decompose antibiotics (e.g., ciprofloxacin, tetracycline), pesticides (e.g., atrazine, imidacloprid), PFAS, dyes, and endocrine disruptors. They frequently surpass their non-carbon equivalents in degradation rate, reusability, and photostability. From the perspectives of reusability and structural stability, carbon scaffolds preserve their structural integrity over repeated applications. They inhibit the aggregation of metal nanoparticles and withstand photo-corrosion. Numerous carbon-based methods (e.g., magnetic biochar-TiO_2) provide straightforward recovery through filtration or magnetic separation (Baptista-Pires et al., 2021). Furthermore, carbon scaffolds provide multifunctional treatment platforms, such as adsorption combined with photocatalysis, Fenton-like reactions paired with photocatalysis, or photoelectrocatalytic systems. This renders them optimal for intricate water matrices and diverse pollutant types.

4.6.1 Development of Carbon Scaffold-based Nanophotocatalysts

Creating carbon scaffold-based nanophotocatalysts for the efficient remediation of enduring contaminants necessitates a methodical strategy encompassing material design, synthesis, functionality, and application. The following are the key considerations throughout their development.

4.6.1.1 Choice of Carbon Scaffold

A. Material Types

- *Graphene/Graphene Oxide (GO)* – These materials exhibit high conductivity and promote tunable surface chemistry.
- *Carbon Nanotubes (CNTs)* – They facilitate excellent electron mobility.
- *Carbon Quantum Dots (CQDs)* – These have tunable optical properties and high photoluminescence.
- *Biochar/Activated Carbon* – They are cost-effective, porous, and eco-friendly.
- *g-C_3N_4 (graphitic carbon nitride)* – It is visible light-active, metal-free photocatalysts.

B. Integration with Semiconductors

Semiconductor Pairing: Combine with materials like TiO_2, ZnO, $BiVO_4$, and g-C_3N_4 to form heterostructures to enhance charge separation, improve light harvesting, and extend catalyst life and recyclability.

C. Bandgap Engineering for Visible-light Activity

- Carbon scaffolds help narrow the bandgap of semiconductors.
- Doping (e.g., N, S, B) or surface modification of carbon materials improves:
  - Light absorption in the visible range
  - Redox potential
  - Photocatalytic activity under natural sunlight

D. Recovery and Reusability
- Incorporate magnetic nanoparticles (e.g., Fe_3O_4) for easy separation posttreatment.
- Ensure stability and activity over multiple photocatalytic cycles.

E. Green and Scalable Synthesis Methods
- Choose eco-friendly methods like:
  - Hydrothermal or solvothermal synthesis
  - Green reduction of GO using plant extracts
  - Biochar from agricultural waste pyrolysis

F. Toxicity and Environmental Safety
- No toxic by-products or nanoparticles leach into the environment
- Materials do not pose cytotoxic or ecotoxic risks.

G. Real-World Application Suitability
Design for water, wastewater, or soil environments, tolerant to variable pH, temperature, and salinity conditions, and target-specific pollutants (e.g., PFAS, antibiotics, pesticides).

H. Stability and Photocorrosion Resistance
- Carbon scaffolds improve **photostability** of traditional semiconductors like ZnO or CdS.
- Long-term exposure to light and contaminants should not degrade performance.

4.6.2 Mechanistic Understanding and Modeling

A mechanistic knowledge enables researchers to customize the catalyst structure for certain contaminants. It can forecast the long-term behavior and stability of the nanophotocatalysts. It can prevent undesirable by-products. It can enable the construction of hybrid systems (e.g., Fenton + photocatalysis) and expedite the transition from laboratory to field scale. Multiple approaches to mechanistic understanding and modeling can be achieved through the following:

4.6.2.1 Advanced Characterization Tools (e.g., XPS, PL, EIS) to Understand

- Charge transfer dynamics
- Surface interactions with pollutants
- Radical generation mechanisms

As seen previously, photocatalysis relies on light-induced excitation of electrons in a semiconductor, creating electrons (e^-) in the conduction band (CB) and holes (h^+) in the valence band (VB). These charge carriers participate in redox reactions, generating reactive oxygen species (ROS) such as •OH (hydroxyl radicals), O_2^-• (superoxide anions), H_2O_2 (hydrogen peroxide). These ROS then oxidize or reduce POPs into nontoxic end products such as CO_2 and H_2O. Carbon-based scaffolds improve and modify this mechanism in several key ways:

- Act as electron sinks → capture and transfer photogenerated electrons
- Suppress recombination → increase the lifetime of charge carriers

- Extend light absorption → due to π–π^* transitions and surface functional groups
- Improve pollutant adsorption → through high surface area and π–π stacking interactions

This enhances the generation of ROS and increases the degradation efficiency of persistent pollutants. To study the mechanism, various characterization and analytical techniques are used.

4.6.2.2 Analytical Techniques

1. **Photoluminescence (PL) Spectroscopy**
   - Electron-hole recombination is analyzed, where lower PL intensity signifies better charge separation.
2. **Electrochemical Impedance Spectroscopy (EIS)**
   - EIS evaluates charge-transfer resistance and assists in understanding the conductivity of the composite.
3. **Transient Photocurrent Response**
   - Determines how efficiently photo-generated charges move through the material
4. **Electron Spin Resonance (ESR)**
   - Detects ROS such as •OH, O_2^-• and confirms radical generation during photocatalysis.
5. **X-ray Photoelectron Spectroscopy (XPS)**
   - Analyzes surface composition and valence states. Also verifies bonding interactions between carbon and metal oxides.

4.6.2.3 Computational Modeling and Theoretical Studies

1. **Density Functional Theory (DFT)**
   - Predicts band structures and electronic properties in the following ways
     - Bandgap tuning
     - Understanding charge-transfer paths
     - Identifying active sites on carbon scaffold
2. **Molecular Dynamics (MD) Simulations**
   These can be used to study pollutant interactions with catalyst surfaces. Furthermore, they also help in understanding better adsorptive and reactive interfaces.

4.6.2.4 Kinetic Modeling

Kinetic modeling can be realized using Langmuir–Hinshelwood or pseudo-first-order kinetics. They quantify degradation rates and mechanisms, and are useful in comparing material efficiency.

4.6.2.5 Real-time Mechanistic Insights

Following cutting-edge methods have been established for real-time reaction monitoring: such as in situ UV-Vis spectroscopy (to track pollutant degradation), in situ

FTIR (to monitor intermediate products and functional group changes and time-resolved spectroscopy (to reveal charge carrier lifetimes).

4.7 Future Directions and Emerging Trends

Creation of intelligent nanophotocatalysts that react to environmental stimuli, including pH, temperature, or type of pollutant is one of the important directions toward a sustainable future. In another instance, catalysts that modulate activity according to contamination levels or light intensity, influencing improved selectivity, enhanced energy efficiency, and regulated pollutant breakdown. To optimize solar spectrum for economical off-grid water and soil treatment, carbon-based materials (e.g., CQDs, doped g-C_3N_4) can be engineered to facilitate broad-spectrum or near-infrared (NIR) light absorption. Furthermore, the incorporation of additional technologies (such as Fenton-like reactions, electrocatalysts, and biophotocatalysts) may enhance the simultaneous removal of diverse pollutant categories (e.g., heavy metals and organics) in a single process. Recent studies indicate that agricultural waste and biomass-derived carbon compounds, such as biochar and hydrochar, have been utilized to create low-cost, sustainable photocatalysts while effectively managing organic waste streams. Moreover, carbon compounds derived from waste frequently exhibit intrinsic porosity and functional groups conducive to photocatalysis. To mitigate secondary pollutants and enhance life cycle economics, magnetic nanostructures (Fe_3O_4, $CoFe_2O_4$) have been integrated into carbon-based catalysts for efficient recovery and reusability. Reports indicate ultra-efficient degradation with little material consumption utilizing single-atom catalysts (SACs) on carbon supports to optimize atom utilization and ensure exact catalytic sites, exemplified by isolated Fe or Co atoms on nitrogen-doped graphene. Numerous reports have documented the real-time detection of contaminants, degradation efficiency, and intermediate products with nanophotocatalyst systems equipped with integrated sensors. Furthermore, to expedite discovery and minimize experimental trial-and-error, machine learning (ML) and computer modeling have been utilized to predict photocatalytic activity, optimize material combinations, and screen dopants and functional groups. Furthermore, to diminish dependence on hazardous solvents, costly precursors, and energy-demanding processes, synthesis methods using hydrothermal, microwave, or sol–gel procedures with natural reductants at ambient temperature have been advocated.

4.8 Conclusion

The advent of carbon scaffold-based nanophotocatalysts has signified a major development in environmental remediation, especially in addressing persistent contaminants. These pollutants—such as medicines, pesticides, industrial dyes, PFAS, and other xenobiotics—constitute a significant hazard to the environment and public health owing to their stability, resistance to natural degradation, and bioaccumulative characteristics. Conventional techniques including incineration, chemical oxidation, and standard photocatalysis have deficiencies in efficiency, selectivity, energy consumption, and environmental sustainability. The incorporation of carbon-based scaffolds into photocatalytic systems has advanced sustainable, high-performance remediation solutions. Carbon materials, including graphene, carbon nanotubes (CNTs), biochar, carbon quantum dots (CQDs), and graphitic carbon nitride (g-C_3N_4), possess distinctive features that render them optimal for facilitating and augmenting the photocatalytic process. These encompass elevated surface area, superior electrical conductivity, chemical stability, adjustable porosity, and functional adaptability.

In summary, carbon scaffold-based nanophotocatalysts constitute a potent, versatile, and sustainable method for the remediation of enduring pollutants. Their capacity to utilize light energy, facilitate intricate reactions, and function in varied environmental circumstances presents a viable resolution to one of the most urgent concerns of our times. Through ongoing study, interdisciplinary collaboration, and innovation, these sophisticated materials can significantly contribute to restoring ecological balance and securing a cleaner, safer environment for future generations.

References

Ahmed, H., Sharif, A., Bakht, S., Javed, F., & Hassan, W. (2021). Persistent organic pollutants and neurological disorders: From exposure to preventive interventions. In M. S. H. Akash & K. Rehman (Eds.), *Environmental Contaminants and Neurological Disorders* (pp. 231–247). Springer International Publishing. https://doi.org/10.1007/978-3-030-66376-6_11

Alharbi, O. M. L., Basheer, A. A., Khattab, R. A., & Ali, I. (2018). Health and environmental effects of persistent organic pollutants. *Journal of Molecular Liquids*, 263, 442–453. https://doi.org/10.1016/j.molliq.2018.05.029

Andriyashina, T. V., Saratovskikh, E. A., Martynova, V. F., Kaz'min, V. M., Chepegin, I. V., & Chizhova, M. A. (2015). Content of radionuclides and persistent organic pollutants in agricultural soils after the Chernobyl nuclear power plant accident. *Russian Journal of Physical Chemistry B*, 9(3), 445–453. https://doi.org/10.1134/S1990793115030148

Ashraf, M. A. (2017). Persistent organic pollutants (POPs): A global issue, a global challenge. *Environmental Science and Pollution Research*, 24(5), 4223–4227. https://doi.org/10.1007/s11356-015-5225-9

Baptista-Pires, L., Norra, G.F., & Radjenovic, J. (2021). Graphene-based sponges for electrochemical degradation of persistent organic contaminants. *Water Research*, *203*, 117492. https://doi.org/10.1016/j.watres.2021.117492

Basel Convention (1989). *Basel Convention on the Control of Transboundary Movements of Hazardous Wastes and their Disposal*. https://www.basel.int/TheConvention/Overview/TextoftheConvention/tabid/1275/Default.aspx

Beyer, A., Mackay, D., Matthies, M., Wania, F., & Webster, E. (2000). Assessing long-range transport potential of persistent organic pollutants. *Environmental Science & Technology*, 34(4), 699–703. https://doi.org/10.1021/es990207w

Boulkhessaim, S., Gacem, A., Khan, S. H. et al. (2022). Emerging trends in the remediation of persistent organic pollutants using nanomaterials and related processes: A review. *Nanomaterials*, 12(13), Article 13. https://doi.org/10.3390/nano12132148

Busquets, R., Ivanov, A. E., Mbundi, L. et al. (2016). Carbon-cryogel hierarchical composites as effective and scalable filters for removal of trace organic pollutants from water. *Journal of Environmental Management*, 182, 141–148. https://doi.org/10.1016/j.jenvman.2016.07.061

Chakraborty, J. & Das, S. (2016). Molecular perspectives and recent advances in microbial remediation of persistent organic pollutants. *Environmental Science and Pollution Research*, 23(17), 16883–16903. https://doi.org/10.1007/s11356-016-6887-7

Chandra, P. & Enespa. (2019). Mycoremediation of environmental pollutants from contaminated soil. In A. Varma & D. K. Choudhary (Eds.), *Mycorrhizosphere and Pedogenesis* (pp. 239–274). Springer. https://doi.org/10.1007/978-981-13-6480-8_15

Christensen, E. R., Wang, Y., Huo, J., & Li, A. (2022). Properties and fate and transport of persistent and mobile polar organic water pollutants: A review. *Journal of Environmental Chemical Engineering*, 10(2), 107201. https://doi.org/10.1016/j.jece.2022.107201

Corsolini, S., Ademollo, N., Romeo, T., Greco, S., & Focardi, S. (2005). Persistent organic pollutants in edible fish: A human and environmental health problem. *Microchemical Journal*, 79(1), 115–123. https://doi.org/10.1016/j.microc.2004.10.006

Ding, Y., Tang, Y., Yang, L. et al. (2016). Porous nitrogen-rich carbon materials from carbon self-repairing g-C_3N_4 assembled with graphene for high-performance supercapacitor. *Journal of Materials Chemistry A*, 4(37), 14307–14315. https://doi.org/10.1039/c6ta05267c

Dirinck, E., Jorens, P. G., Covaci, A. et al. (2011). Obesity and persistent organic pollutants: Possible obesogenic effect of organochlorine pesticides and polychlorinated biphenyls. *Obesity*, 19(4), 709–714. https://doi.org/10.1038/oby.2010.133

Dong, H., Zeng, G., Tang, L. et al. (2015). An overview on limitations of TiO_2-based particles for photocatalytic degradation of organic pollutants and the corresponding countermeasures. *Water Research*, 79, 128–146. https://doi.org/10.1016/j.watres.2015.04.038

Doula, M. K., Kavvadias, V. A., & Elaiopoulos, K. (2012). Zeolites in soil remediation processes. In *Handbook of Natural Zeolites* (pp. 519–568). Bentham Science Publishers. https://www.benthamdirect.com/content/books/9781608052615.chapter-22

El-Shahawi, M. S., Hamza, A., Bashammakh, A. S., & Al-Saggaf, W. T. (2010). An overview on the accumulation, distribution, transformations, toxicity and analytical methods for the monitoring of persistent organic pollutants. *Talanta*, 80(5), 1587–1597. https://doi.org/10.1016/j.talanta.2009.09.055

Ewis, D., Ba-Abbad, M. M., Benamor, A., & El-Naas, M. H. (2022). Adsorption of organic water pollutants by clays and clay minerals composites: A comprehensive review. *Applied Clay Science*, 229, 106686. https://doi.org/10.1016/j.clay.2022.106686

Fei, L., Bilal, M., Qamar, S. A. et al. (2022). Nano-remediation technologies for the sustainable mitigation of persistent organic pollutants. *Environmental Research*, 211, 113060. https://doi.org/10.1016/j.envres.2022.113060

Feng, B., Onda, Y., Wakiyama, Y., Taniguchi, K., Hashimoto, A., & Zhang, Y. (2022). Persistent impact of Fukushima decontamination on soil erosion and suspended sediment. *Nature Sustainability*, 5(10), 879–889. https://doi.org/10.1038/s41893-022-00924-6

Gaur, N., Dutta, D., Singh, A., Dubey, R., & Kamboj, D. V. (2022). Recent advances in the elimination of persistent organic pollutants by photocatalysis. *Frontiers in Environmental Science*, 10, 22076. https://doi.org/10.3389/fenvs.2022.872514

Gill, D. A. & Mix, T. L. (2020). Chapter 25 - Love canal: A classic case study of a contaminated community. In C. N. Pope & J. Liu (Eds.), *An Introduction to Interdisciplinary Toxicology* (pp. 341–352). Academic Press. https://doi.org/10.1016/B978-0-12-813602-7.00025-9

Giroux, M. S., Siddiqui, S., & Brander, S. M. (2024). Classic contaminants in aquatic ecosystems: POPs, PFAS, heavy metals, and microplastics. In S. Siddiqui & S. M. Brander (Eds.), *Aquatic Ecotoxicology: Understanding Pollutants, Aquatic Organisms, and their Environments* (pp. 43–58). Springer International Publishing. https://doi.org/10.1007/978-3-031-53130-9_4

Gong, P., Xu, H., Wang, C., Chen, Y., Guo, L., & Wang, X. (2021). Persistent organic pollutant cycling in forests. *Nature Reviews Earth & Environment*, 2(3), 182–197. https://doi.org/10.1038/s43017-020-00137-5

Hao, Y., Zheng, S., Wang, P. et al. (2021). Ecotoxicology of persistent organic pollutants in birds. *Environmental Science: Processes & Impacts*, 23(3), 400–416. https://doi.org/10.1039/D0EM00451K

Helmer, R. W., Reeves, D. M., & Cassidy, D. P. (2022). Per- and polyfluorinated alkyl substances (PFAS) cycling within Michigan: Contaminated sites, landfills and wastewater treatment plants. *Water Research*, 210, 117983. https://doi.org/10.1016/j.watres.2021.117983

Islam, Md. M. (2025). A thorough examination of photocatalysis for addressing wastewater pollution. *ChemistrySelect*, 10(14), e202500274. https://doi.org/10.1002/slct.202500274

Jagannath, A., Shore, R. F., Walker, L. A., Ferns, P. N., & Gosler, A. G. (2008). Eggshell pigmentation indicates pesticide contamination. *Journal of Applied Ecology*, 45(1), 133–140. https://doi.org/10.1111/j.1365-2664.2007.01386.x

Jang, Y.-C. & Townsend, T. G. (2001). Occurrence of organic pollutants in recovered soil fines from construction and demolition waste. *Waste Management*, 21(8), 703–715. https://doi.org/10.1016/S0956-053X(01)00007-1

Jeirani, Z., Niu, C. H., & Soltan, J. (2017). Adsorption of emerging pollutants on activated carbon. *Reviews in Chemical Engineering*, 33(5), 491–522. https://doi.org/10.1515/revce-2016-0027

Jitae, K., Taeyoung, K., & Thi-Huong, P. (2025). Degradation of nonpoint source pollutants derived from plastic waste using WO_3/g-C_3N_4 heterojunction photocatalyst. *Process Safety and Environmental Protection*, 195, 106704. https://doi.org/10.1016/j.psep.2024.12.085

Jones, K. C. & de Voogt, P. (1999). Persistent organic pollutants (POPs): State of the science. *Environmental Pollution*, 100(1), 209–221. https://doi.org/10.1016/S0269-7491(99)00098-6

Khan, R., Inam, M. A., Khan, S., Park, D. R., & Yeom, I. T. (2019). Interaction between persistent organic pollutants and ZnO NPs in synthetic and natural waters. *Nanomaterials*, 9(3), Article 3. https://doi.org/10.3390/nano9030472

Lallas, P. L. (2001). The Stockholm convention on persistent organic pollutants. *American Journal of International Law*, 95(3), 692–708. https://doi.org/10.2307/2668517

Lin, F., Wang, Z., Zhang, Z. et al. (2020). Flue gas treatment with ozone oxidation: An overview on NO*x*, organic pollutants, and mercury. *Chemical Engineering Journal*, 382, 123030. https://doi.org/10.1016/j.cej.2019.123030

Liu, Z., Ren, X., Duan, X., Sarmah, A. K., & Zhao, X. (2023). Remediation of environmentally persistent organic pollutants (POPs) by persulfates oxidation system (PS): A review. *Science of The Total Environment*, 863, 160818. https://doi.org/10.1016/j.scitotenv.2022.160818

Maldani, M., Nassiri, L., & Ibijbijen, J. (2022). Biodegradation and remediation of pesticides in contaminated agroecosystems: Special reference to Glyphosate and Paraquat. In N. K. Arora & B. Bouizgarne (Eds.), *Microbial BioTechnology for Sustainable Agriculture Volume 1* (pp. 489–545). Springer Nature. https://doi.org/10.1007/978-981-16-4843-4_17

Masih, A., Lal, J. K., & Patel, D. K. (2014). Contamination and exposure profiles of persistent organic pollutants (PAHs and OCPs) in groundwater at a terai belt of North India. *Water Quality, Exposure and Health*, 6(4), 187–198. https://doi.org/10.1007/s12403-014-0126-6

Matthies, M., Solomon, K., Vighi, M., Gilman, A., & Tarazona, J. V. (2016). The origin and evolution of assessment criteria for persistent, bioaccumulative and toxic (PBT) chemicals and persistent organic pollutants (POPs). *Environmental Science: Processes & Impacts*, 18(9), 1114–1128. https://doi.org/10.1039/C6EM00311G

Meylemans, H. A., Groshens, T. J., & Harvey, B. G. (2012). Synthesis of renewable bisphenols from creosol. *ChemSusChem*, 5(1), 206–210. https://doi.org/10.1002/cssc.201100402

Milic, J., Curcic, M., Brnjas, Z. et al. (2019). The socio-economic impact timeline in Serbia for persistent organic pollutants (POPs). *Science of The Total Environment*, 688, 486–493. https://doi.org/10.1016/j.scitotenv.2019.06.161

Muir, D. C. G. & Howard, P. H. (2006). Are there other persistent organic pollutants? A challenge for environmental chemists. *Environmental Science & Technology*, 40(23), 7157–7166. https://doi.org/10.1021/es061677a

Naghdi, S., Shahrestani, M. M., Zendehbad, M., Djahaniani, H., Kazemian, H., & Eder, D. (2023). Recent advances in application of metal-organic frameworks (MOFs) as adsorbent and catalyst in removal of persistent organic pollutants (POPs). *Journal of Hazardous Materials*, 442, 130127. https://doi.org/10.1016/j.jhazmat.2022.130127

Nassri, I., khattabi rifi, S., Sayerh, F., & Souabi, S. (2023). Occurrence, pollution sources, and mitigation prospects of Antibiotics, anti-inflammatories, and endocrine disruptors in the aquatic environment. *Environmental Nanotechnology, Monitoring & Management*, 20, 100878. https://doi.org/10.1016/j.enmm.2023.100878

Ni, N., Kong, D., Wu, W., He, J., Shan, Z., Li, J. et al. (2020). The role of biochar in reducing the bioavailability and migration of persistent organic pollutants in soil–plant systems: A review. *Bulletin of Environmental Contamination and Toxicology*, 104(2), 157–165. https://doi.org/10.1007/s00128-019-02779-8

Noor, S., Ashar, A., Taj, M. B., & Bhutta, Z. A. (2022). Advanced oxidation processes for remediation of persistent organic pollutants. In *Advanced Oxidation Processes for Wastewater Treatment*. CRC Press.

Novikau, R. & Lujaniene, G. (2022). Adsorption behaviour of pollutants: Heavy metals, radionuclides, organic pollutants, on clays and their minerals (raw, modified and treated): A review. *Journal of Environmental Management*, 309, 114685. https://doi.org/10.1016/j.jenvman.2022.114685

O'Connor, D., Hou, D., Ok, Y. S. et al. (2018). Sustainable in situ remediation of recalcitrant organic pollutants in groundwater with controlled release materials: A review. *Journal of Controlled Release*, 283, 200–213. https://doi.org/10.1016/j.jconrel.2018.06.007

Organic Pollutants in Leachates from Landfill Sites—Sawhney—1984—Journal of Environmental Quality—Wiley Online Library. (n.d.). Retrieved April 20, 2025, from https://acsess.onlinelibrary.wiley.com/doi/abs/10.2134/jeq1984.00472425001300030005x

Pariatamby, A. & Kee, Y. L. (2016). Persistent organic pollutants management and remediation. *Procedia Environmental Sciences*, 31, 842–848. https://doi.org/10.1016/j.proenv.2016.02.093

Pierri, D. (2021). Actual decay of tetrachloroethene (PCE) and trichloroethene (TCE) in a highly contaminated shallow groundwater system. *Environmental Advances*, 5, 100090. https://doi.org/10.1016/j.envadv.2021.100090

Tahir, M. B., Nawaz, T., Nabi, G., Sagir, M., Isa Khan, M., & Malik, N. (2022). Role of nanophotocatalysts for the treatment of hazardous organic and inorganic pollutants in wastewater. *International Journal of Environmental Analytical Chemistry*, 102(2), 491–515. https://doi.org/10.1080/03067319.2020.1723570

Titchou, F. E., Zazou, H., Afanga, H., El Gaayda, J., Akbour, R. A., & Hamdani, M. (2021). Removal of Persistent Organic Pollutants (POPs) from water and wastewater by adsorption and electrocoagulation process. *Groundwater for Sustainable Development*, 13, 100575. https://doi.org/10.1016/j.gsd.2021.100575

Trivedi, J. & Majumdar, D. (2013). Memory effect driven emissions of persistent organic pollutants from industrial thermal processes, their implications and management: A review. *Journal of Environmental Management*, 119, 111–120. https://doi.org/10.1016/j.jenvman.2013.01.026

UN Treaty Collection (2013). *Chapter XXVII – Environment, 17. Minamata Convention on Mercury*. https://treaties.un.org/Pages/ViewDetails.aspx?src=TREATY&mtdsg_no=XXVII-17&chapter=27&clang=_en

Upadhyay, S., Sharma, P. K., Dogra, K. et al. (2024). Microplastics in freshwater: Unveiling sources, fate, and removal strategies. *Groundwater for Sustainable Development*, 26, 101185. https://doi.org/10.1016/j.gsd.2024.101185

Venny, Gan, S., & Ng, H. K. (2012). Current status and prospects of Fenton oxidation for the decontamination of persistent organic pollutants (POPs) in soils. *Chemical Engineering Journal*, 213, 295–317. https://doi.org/10.1016/j.cej.2012.10.005

Wagner, M., Andrew Lin, K.-Y., Oh, W.-D., & Lisak, G. (2021). Metal-organic frameworks for pesticidal persistent organic pollutants detection and adsorption – A mini review. *Journal of Hazardous Materials*, 413, 125325. https://doi.org/10.1016/j.jhazmat.2021.125325

Wang, H., Xu, Q., Zhou, X. et al. (2022). Highly efficient adsorptive removal of persistent organic pollutants using NPD-acid combined modified NaY zeolites. *Chemical Engineering Journal*, 431, 133858. https://doi.org/10.1016/j.cej.2021.133858

Yadav, I. C., Devi, N. L., Syed, J. H. et al. (2015). Current status of persistent organic pesticides residues in air, water, and soil, and their possible effect on neighboring countries: A comprehensive review of India. *Science of The Total Environment*, 511, 123–137. https://doi.org/10.1016/j.scitotenv.2014.12.041

Yang, L., Wang, S., Peng, X. et al. (2019). Gas chromatography-orbitrap mass spectrometry screening of organic chemicals in fly ash samples from industrial sources and implications for understanding the formation mechanisms of unintentional persistent organic pollutants. *Science of The Total Environment*, 664, 107–115. https://doi.org/10.1016/j.scitotenv.2019.02.001

Yassin, M. T., Negi, A., Ringwal, S., Pandey, M., Sati, P., & Tripathi, P. K. (2024). Enhancing wastewater treatment: DFT and experimental insights into ZnO-implanted TiO_2 Z-scheme photocatalyst mediated via *Smilax aspera*. *Optical Materials*, 155, 115832. https://doi.org/10.1016/j.optmat.2024.115832

Yu, J.-G., Zou, J., Liu, L.-L., Jiang, X.-Y., Jiao, F.-P., & Chen, X.-Q. (2017). Preparation of TiO_2-based photocatalysts and their photocatalytic degradation properties for methylene blue, rhodamine B and methyl orange. *Desalination and Water Treatment*, 81, 282–290. https://doi.org/10.5004/dwt.2017.21109

Zhang, S., Li, B., Wang, X. et al. (2020). Recent developments of two-dimensional graphene-based composites in visible-light photocatalysis for eliminating persistent organic pollutants from wastewater. *Chemical Engineering Journal*, 390, 124642. https://doi.org/10.1016/j.cej.2020.124642

Section 3

Industrial Implications and Energy Economics

5

Carbon Scaffold-based Nanophotocatalysts in Semiconductors: Fabrication, Performance, and Prospects

*Md. Iftikar Hussain[1], Gyandeep Saikia[1], Ranjeeta Singh[1], Unmilita Phukan[1], Dipjyoti Dutta[2], Dhrubajit Borah[3], Himashree Mahanta[1], and Rekha Rani Dutta[1]**

[1] *Department of Chemistry, School of Basic Sciences, The Assam Kaziranga University, Jorhat, Assam*
[2] *Material sciences and Technology Division (MSTD), CSIR-North East Institute of Science and Technology (NEIST), Jorhat, Assam, India*
[3] *Department of Chemistry, N.N. Saikia College, Titabar, Jorhat, Assam, India*

*Corresponding author: rekha@kazirangauniversity.in

5.1 Introduction

The technologies for semiconductors are continually evolving. Finding long-lasting, effective, and adaptable photocatalysts is crucial for progress in a number of scientific and technological domains. Owing to the special qualities of carbon scaffold-based materials, such as their large surface area, adjustable electrical structure, and remarkable chemical stability, they have emerged as viable options for a variety of uses. Because of their unique characteristics, such as small size, increased adsorption and active sites, bandgap narrowing, co-catalysis, photosensitization, and photocatalysis, carbon scaffold-based nanomaterials have previously shown a great deal of promise to increase the efficiency of semiconductor photocatalysts. Nanosized photocatalysts have multipurpose properties and can therefore be used successfully in various applications including energy, environment and medical fields (Ahmed et al., 2023). The adsorption site of the reagent's active center can be improved by combining carbon scaffold-based nanomaterial with a photocatalyst. Examples include cocatalyst, photosensitizer, photocatalyst, and E_g narrowing effect; photocatalytic water splitting of semiconductor photocatalysts can be enhanced by employing carbon scaffold-based nanomaterials as catalyst support, an electron acceptor, and a transference canal booster. Furthermore, semiconductor nanomaterials can also be

Photocatalysts based on Carbon Scaffolds: Environmental and Industrial Applications, First Edition. Edited by Ramesh Oraon et al.
Published 2026 by Wiley-VCH GmbH.

used as photocatalyst for the acceleration of redox reactions by direct water splitting to generate chemical fuel or electricity Rasool et al. (2022). For water splitting, carbon scaffold-based nanomaterials might be a better option due to their many benefits, which include inexpensive costs, large reserves, favorable qualities, and the possibility of reducing their environmental impact. Many methods have emerged, however there are still certain drawbacks, such poor hydrogen storage and high manufacturing costs. Recent developments in heterostructures and hybrid applications have increased the efficiency of H_2 production significantly. In this aspect, carbon-based nanomaterials and their modifications have successfully removed the existing limitations that have an adverse effect on photocatalytic performance (Ahmed et al., 2023). Techniques like the hydrothermal method, electrospinning, self-assembly, covalent reaction, sol–gel method electrostatic self-assembly, phase inversion method, noncovalent reaction, solution phase reduction, photochemical reaction, electrostatic interaction, irradiation technique, sputtering, mechanochemical synthesis, calcination, ultrasonic method, a solvothermal technique, and deposition method can be used to fabricate the carbon scaffold-based nanophotocatalyst mediated semiconductors. The performance of such semiconductors can be enhanced using various processes such as doping, heterojunction, crystal phase engineering, cocatalyst loading, etc. The following paragraph summarizes the various fabrication methods of carbon scaffold-based nanophotocatalyst mediated semiconductor material.

5.2 Fabrication Method

5.2.1 Hydrothermal Method

The hydrothermal method uses an autoclave to heat the precursors in a water solvent at a high pressure and temperature. Generally, heating occurs at a temperature that is higher than 100 °C but below 300 °C. The closed system's high temperature creates a high-pressure environment that promotes reactant solubility. The two main phases in the crystallization process used in this hydrothermal approach are nucleation and crystal growth. This technique is most commonly utilized for different nanomaterial preparations because of how easy and straightforward it is to use. Temperature affects the product greatly because it is the sole direct controllable parameter. For instance, in standard ZnO nanorod production, the temperature closely correlated with the nanorods' diameters. Moreover, the residency period might be advantageous when the nanorod width grows with an increase in reaction time. In the same investigation, pH also affected the structural morphology of ZnO synthesis; that is, an acidic pH promoted nanorod morphology whereas an alkaline pH favored a structure resembling a flower. Numerous nanomaterials, including heterostructures like $g\text{-}C_3N_4/Bi_2WO_6$, were created by this technique. The resulting $g\text{-}C_3N_4$ was morphologically flake-like, while the morphology of Bi_2WO_6 was aggregated laminar. Similarly, a photocatalyst with an uneven sheet shape was also synthesized using this method. Irregular fragments, spheres like flowers, spherical nanoparticles, and hexagonal morphology (Rani et al., 2018). Additional instances of creating carbon scaffold-based nanophotocatalysts for semiconductor applications include creating the core–shell structure of a TiO_2-carbon sphere, in which TiO_2 is anchored on carbon spheres through hydrothermal polymerization of sucrose to bathochromically increase its light-response area; creating a carbon–ZnO nanocomposite for photocatalytically breaking down Rhodamine B dyes; creating a graphene–ZnO–Au nanocomposite for breaking down nitrobenzene; creating a graphene–TiO_2–Au nanocomposite for breaking down 2,4-dichlorophenol; creating a graphdiyne–ZnO nanocomposite for photodegradation of azo dyes; and creating carbon–indium oxide (In_2O_3) nanobelts for photocatalytically breaking down CO_2 into CO and CH_4 with water (Mondal & Gupta, 2018). Additional research in this area includes the synthesis of CQDs/ZnO nanocomposite (~20–30 nm) using a hydrothermal one-step method, which was then used to photodegrade toxic methanol and benzene (>80% degradation efficiency after 24 hours) in air at room temperature under visible light; the evaluation of the synthesis of CQDs/CdS photocatalysts toward the photodegradation of RhB; and the photodegradation of Alizarin red S (ARS) dye using CQDs/ZnS photo/nanocatalysts under similar visible-light irradiation conditions. Recent literature also reviews the fabrication of polymer dots from polymers whose core contains carbon dots using hydrothermal techniques. The suitability of the hydrothermal method is demonstrated by works such as the synthesis of CDs/PDs–TiO_2 nanocomposites with Ti—O—C bonds, which were made by grafting CDs/PDs with TiO_2 nanoparticles via a simple hydrothermal thermal treatment and their application toward MO dye degradation in aqueous solutions (Figure 5.1); the hydrothermal fabrication of Fe_2O_3/CQDs nanocomposites as efficient photocatalysts for the degradation of noxious gases, namely benzene or methanol in vis-light conditions; the synthesis of N-GQDs/TiO_2 nanocomposite via a hydrothermal procedure and its deployment as a photo/nanocatalyst for the photodegradation of aqueous MB (Nasrollahzadeh et al., 2021).

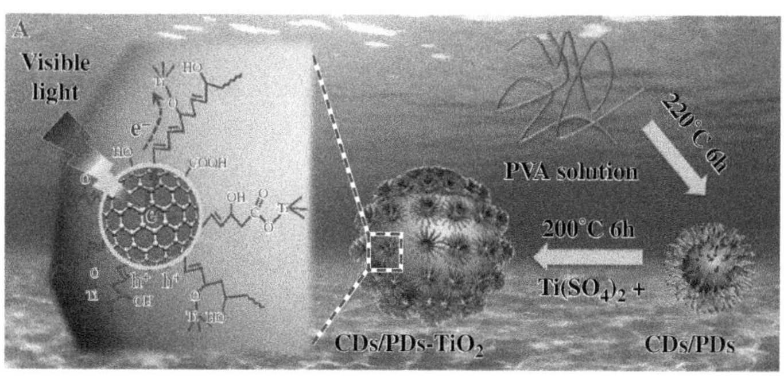

Figure 5.1 Mechanistic photo catalytic illustration for CDs/PDs-TiO_2 under Vis-light irradiation. *Source*: Nasrollahzadeh et al. (2021) / with permission of Elsevier.

5.2.2 Electro Spinning Method

Many attempts are currently being made to improve the qualities of nanofibers and scale up production. One of the more promising fabrication techniques is electro spinning, which allows one to create nanofibers with diverse materials in a range of fibrous configurations. In recent years, there has been a significant increase in interest in electrospinning among academics and industry because of its affordability, ease of use, and versatility with a broad range of materials. A dry spinning method called electrospinning is utilized to create continuous nanofibers. Electrostatic force is used to extract fibers from a molten or liquid polymer solution, and nanofiber networks are created all at once. It is regarded as an appropriate, adaptable method for mass-producing electrostatic fibers. With this technique, it is easy to regulate the fiber size from nanometer to micrometer.

Because this method may provide nanofibers with a high surface-to-volume ratio and a structure mimicking the extracellular matrix (ECM), it has attracted a lot of interest for biomedical research. Many different types of nanofibers, such as smooth, branching, core–shell, ribbon-like, porous, and having fractal surface features, can be produced using electrospinning. A mass manufacturing infrastructure for associated equipment has been established by the widespread usage of electrospinning in various industries and high-tech domains. This equipment is now widely available alongside nanofibrous material. Different electrospinning-based methods can be used to create 2D and 3D hybrid nanofiber structures for varied uses. Along with the main 3D electrospinning techniques, such as multi-layer electrospinning, wet electrospinning, template-assisted electrospinning, electrospinning with postprocessing, self-assembly electrospinning, and combining 3D printing with electrospinning, the conventional methods used for 2D nanostructures include blend electrospinning, coaxial electrospinning, emulsion electrospinning, and side by side electrospinning. The benefit of the spinning techniques is their ease of use in producing substantial quantities of carbon fiber, both in lab and industrial settings. Different carbon nanofibers obtained by electrospinning including carbon nanofibers from modified single-nozzle electrospinning, porous carbon nanofibers, metal-oxide-decorated carbon nanofibers and carbon nanofibers from modified dual-nozzle electrospinning (Abadi et al., 2022).

5.2.3 Self-assembly Method

A straightforward and economical method is the self-assembly approach-based "bottom-up" technology. Using the benefit of molecular self-interaction, self-assembly is one of the most crucial "molecular engineering" techniques for creating complex functional structures at the micro- and molecular levels. Designer molecules and supramolecular entities are used in the nanofabrication process of "molecular self-assembly," which allows molecules to spontaneously group into desired forms. This technique simplifies a lot of challenging nanofabrication tasks, especially those that need for atomic level structural alterations. Furthermore, molecule self-assembly tends to produce highly defect-free and self-healing structures since the target structures are selective and have thermodynamically stable assembly between the possible configurations. A number of self-assembly techniques, including liquid crystals, zero gels, phase-separation of copolymers, pore generation in alumina, and polymer spheres templating, have been developed recently (Pradeesh et al., 2013).

Other instances of carbon-scaffold based nano-photocatalysts in semiconductors made by self-assembly methods include the synthesis of Ti_3C_2/g-C_3N_4 nanophotocatalyst by electrostatic self-assembly approach for photocatalytic H_2 evolution, the self-assembly of uniform mesoporous anatase TiO_2 on graphene sheets (Radhika et al., 2019), Pt-TiO_2 nano-photocatalysts supported on various carbonaceous material (MWCNT, SWCNT, rGO, and AC) prepared by the electrostatic self-assembly method to use in CO_2 photoreduction under visible-light irradiation for CH_4 production (Larimi et al., 2020), and synthesis of Ti_3C_2/g-C_3N_4 nanophotocatalyst by electrostatic self-assembly approach for photocatalytic H_2 evolution. The synthesis of WO_3/g-C_3N_4 involved electrostatic self-assembly, WO_3 nanosheets, and g-C_3N_4 were used together for photocatalytic H_2 evolution (Kumar et al., 2021). O-doped g-C_3N_4 nanosheets and Ti_3C_2 MXene nanosheets were used together for photocatalytic H_2 evolution.

5.2.4 Covalent Reaction

By covalently adding polyurethane (PU) prepolymers to trace amounts of functionalized CNTs $[OH]_n$-SWCNTs using an effective method with mild chemicals, polymer nanocomposites have been created as transparent films. These PU nanocomposites exhibit improved mechanical, thermal, and conductivity (10^{-4} S cm^{-1}) capabilities because SWCNTs were uniformly distributed throughout them via covalent interaction between SWCNTs and the polyurethane network. The more covalent nature of bonding produces semiconductors with comparatively high carrier mobilities for closed-shell oxides based on cations with relatively high electronegativity, such as in ZnO and SnO_2 (Banerjee et al., 2011).

5.2.5 Sol–gel Method

Making a sol that disperses the target material's precursor in a polymeric liquid is the first stage in the sol–gel process, which is then followed by gelation. Because of its inherent benefits of simplicity and low cost, the sol–gel technique has been widely employed in the construction of heterojunctions. The included polymer can be removed by calcination. The core–shell structure is uniquely designed using sol–gel methods. This is one method of creating metal oxide nanoparticles using wet chemistry. This approach has gained popularity specifically because it is less expensive, uses a low-temperature procedure, and blends molecules homogeneously (Pandikumar & Jothivenkatachalam, 2019). The sol–gel procedure is one way to create nanocrystalline semiconductor metal oxide from an organometal or metal alkoxide.

A sol is created when graphene is mixed with a metal alkoxide or an organometal and magnetically agitated in the presence of an ethyl alcohol-based dispersion medium. Following a sequence of polycondensation and phase transitions brought on by heating and drying, semiconductor nanoparticles in the sol become uniformly attached on grapheme sheets. Using ethanol as a solvent and titanium (IV) butoxide as the Ti source, the procedure was used to create the TiO_2–GR nanocomposite. Additionally, it is necessary to add processes to reduce GO in all circumstances when it is not reduced in situ while the semiconductor's nanocomposite is being prepared. Usually, heating or chemical reduction are used to accomplish this. Thus, the most successfully produced nanocomposites are semiconductor-reduced graphene oxide (RGO) composites. If reduction is not performed, GO's comparatively low electrical conductivity will prevent it from extending the lifetime of generated electrons and holes to a point where they can be beneficial. After graphene is obtained, the semiconductor is frequently added to the nanocomposite using sol–gel, solvothermal, in situ growth, or solution-mixing techniques (Radhika et al., 2019). Some examples of fabrication of carbon-scaffold based nano-photocatalysts in semiconductors by sol–gel methods include the synthesis of TiO_2/AC nanocomposite by sol–gel method for Oxidation of propene. synthesis of TiO_2/MWCNTs and TiO_2/AC nanocomposites by sol–gel method for degradation of Acid Blue 92, synthesis of CNTs/TiO_2 nanocomposite by sol–gel method for degradation of MB, synthesis of GO-TiO_2 NFs by sol–gel method for photocatalytic H_2 evolution and dye-sensitized H_2 evolution, synthesis of $LaFeO_3$–rGO nanocomposite by high temperature sol–gel method for oxidation of MB or RhB (Han et al., 2017).

5.2.6 Phase Inversion Method

Arguably, the most straightforward method used to create polymeric films – primarily membranes – is casting. Phase inversion is the process by which a forming solution is mechanically distributed to generate polymeric films. Thicker films (micrometer range) are typically produced by this method, and porosity is necessary to attain the necessary characteristics for water treatment like adsorption, catalysis, or ultrafiltration. To impart the active characteristics, porosity, and expanded surface area of the membrane, additives and active materials can be added to the forming fluid. As an alternative, the selection of the polymer and the conditions of film creation can cause a porous network to self-assemble. Phase inversion refers to the separation of the components of the forming solution into their respective phases. This can be achieved by either chilling the solution or by introducing an anti-solvent to force the polymeric phase to precipitate. After this procedure, the films are cleaned of the additive or templating phase to create a porous structure with a large surface area. Through manipulation of the forming solution and the phase separation of its constituent parts, various porous nanostructures can be produced. Typically, a blend of polymers that are recognized for forming domains with the appropriate pore size is utilized (Karchiyappan et al., 2022). Multiwalled CNT/PSf-mixed matrix membranes were cast by Choi et al. using nonsolvent-induced phase inversion. Acid is applied beforehand to the nanotubes to help with their dispersion in the solvent. Because carboxylic acid groups accumulate on CNT surfaces after acid pretreatment, the presence of CNTs boosts the membranes' surface hydrophilicity. By spreading multiwalled CNTs (4 wt%) throughout a PSf/PVP polymer matrix by phase inversion, Brunet et al. created nanotube/polymer membranes. It seems that PVP helped the CNTs disperse throughout the membrane casting solution (Pendergast & Hoek, 2011).

5.2.7 Noncovalent Reaction

The potential for noncovalent interactions between CNTs and polymers has been studied by numerous teams. The field of study on noncovalent polymer-CNT composites has advanced significantly due to a number of breakthroughs. On the other hand, the covalent bonds between polymers and CNTs are equally important since the lengthy polymeric chains let the CNTs dissolve into a wide variety of media even with minimal modification. To date, two main strategies referred to as "grafting from" and "grafting to" approaches have been used to generate covalent links between polymeric chains and the surface graphitic structures of CNTs (Figure 5.3). The "grafting from" method focuses on immobilizing the polymer precursors onto the central helical filaments of CNTs, followed by polymerization processes and polymer chain elongation in the presence of monomers. This process generates the macromolecules on the exterior of the CNTs. For example, Qin et al. reported polystyrene-sulfonate grafting on CNT surfaces via in-situ free radical polymerization. An analogous design was used by Viswanathan et al. Using in-situ anionic polymerization, they have incorporated the polystyrene chains into the raw CNTs. The pristine CNTs were coated with secondary-butyllithium to bring carbanion moieties to the surface and start the bundles' exfoliation process. After the delivery of monomers, the CNT-carbanions start the polymerization process, which leads to the covalent grafting of the polystyrene (Tiwari et al., 2023).

The "grafting to" technique depends on the formation of a reactive macromolecule and the coupling reactions that follow, which result in the formation of a covalent connection between the synthesized polymer chain and the CNT surface. For example, the ultrasonication approach is used to react poly (methyl methacrylate) (pMMa) with CNTs in a solution containing monochlorobenzene. Another example involves Wu et al. using poly (vinylcarbozole) or poly(butadiene) and treating them with Bu-Li or NaOH to produce the macromolecular anions prior to grafting CNTs on top of them. As an alternative, the Blau group suggested attaching halogenated polymers to CNTs after functionalizing them using n-Bu-Li. Additionally, a number of other techniques, including radical coupling and cycloaddition, can be used to graft the chains of certain macromolecules onto the surface of CNTs. This noncovalent functionalization is primarily caused by a variety of physical interactions, including physical adsorption, electrostatic forces, hydrophobic contact, and van der Waals force. Consequently, this functionalization is accomplished by means of polymer wrapping, the adsorption of surfactants or small aromatic molecules, and interaction with porphyrins or biomolecules such as peptides and deoxyribonucleic acid (DNA). In general, the creation of surface functionalized carbon-based nanomaterials is where this approach is most helpful. This method was mostly employed in the past to surface functionalizing CNTs' sp^2 structure. According to recent research, graphene can be treated with the same expertise by employing a variety of organic modifiers (Tiwari et al., 2023).

5.2.8 Photochemical Reaction

The improved photochemical conversion activity of semiconductor photocatalysts has prompted substantial research into their combination with carbon nanostructures. Even though many methods for synthesizing nanocomposite materials have been reported, the scientific community is growing increasingly intrigued by the innovative use of photocatalytic processes in the creation of these carbon-based semiconductor composites and even pure carbon nanomaterials. These carbon-based nanomaterials fall under the category of photocatalysis and are created by using a photocatalyst's valence band hole or conduction band electron. Techniques for using photocatalysis to create carbon and carbon-based nanocomposites have been developed. A delicate understanding of the process is needed to exploit the redox capabilities of a photocatalyst to start the production of carbon nanostructures or carbon precursors from the reactions. These days, it has emerged as a useful alternative process for producing carbon-based nanomaterials because it provides opportunities to modify the product's properties, including shape, porosity, hydrophilicity, and hydrophobicity of composites, electrical conductivity, and storage capacity. Each of these factors holds significance in numerous applications. It's interesting to note that many of these carbon nanostructures and carbon-based nanomaterials that were produced through photocatalytic reactions were employed in light energy-related processes, including organic degradation catalyzed by photocatalysis, hydrogen production via water splitting, photovoltaic and photoelectrochemical cells, and antibacterial applications. In addition to its traditional uses, photocatalysts are finding vital use in the synthesis of pure carbon and carbon-based composites. This viewpoint critically explores the principles of the photocatalytic method in this regard. The basic theories fall into one of two categories: either the use of the semiconductor photocatalysts' (i) valance band or (ii) conduction band, where the potential of each band has a significant impact on the characteristics of the resulting carbon-based nanomaterials. For instance, the production of porous carbon and graphene/RGO through photocatalytic reaction based on $H_3PW_{12}O_{40}$, ZnO, CdS, TiO_2, and photocatalytic reduction. TiO_2, ZnO, based on photocatalytic oxidation (Ng et al., 2012).

5.2.9 Electrostatic Interaction

Supercapacitors are energy-storage devices that fall into two categories: pseudocapacitors and electrochemical double-layer capacitors (EDLC). In the first case, electrostatic adsorption of electrolyte ions onto the surface area of electrically conducting porous electrodes causes capacitance. Second, reversible redox reactions store energy at the electrode/electrolyte interface. Charge separation in the electrical double layer is one of the two primary methods by which these devices function, and high surface area electrodes are where many of the duties are situated. Since carbon-based materials (CBMs) are easily modulable to acquire the appropriate physical-chemical properties, their use for that purpose is particularly intriguing. These characteristics directly affect how charges from an electrolyte solution will travel through these carbonaceous materials and how ions will flow through the electrode surfaces. Carbon-based electrodes should possess the following characteristics in addition to their textural qualities: High conductivity, corrosion resistance, thermal stability, controlled pore structure, compatibility with composite materials, and low cost are just a few of the benefits (Tiwari et al., 2023).

The physicochemical features, high electrical conductivity, tunable porosity, ease of processing, and variety of geometries of CBMs make them widely used (Deng et al., 2016). Supercapacitors use carbon compounds generated from biomass as electrodes, and these materials may perform better than nonbiomass materials. Direct pyrolysis or carbonization of the biomass is the most popular synthesis technique for this kind of material; however, it provides little control over the physicochemical characteristics of CBMs (Tiwari et al., 2023).

Carbon materials' electrical properties can be directly impacted by their structure, particularly their resistivity and conductivity, which are influenced by the activation or preparation process used to create them. In this way, the temperature at which carbon-rich precursors are converted affects the conductivity of the material. This conversion in sp^2 hybridization may produce large amounts of conjugated carbon, which is crucial for the electrochemical properties (as more efficient charge carriers). Extremely disordered materials require a high degree of graphitization, which is highly dependent on the temperature of thermal treatment, due to their weak conductivity. In addition to their intrinsic resistivity, carbon particles' electrical resistance must also be considered. Particle contact resistance is a crucial component of the powdered carbon aggregates. This resistance is dependent on the shape of the carbon particles and the total pressure that is applied to the material that is used in the supercapacitor; the pressure is higher for smaller carbon particles (Tiwari et al., 2023).

Surface oxygen groups must exist for a carbon-based super capacitor to function satisfactorily. These materials' high oxygen concentration raises the obstacle to electron transmission between the microcrystalline components. Nonetheless, there needs to be agreement on the fact that these groupings are ultimately harmful to a super capacitor's ability to function because their business is still being contested in the specialized literature. It's an issue that needs to be resolved since processing carbon precursors in an oxygen-rich environment results in a material with a high concentration of this heteroatom and a super capacitor with a high electrical resistance. These surface groups play a crucial role in the carbon material's overall behavior (Tiwari et al., 2023). They might affect the point of zero charge, wettability, ion adsorption, self-discharge, and electrical contact resistance. These materials also include other heteroatoms that are integrated as a result of thermal treatments, such as hydrogen, halogens, nitrogen, and sulfur, which come from the raw materials that were used to make them. However, as Figure 5.1 shows, the surface oxides with regard to oxygenated carbons can be classified as basic, acidic, or neutral. Acidic functionalities usually result in large rates of self-discharges, which makes them undesirable. Conversely, the enhanced wettability necessitates basic and neutral oxygenated functionalities, which raises specific capacitance (Tiwari et al., 2023). In this context, electrostatic interaction can be used to create modified mesoporous-C_3N_4 heterojunctions, or GQDs (Yu et al., 2017). As seen in Figure 5.5, this enables the nanocomposite to produce more superoxide radical species and tiny holes, which aid in the efficient photodegradation of RhB and TC. On the other hand, Liu et al. (Nasrollahzadeh et al., 2021) examined the GQD/zinc porphyrin heterojunction and evaluated how well it worked to speed up the visible light-driven photodegradation processes in aqueous MB (Figure 5.2).

5.2.10 Irradiation Technique

Using the microwave-assisted solvothermal approach, the RGO/CdS/TiO S-scheme heterojunction photocatalysts with an exposed high energy {001} facet of TiO were synthesized. Using methylene blue (MB) catalyzed by visible light, XRD, SEM, TEM-EDS, UV–vis, FT-IR, photocurrent measurements, and photocatalytic properties were re-examined.

These techniques were used to characterize the morphology and structure of the as-prepared samples. The effect of TiO loading on the photocatalytic properties of RGO/CdS/TiO samples was also examined. As expected, the generated composite nanomaterials showed higher photocatalytic activity in the presence of visible light. Moreover, PL, XPS, and free radical capture studies show that the generated RGO/CdS/TiO photocatalyst follows the S-scheme photocatalytic enhancement mechanism. In a two-step, microwave-assisted solvothermal method, the exposed {001} facets of TiO in the RGO/CdS/TiO core–shell heterojunctions were com-pounded.

GO (10 mg) and cadmium chloride (342 mg) were first sonicated for 30 minutes in 10 mL of ethylene glycol to form a solution of GO nanosheets. After dissolving

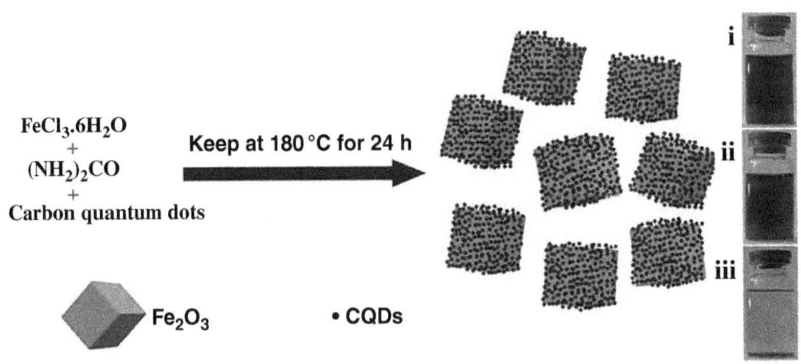

Figure 5.2 The photos show the hydrothermal method used to fabricate Fe_2O_3/CQD nanocomposites. They show the (i) aqueous dispersion of CQDs, (ii) solution containing CQDs, $(NH_2)_2CO$, and $FeCl_3$, and (iii) end result. *Source*: Reproduced with permission from Nasrollahzadeh et al. (2021) / Elsevier.

342 mg of thiourea and 100 mg of urea in an additional 10 mL of ethylene glycol, the mixture was progressively added to the GO mixed solution using rate-controlled titration (5 mL/30 min) while being sonicated for 30 minutes. After that, the mixture was poured into a 35 mL microwave reaction tube, which was cooked in the microwave for 40 minutes at 120 °C inside an automated focused microwave reaction system. After the reaction, the product was washed several times with ethanol and deionized water before being filtered through a microporous membrane. Ultimately, the sample was dried in a vacuum oven at 60 °C for 12 hours, yielding the RGO/CdS (GC) composite. According to our earlier research, the RGO/CdS system with a weight content of 5 wt% GO demonstrated the best photocatalytic activity when compared to other GO contents (Ma et al., 2020) (Figure 5.3).

The RGO/CdS sample (0.4 g) and a predetermined amount of titanium sulfate were dissolved in 10 mL of ethanol for 30 minutes in the second phase, which involved ultrasonication. Then, 18.5 mg of ammonium fluoride was dissolved in 5 mL of deionized water to generate a homogeneous solution, which was then injected into the RGO/CdS mixture using rate-controlled titration (5 mL/30 min) and ultrasonicating the mixture for 30 minutes. The combined solution was then put inside the microwave reaction tube and heated for four hours to 120 °C. In order to produce RGO/CdS/TiO_2, the precipitate was ultimately recovered by centrifugation, dried under vacuum for 12 hours at 60 °C, and filtered through a microporous membrane. It was then repeatedly cleaned with 100% ethanol and deionized water. Figure 5.6 shows the synthesis route for the RGO/CdS/TiO photocatalysts. The as-produced RGO/CdS/TiO samples with 8, 10, 12, and 14 wt% TiO are denoted as GCT-8, GCT-10, GCT-12, and GCT-14, respectively. The samples prepared using RGO/CdS are designated as GC.

N-TiO_2/g-C_3N_4 composites can also be made using an in situ microwave-assisted synthesis method that has been devised. Using cheap H_2TiO_3 as the reactant and NH_3H_2O as the N-doping source, N-doped TiO_2 was created without the need of surfactants or a particular gas environment.

5.2.11 Sputtering Method

For many years, researchers have studied the exceptional mechanical properties of metal containing amorphous carbon (Me–aC) coatings. Graphite and metallic targets are often sputtered simultaneously to provide Me–aC coatings. The close

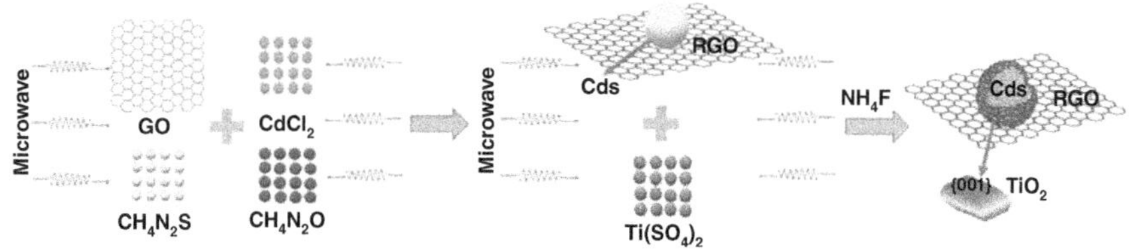

Figure 5.3 Synthetic route of GCT photo catalysts. *Source*: Ma et al. (2020) / Royal Society of Chemistry.

field unbalanced DC magnetron sputtering method, a rapidly scaled-up deposition process, was used to create the Me–aC coatings. This setup has four targets fixed to the side walls of the deposition chamber. The magnetic pole arrangement of the cathodes is designed to create a closed magnetic field configuration, which extends the plasma to the substrate and enhances ion plating. However, the reproducibility, adhesion, and uniformity of the coatings, as well as the stability of the sputtering process, are not sufficient when such systems are employed to coat complex work pieces.

A new middle frequency (MF) powered twin magnetron cathode with increased plasma density and target utilization has been created recently (Yang et al., 2005). Recently developed sputtering tool MF technology has been proven to be successful in Low-e and antireflection/anti-static (ARAS) coating mass production systems. It offers better coating quality for both optical and cutting tool coatings and has higher deposition speeds. In all commercial twin-magnet coating systems, the two identical rectangular magnetron targets are positioned near to and apart from one another in the sputter chamber at angles of 120° or 180°. To produce hard coatings on machining tools in higher quantities, a high plasma density needs to be kept in the deposition chamber. Unbalanced magnetron (UBM) sputtering of graphite and Cr metal targets has produced nanoscale hydrogen-free C/Cr coatings for tribological purposes. The microstructure and characteristics of the coating are significantly influenced by the substrate bias voltage (Zhou et al., 2011).

5.2.12 Mechanochemical Synthesis

Chemical bond changes that cannot be triggered by other known physical stimuli, like heat or light, can be triggered by mechanical force. Mechanochemistry, a rapidly expanding and versatile technology, was chosen as one of ten game-changing discoveries. Thus, it was anticipated that carbon-rich nanomaterials would be synthesized and activated using mechanochemistry; one of the earliest examples dates to the late 1990s. Suslick categorizes mechanochemistry into four independent but linked areas: polymer mechanochemistry, tribochemistry, trituration, and sonochemistry (Xuan et al., 2022). Although several parts of these subfields overlap, polymer and trituration mechanochemistry are the primary sources of methodologies employed for specific synthetic goals.

Research on the development of heteroatom-doped graphitic carbon nitrides (g-CN) has been rapidly developing since its first publication toward water splitting in 2009. Since then, it has been demonstrated that adding elements like boron, phosphorus, sulfur, and oxygen can help reduce the bandgap of these metal-free photocatalysts and increase their overall stability. The incorporation of phosphorus has traditionally been achieved using high-temperature or microwave synthesis. These techniques usually include inserting a phosphorus atom into the g-CN building block, the heptazine ring, as opposed to a connecting point. Hartley and Martsinovich have computationally investigated the effect of various linkers, such as phosphorus atoms, on the structure and optical behavior of heptazine-based graphitic carbon nitrides. However, there aren't many examples of carbon nitride materials that have phosphorus atoms bonded to them. This is probably because it can be difficult to regulate phosphorus atom insertion as a linker under high energy circumstances has shown promise in the synthesis of various polymers, nanomaterials crystal engineering, inorganic material synthesis, and organic small molecule synthesis. The benefits of employing such approaches lie not only in their capacity to circumvent bulk solvent and mild reaction conditions, but also in their provision of favorable conditions for the emergence of novel materials and new relativities (Fiss et al., 2022).

5.2.13 Calcination

Titania and 3D graphene foam hybrids that were self-sustained and supported by a nickel scaffold were able to serve as powerful photocatalysts when exposed to direct sunshine. The hybrid was produced by combining the template-assisted CVD development of a few-layer graphene film on nickel foam with the sol–gel technique for the subsequent deposition of the anatase TiO_2 layer. The advantages of sol–gel technique include reproducible low-temperature synthesis, low infrastructure needs, and inexpensive prices for industrial application. The graphene surface was treated with UV ozone prior to titania deposition in order to change its hydrophobicity and introduce oxygen-containing groups for efficient metal oxide-graphene interaction. Because there is very little defect density, this approach alters the surface energy of graphene while maintaining its superb transport capabilities (Mihai et al., 2022). The nickel foam provides the photocatalyst with a larger surface area than planar substrates and the resilience to be easily and reasonably reused in successive treatments, in contrast to titania-based nanoparticle powder. Because a wide surface area has many active sites, it can also promote high pollutant adsorption and photocatalysis.

In photocatalytic activity, the anatase titania polymorph is preferred even though it has a larger bandgap than rutile. This is a result of its reduced charge recombination rate and improved adsorptive affinity for dyes. The ideal heat treatment for TiO formation of a dominant anatase phase was determined by testing a wide range of calcination settings, including temperature and time changes. As a result, TiO-coated FTO films were air-annealed at 450 and 550 °C for two and a half hours. On the other hand, the anatase bandgap was shown to grow with increasing calcination time and temperature based on the results of the diffuse reflectance tests performed to estimate the bandgap. As a result, when the calcination period was increased from two to three hours, there was a noticeable rise in energy from 3.17 to 3.41 eV at 450 °C and just a minor increase from 3.35 to 3.4 eV at 550 °C. These findings support the widely held belief that the indirect transition-related absorption edge of anatase is located at approximately 3.2 eV.

In light of structural characterization, which revealed only the presence of a well-defined anatase phase without significant changes in crystallinity with increasing the calcination temperature or time at 550 °C or three hours, and the lowest bandgap of 3.17 eV observed for the 450 °C and two hours, the TiO/3D-GF/Ni photocatalyst synthesis was decided to be completed at a calcination temperature of 450 °C for two hours.

Utilizing scanning electron microscopy, the surface morphology of the 3DGF/Ni was evaluated. The outcomes demonstrate the construction of a continuous, linked graphene network (Figure 5.4a) with the characteristic wrinkled and corrugated graphene layers (Figure 5.4b), exactly replicating the architecture of the Ni template. The thin TiO film displays a flower-like morphology (Figure 5.4d) and adheres to the architecture of the Ni foam (Figure 5.4c) after the TiO sol deposition and two hours of calcination at 450 °C. It is noteworthy that following four rounds of deposition, the graphene skeleton's surface is uniformly covered with TiO coating, exhibiting minimal fractures.

5.2.14 Ultrasonic Method

Nanoscale zero-valent iron (nZVI) particles have been widely used as a new tool for the treatment of wastewater contaminated with various pollutants in recent years due to its small particle size, large specific surface area, and strong reactivity. Unfortunately, there are still some technical issues with nZVI. On the one hand, the interparticle Van der Waals and magnetic interactions that cause nZVI particles to clump together greatly limit their dispersibility. However, nanoparticles

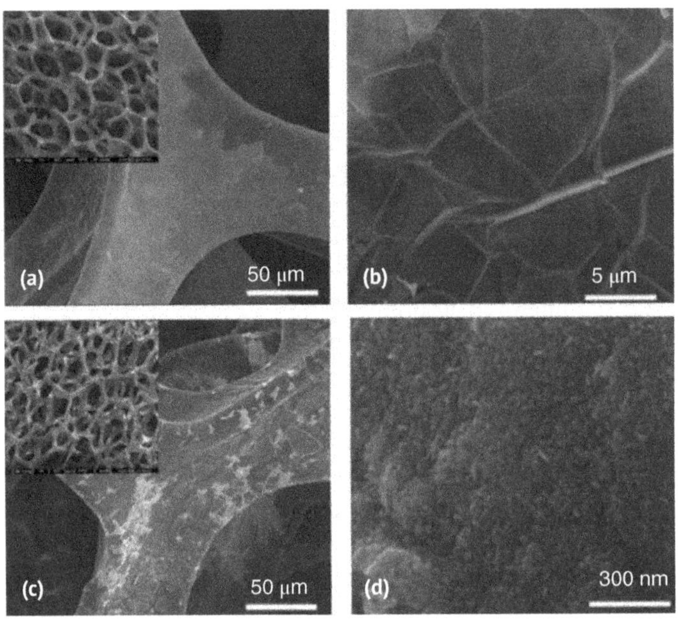

Figure 5.4 Morphological characterization of the 3D-GF/Ni and TiO/3D-GF/Ni hybrids (a, b) and (c, d) shows how the TiO film completely coats the 3D-GF/Ni skeleton with a flower-like morphology, while the graphene network (insets) displays the uniformity of graphene with its microscopic wrinkled and corrugated morphology. *Source*: Mihai et al. (2022) / MDPI / CC BY 4.0.

have a propensity to oxidize, and the easy creation of oxide coating blocks the useful active surface sites, ultimately reducing reactivity. One way to eliminate these drawbacks is to support nZVI particles. The bulk of studies conducted in the last few years have loaded nZVI particles onto a variety of porous materials as a carrier, such as graphene, activated carbon mesoporous carbon, activated carbon bentonite, mesoporous silica, titanium oxide pumice, and graphene, using the liquid phase reduction approach. However, there are still several issues with the technology, including high production costs, limited production efficiency, and significant hydrogen generation during the preparation phase. Large-scale application limits are caused by all of these issues. The compound substance used in this investigation was made in two processes. First, pulse electrodeposition was used to directly obtain the nZVI particles from steel scrap. Second, the nZVI particles were quickly supported by the biomass activated carbon (BC) in the presence of mechanical agitation. Methyl orange (MO) removal was used to characterize and assess the BC-nZVI material's decontamination capabilities (Homaeigohar, 2020) (Figure 5.5).

5.2.15 Solvothermal Method

Photocatalysts based on reduced GO nanocomposites can be made in a number of ways. One popular method for producing these nanophotocatalysts is to introduce presynthesized photocatalyst nanoparticles and make direct contact between them and GO flakes. Furthermore, the sonication energy is employed to improve dispersion and encourage the inclusion of nanoparticles on the rGO during the solvothermal process that lowers GOES to rGO nanosheets. The nano-heterostructure was synthesized using different weighted quantities of GO and a mixed phase of 60% BiOCl and 40% $Bi_{24}O_{31}C_{110}$ and rGO. In order to create these nanocomposites,

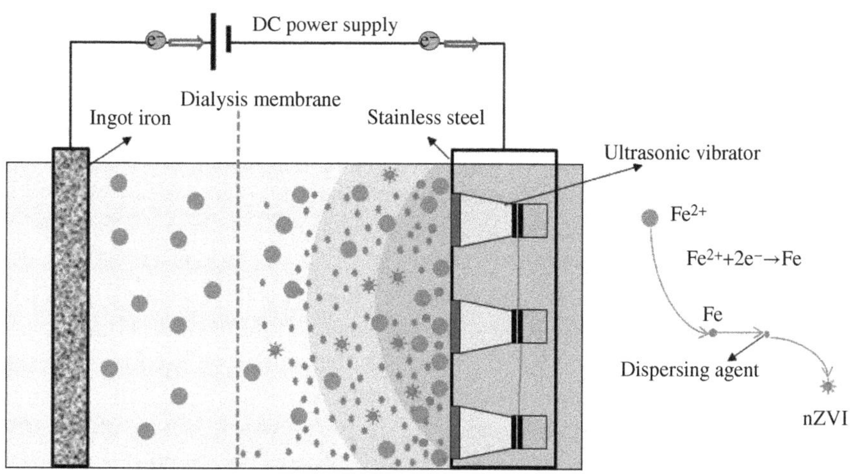

Figure 5.5 Electrolytic cell for the production of zero-valent iron at the nanoscale (nZVI). *Source*: Homaeigohar (2020) / MDPI / CC BY 4.0.

an easy sorbitol-nitrate self-combustion approach was utilized to create a mixed-phase of BiOCl- $Bi_{24}O_{31}C_{110}$, and the rGO was subsequently included via a sono-solvothermal process. The generated nanocomposites were characterized and their photocatalytic efficiency in the photodegradation of the antibiotic ofloxacin (OFL) evaluated under artificial sunlight illumination. Compared to the other samples, the BiOCl-$Bi_{24}O_{31}C_{110}$/rGO nano-heterostructure, which contained 0.5 wt% reduced GO, had the highest photocatalytic efficiency and the highest kinetic rate. To learn more about nanocomposite's photocatalytic performance and capabilities, it was also used to photodegrade other FQs antibiotics, like levofloxacin and ciprofloxacin. Furthermore, over the BiOCl- $Bi_{24}O_{31}C_{110}$/rGO system, a degradation mechanism for the elimination of fluoroquinolone antibiotics under simulated solar light was proposed (Shabani et al., 2019).

Using a solvothermal technique, Shen et al. created a TiO_2 (TiO_2@CF) monolith. The artificial TiO_2@CF monolith has demonstrated a sizable surface area and a distinct three-dimensional (3D) structure. The methyl orange dye degradation potential of the TiO_2@CF monolith was assessed using photocatalytic activity. These findings demonstrate the strong post depollution, recyclability, and photocatalytic potential of TiO_2@CF monolith nanomaterials. Using a hydrothermal process, Qiu et al. synthesized 3D-structured TiO_2/GA nanocomposites out of glucose. The huge surface area, hydrophobic nature, massive appearance, and great recyclability of these materials contributed to their observed remarkable photocatalytic effectiveness to breakdown methyl orange dye. Yang and coworkers synthesized graphene microspheres covered with colloidal silicon dioxide (TiO_2), which they employed as photocatalysts to break down methylene blue dye (MB). The generated graphene microsphere colloidal-coated TiO_2 has demonstrated strong visible light absorption, increased adsorption capacity, and quick charge transfer, resulting in excellent photocatalytic efficiency for the breakdown of methylene blue dye. The production of polyhydroxy fullerenes (PHF)-modified TiO_2 was described by Krishna et al. This modification improved TiO_2's photocatalytic performance to breakdown procion red dye. For instance, Liu et al. found that the azo dyes in water may be broken down by the hollow structured {001} facet TiO_2 nanocomposite with adjustable photocatalytic effectiveness. Using a hydrothermal process, Chen et al. were also successful in synthesizing 100% exposed {001} faceted ultrathin anatase TiO_2 nanosheets. For photocatalytic applications, they employed them. The TiO_2/carbon black (CB) nanocomposite catalyst, prepared by Mao and Weng et al., is more active than the TiO_2/activated carbon composite and bare TiO_2 catalysts. The sol–gel process was used to generate the TiO_2/CB nanocomposite, which was then calcined. The TiO_2/CB nanocomposite that was synthesized was examined using electron microscopy methods and surface area analysis, among other methods. By preventing the rutile TiO_2 phase from forming, the CB decreased the amount of photogenerated electrons and holes that were recombined, increasing the photocatalytic activity of TiO_2/CB. Testing has been done on the photocatalytic activity on methyl orange. According to the authors, the CB increased TiO_2's photocatalytic activity. Li and his group have documented the enhanced photocatalytic performance of their TiO_2/porous carbon nanofibers (PCNFs). The PCNFs were created using an electrospinning method, yielding a one-dimensional (1D) PCNF structure. After applying $TiCl_4$ to the synthesized PCNFs using a solvothermal technique, highly porous PCNF-coated TiO_2 nanorods were the end result. Next, this nanomaterial was employed to cause the methyl orange dye to photodegrade in the presence of visible light (Pandikumar & Jothivenkatachalam, 2019).

5.2.16 Deposition Method

Electrophoretic deposition (EPD) is the process where charged particles from a colloidal dispersion move toward an electrode under continuous application of an electric field. During the EPD operations, charged CNTs build up at the electrode to create a uniform, stiff film. The fact that EPD can only be deposited on conductive surfaces is one of its main drawbacks. Du et al.'s paper was the first to describe the EPD method of depositing CNTs. After layering MWCNTs from ethanol/acetone dispersion onto a metallic substrate, which produced a porous structure for the film, they observed hydrogen evolution on the cathode. After CNTs are deposited, more EPD can be used to create a new layer or to deposit nanoparticles into the CNTs film structure. As an alternative, stable dispersions including two or more components can be used to create CNT composite films via EPD. Preassembling the parts could also create building blocks in dispersion prior to EPD.

Pulse electrodeposition involves rapidly alternating potential or current to produce a succession of pulses with similar amplitude, length, and polarity that are separated by zero current. The negatively charged layer is discharged during the pulse electrodeposition process, and the ions are let permeate toward the substrate during the off time. Through the adjustment of pulse parameters, this technique enables the synthesis of coatings with regulated composition and thickness. Additionally, it may co-deposit metal particles on pre-existing films that are both nano and micron in size. Using the pulse electro deposition method, CNTs have been combined with a variety of metals and metallic oxides to create composite films. MWCNTs/Co films might be produced by pulse reverse current (PRC) electro deposition from an aqueous bath comprising MWCNTs and cobalt sulfate, as Su et al. (Fu & Yu, 2014) showed.

One chemical method for growing thin films is chemical vapor deposition (CVD). The substrate is treated with one or more vaporized compounds, and the vapor is either heated to a low temperature or forms a thin coating by reacting with other gases on the substrate surface. CVD may be used to create CNT thin films on a range of substrates. Spray pyrolysis, a CVD technique, allows for the simultaneous provision of hydrocarbon and organometallic chemicals, enabling the growth of CNT thin films on substrate. In order to facilitate the reaction, catalysts are typically added during the growth phase of the CVD process. Carbon resources that are frequently employed are cyclohexane and ferrocene ($FeCp_2$) dissolved in benzene. The CVD process can be applied to numerous or single depositions. Various CVD techniques have been used to create various CNT-based composite films. There is a study on a simple CVD method that uses pyrolysis of $FeCp_2$ and acetonitrile (ACN) to effectively create N_2-encapsulated CNTs film. The catalysts' batch-type pyrolysis resulted in a rapid extrusion of N_2 along the conical catalyst particles, which changed the reaction gases' flow. As a result, branched CNTs formed, encasing N_2 within them. Dong et al. created a one-step procedure for using a CVD technique to generate a hybrid thin film of CNTs and graphene over copper foil that has been coated with silicon nanoparticles. On the copper foil, a graphene sheet forms consistently, and on top of it, CNTs with a multiple-wall structure resembling bamboo emerge from the Si NPs to form a network. On the other hand, repeated depositions using CVD is another method for combining CNTs with other materials after the creation of CNT-based composite films. By using hot filament CVD (HFCVD) at 650 °C, it is possible to successfully synthesize CNTs/diamond thin films that are coated on stainless steels (Fu & Yu, 2014).

5.3 Performance

In semiconductor applications, carbon scaffold-based nanophotocatalysts have shown potential performance for several kinds of reasons:

a. **Large Surface Area-** The large surface area of carbon scaffold structures ensures the availability of numerous active sites for catalytic reactions and efficient photon utilization (Kumar et al., 2020).
b. **Enhanced Charge Separation-** Carbon scaffolds' special construction enables effective charge transfer and separation, which lowers recombination rates and boosts photocatalytic performance (Zhang et al., 2020).
c. **Tunable Properties-** By engineering carbon materials with certain characteristics such as pore size, surface functional groups, and electronic structure, photocatalytic performance can be maximized (Assayehegn et al., 2020).
d. **Chemical Stability-** Excellent chemical stability is exhibited by carbon-based scaffolds, which assures sustained catalytic activity even in severe reaction conditions (Kumar et al., 2019).
e. **Environmental Friendliness-** Compared to some other semiconductor materials, carbon is abundant, nontoxic, and ecologically benign, making carbon scaffold-based photocatalysts attractive for sustainable applications (Kumar et al., 2020).

There are also scopes for increasing the performance of carbon scaffold-based photocatalysts in semiconductors. Some of them are explained.

1. **Doping-** The electronic structure of the carbon scaffold may be changed by adding dopant atoms, which can enhance charge separation and light absorption. Additionally, doping can help produce active sites for catalytic reactions. The catalytic activities of carbon scaffold-based nanophotocatalysts are influenced by various types of doping, including heteroatom doping, charge-transfer doping, and defective doping, to modulate the charge distribution of the carbon structure and create synergistic effects (Miao et al., 2018; Mao et al., 2018; Wang et al., 2018; Gui et al., 2014; Yan et al., 2017).

2. **Heterojunction-** Charge separation and photocatalytic activity can be enhanced by forming heterojunctions between various semiconductor materials or between a semiconductor and a carbon-scaffold. Heterojunctions can lessen recombination losses by facilitating the flow of photogenerated electrons and holes.

   Liu et al. synthesized a composite of nitrogen-doped carbon quantum dots (NCQDs) and $BiVO_4$, which exhibited excellent photocatalytic performance. Enhanced photocatalytic efficiency can be attributed to the formation of a heterojunction, which enhances light-harvesting efficiency and facilitates the separation and migration of photogenerated carriers (Liu et al., 2021a).
3. **Crystal Phase Engineering-** Crystal phase engineering is a crucial aspect that significantly evaluates the efficiency of the nanophotocatalyst. This technique involves controlling the crystalline structure of the catalyst at the nanoscale level to optimize its properties. Furthermore, because different crystalline phases typically exhibit distinct physical and chemical properties, a semiconductor's photocatalytic efficiency is strongly correlated with its crystalline phase Liang et al. (2020). The efficiency and selectivity of nanophotocatalysts in photocatalytic reactions can be improved by customizing their electronic band structure, surface area, and surface energy through crystal phase engineering. Different carbon scaffold-based compounds have different crystal phases and characteristics, which makes them good candidates for particular applications. Cai et al. utilized crystal phase engineering on $BiVO_4$ by incorporating N-doped carbon quantum dots (NCDs), leading to a transformation of $BiVO_4$ from its monoclinic phase to a tetragonal phase, thereby creating a self-heterostructure Cai et al. (2024). Through crystal phase engineering, various techniques are used to customize characteristics of carbon scaffold-based nanomaterials for specific applications like photocatalysis, biomedical, sensing, energy storage, etc.
4. **Quantum Confinement-** Quantum confinement involves the alteration of a material's properties when its dimensions are reduced to nanoscale. Reducing the dimensionality from 2D to 1D or from 1D to 0D, the semiconductor nanomaterials exhibit intriguing features. Quantum confinement in carbon scaffold-based nanophotocatalysts can be accomplished by controlling the size and morphological regulation of the carbon nanostructures. Examples of confinement for 3D nanomaterials: nanostructured materials, 2D nanomaterials: thin films, 1D nanomaterial: quantum wires, and 0D nanomaterial: quantum dots. In the instance of carbon scaffold-based materials, such as graphene quantum dots (GQDs) or carbon nanodots (CNDs), an adjustable bandgap can result from quantum confinement Sivaranjani et al. (2022). This increases photocatalytic activity by permitting the absorption of a wider spectrum of light wavelengths Kumar et al. (2019). Size-dependent fluorescence and strong light absorption are two optical characteristics of quantum-confined carbon nanostructures. These properties can be utilized for applications such as photoluminescence-based sensing or light harvesting in solar cells.

5. **Vacancy Engineering-** Vacancy engineering is a technique used to manipulate the atomic structure of semiconductor materials by intentionally creating vacancies, or missing atoms, within the lattice structure. Vacancy engineering effectively manipulates photocatalytic processes by extending the light absorption range, improving photo-generated charge separation and transfer, and optimizing photocatalytic reactions at the surface of catalysts (Wang et al., 2021). In the context of carbon scaffold-based nanophotocatalyst, we can take the example of graphitic carbon nitride (g-C_3N_4), which has a poor charge transfer efficiency, resulting in its lower photocatalytic activity (Zhao et al., 2019). However, incorporating vacancies in g-C_3N_4 can improve the transfer and separation of charge carriers (Zhao et al., 2019).
6. **Crystal Facet Engineering-** This crystal facet engineering has been chosen as an excellent approach for improving photocatalytic performance. In recent development, it has been used for controlling the single-component semiconductor system and for designing the surface and inter-surface structures of multicomponent heterostructures for photocatalytic applications Bai et al. (2016). Presynthesized components provide more flexibility for facet engineering as they can control on its own in their procedures Yamada et al. (2011). It has achieved crystal phase engineering and significantly achieved K+ storage performance, and its structural morphologies are determined by carbon scaffold. The interconnected nanorods have given rise to a robust matrix to increase electrical conductivity and ion transport Chang et al. (2022). Recently, researchers have focused on new photoanodes in crystal facet engineering. There is a high performance of facets in water which confirms the importance of crystal facet engineering.
7. **Cocatalyst Loading-** The efficiency of hydrogen production in photocatalytic devices that use a carbon-scaffold can be improved through the use of cocatalysts. Metals such as Pt, Au, and Pd work well as cocatalysts due to their higher work functions compared to semiconductors. Cocatalysts can help with charge separation, provide catalytic sites, and lower the activated energy required for hydrogen production. Employing suitable cocatalysts-particularly dual cocatalysts for oxidation and reduction, the efficiency of the water-splitting process in semiconductor photocatalytic systems can be greatly improved. Cocatalysts can speed up oxidation and reduction reaction sites, prevent charge recombination, and stop reverse reactions. Cocatalysts are positioned on the electrode materials in water-splitting systems to serve as spatially separated active/reaction sites, similar to natural photosynthesis. The characteristics of the cocatalysts and their interaction with the semiconductor at the interface are crucial in both cases Bimetallic cocatalysts have shown cooperative phenomena that could enhance the synergistic effect of improving catalytic activity and stability (Liu et al., 2021b). Carbon scaffolds are well suited to support and enhance the photocatalytic area of the photocatalyst (Figure 5.6).

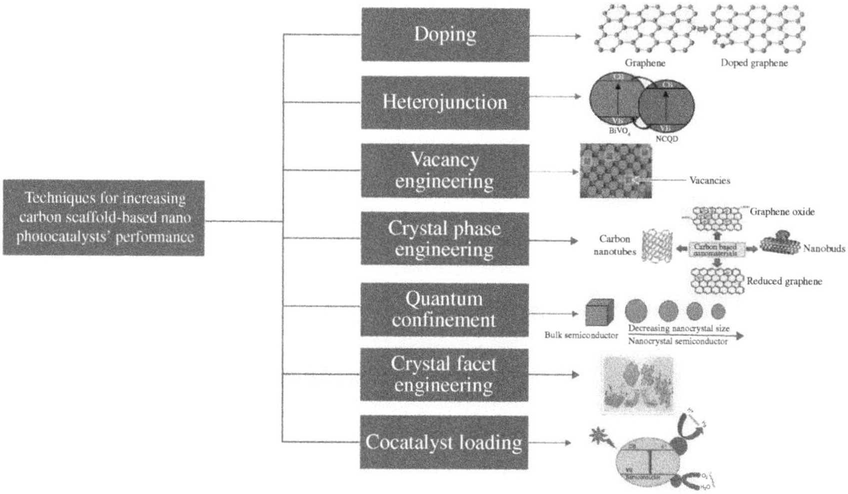

Figure 5.6 Different ways to increase the performance of carbon scaffold-based nanophotocatalyst semiconductors. (Self-prepared.)

5.4 Summary and Future Prospects

This chapter has provided a concise overview of the importance of carbon scaffold-based nanophotocatalyst mediated semiconducting materials and their different modes of fabrications. This chapter also highlights the various scopes of carbon scaffold-based photocatalysts in semiconductors. Carbon nanomaterial-based scaffolds are possessed various advantages like biocompatibility, large surface area, highly porous behavior, mechanical stability, and easily availability, which make these material more suitable for various applications like polluted water treatment, dye degradation, organic reactions catalysis, drug delivery, bio sensing, tissue engineering, cancer therapeutics, biomedical waste management, etc. Due to commercial availability and simple fabrication methodologies, investigations into these materials are expanding day by day, and in upcoming research, it will definitely be the most appropriate applicant in versatile fields.

References

Abadi, B., Goshtasbi, N., Bolourian, S., Tahsili, J., Adeli-Sardou, M., & Forootanfar, H. (2022). Electrospun hybrid nanofibers: Fabrication, characterization, and biomedical applications. *Frontiers in Bioengineering and Biotechnology*, 10, 986975.

Ahmed, S. F., Senthil Kumar, P., Ahmed, B. et al. (2023). Carbon-based nanomaterials: Characteristics, dimensions, advances and challenges in enhancing photocatalytic hydrogen production. *International Journal of Hydrogen Energy*, 52, 424–442.

Assayehegn, E., Solaiappan, A., Chebude, Y., & Alemayehu, E. (2020). Fabrication of tunable anatase/rutile heterojunction N/TiO_2 nanophotocatalyst for enhanced visible light degradation activity. *Applied Surface Science*, 515, 145966.
Bai, S., Wang, L., Li, Z., & Xiong, Y. (2016). Facet-engineered surface and interface design of photocatalytic materials. *Advanced Science*, 4(1), 1600216.
Banerjee, A. N. (2011). The design, fabrication, and photocatalytic utility of nanostructured semiconductors: Focus on TiO_2-based nanostructures. *Nanotechnology, Science and Applications*, 4, 35–65.
Cai, W., Fu, J., Hu, C. et al. (2024). Crystal phase engineering of $BiVO_4$ induced by N-doped carbon quantum dots for photocatalytic application. *Journal of Materials Science*, 59(23), 4118–4135.
Chang, C.-H., Chen, K. T., Hsieh, Y. Y., Chang, C. B., & Tuan, H. Y. (2022). Crystal facet and architecture engineering of metal oxide nanonetwork anodes for high-performance potassium ion batteries and hybrid capacitors. *ACS Nano*, 16(1), 1486–1501.
Deng, J., Li, M., & Wang, Y. (2016). Biomass-derived carbon: synthesis and applications in energy storage and conversion. *Green chemistry*, 18(18), 4824–4854.
Fiss, B. G., Douglas, G., Ferguson, M. et al. (2022). Mechanochemical bottom-up synthesis of phosphorus-linked, heptazine-based carbon nitrides using sodium phosphide. *Beilstein Journal of Organic Chemistry*, 18(1), 1203–1209.
Fu, L. & Yu, A. M. (2014). Carbon nanotubes based thin films: Fabrication, characterization and applications. *Reviews on Advanced Materials Science*, 36(1), 40–61.
Gui, X., Zeng, Z., Zhu, Y. et al. (2014). Three-dimensional carbon nanotube sponge-array architectures with high energy dissipation. *Advanced Materials*, 26(8), 1248–1253.
Han, W., Li, Z., Li, Y. et al. (2017). The promoting role of different carbon allotropes cocatalysts for semiconductors in photocatalytic energy generation and pollutants degradation. *Frontiers in Chemistry*, 5, 84.
Homaeigohar, S. (2020). Water treatment with new nanomaterials. *Water*, 12(5), 1507.
Karchiyappan, T., Karri, R. R., & Dehghani, M. H. (Eds.). (2022). *Industrial Wastewater Treatment: Emerging Technologies for Sustainability*. Springer.
Kumar, A., Kumari, A., Sharma, G., Du, B., Naushad, M., & Stadler, F. J. (2020). Carbon quantum dots and reduced graphene oxide modified self-assembled $S@C_3N_4/B@C_3N_4$ metal-free nano-photocatalyst for high performance degradation of chloramphenicol. *Journal of Molecular Liquids*, 300, 112356.
Kumar, A., Thakur, P. R., Sharma, G. et al. (2019). Carbon nitride, metal nitrides, phosphides, chalcogenides, perovskites and carbides nanophotocatalysts for environmental applications. *Environmental Chemistry Letters*, 17, 655–682.
Kumar, P., Laishram, D., Sharma, R. K., Vinu, A., Hu, J., & Kibria, M. G. (2021). Boosting photocatalytic activity using carbon nitride-based 2D/2D van der Waals heterojunctions. *Chemistry of Materials*, 33(23), 9012–9092.
Larimi, A., Rahimi, M., & Khorasheh, F. (2020). Carbonaceous supports decorated with Pt–TiO_2 nanoparticles using electrostatic self-assembly method as a highly visible-light active photocatalyst for CO_2 photoreduction. *Renewable Energy*, 145, 1862–1869.

Liang, J., Li, Y., Zhang, J. et al. (2020). Crystalline phase engineering in WO_3/g-C_3N_4 composites for improved photocatalytic performance under visible light. *Materials Research Express*, 7(6), 065503.

Liu, Y., Liu, C., Shi, C. et al. (2021a). Carbon-based quantum dots (QDs) modified ms/tz-$BiVO_4$ heterojunction with enhanced photocatalytic performance for water purification. *Journal of Alloys and Compounds*, 881, 160437.

Liu, Y., Sun, Z., & Hu, Y. H. (2021b). Bimetallic cocatalysts for photocatalytic hydrogen production from water. *Chemical Engineering Journal*, 409, 128250.

Ma, Z., Shi, L., Qu, W. et al. (2020). Microwave-assisted synthesis of an RGO/CdS/TiO_2 step-scheme with exposed TiO_2 {001} facets and enhanced visible photocatalytic activity. *RSC Advances*, 10(71), 43447–43458.

Mao, J., Iocozzia, J., Huang, J., Meng, K., Lai, Y., & Lin, Z. (2018). Graphene aerogels for efficient energy storage and conversion. *Energy & Environmental Science*, 11(4), 772–799.

Miao, X., Qu, D., Yang, D. et al. (2018). Synthesis of carbon dots with multiple colour emission by controlled graphitization and surface functionalization. *Advanced Materials*, 30(1), 1704740.

Mihai, E. M., Mihalache, I., Istrate, A. I. et al. (2022). Self-sustained three-dimensional macroporous TiO_2-graphene photocatalyst for sunlight decolorization of methyl orange. *Nanomaterials*, 12(24), 4393.

Mondal, K. & Gupta, A. (2018). Recent advances in carbon–semiconductor nanocomposites for water remediation. In S. Bhattacharya, A. Gupta, & Pandey, A (Eds.), *Water Remediation. Energy, Environment, and Sustainability*. Springer, Singapore.

Nasrollahzadeh, M., Sajjadi, M., Iravani, S., & Varma, R. S. (2021). Carbon-based sustainable nanomaterials for water treatment: State-of-art and future perspectives. *Chemosphere*, 263, 128005.

Ng, Y. H., Ikeda, S., Matsumura, M., & Amal, R. (2012). A perspective on fabricating carbon-based nanomaterials by photocatalysis and their applications. *Energy & Environmental Science*, 5(11), 9307–9318.

Pandikumar, A. & Jothivenkatachalam, K. (Eds.). (2019). *Photocatalytic Functional Materials for Environmental Remediation*. John Wiley & Sons.

Pendergast, M. M. & Hoek, E. M. (2011). A review of water treatment membrane nanotechnologies. *Energy & Environmental Science*, 4(6), 1946–1971.

Pradeesh, K., Rao Kotla, N., Ahmad, S., Dwivedi, V. K., & Prakash, G. V. (2013). Naturally self-assembled nanosystems and their templated structures for photonic applications. *Journal of Nanoparticles*, 2013(1), 531871.

Radhika, N. P., Selvin, R., Kakkar, R., & Umar, A. (2019). Recent advances in nano-photocatalysts for organic synthesis. *Arabian Journal of Chemistry*, 12(8), 4550–4578.

Rani, A., Reddy, R., Sharma, U. et al. (2018). A review on the progress of nanostructure materials for energy harnessing and environmental remediation. *Journal of Nanostructure in Chemistry*, 8, 255–291.

Rasool, M. A., Sattar, R., & Anum, A. (2022). An insight into carbon nanomaterial-based photocatalytic water splitting for green hydrogen production. *Catalysts*, 13(1), 66.

Shabani, M., Haghighi, M., Kahforoushan, D., & Haghighi, A. (2019). Sono-solvothermal hybrid fabrication of BiOCl-$Bi_{24}O_{31}Cl_{10}$/rGO nano-heterostructure photocatalyst with efficient solar-light-driven performance in degradation of fluoroquinolone antibiotics. *Solar Energy Materials and Solar Cells*, 193, 335–350.

Sivaranjani, P. R., Janani, B., Thomas, A. M., & Raju, L. L. (2022). Recent development in MoS_2-based nano-photocatalyst for the degradation of pharmaceutically active compounds. *Journal of Cleaner Production*. 352(5), 131506.

Tiwari, S. K., Bystrzejewski, M., & Kumar, V. (Eds.). (2023). *Biomass-Based Functional Carbon Nanostructures for Supercapacitors*. Springer Nature.

Wang, B., Liu, J., Yao, S. et al. (2021). Vacancy engineering in nanostructured semiconductors for enhancing photocatalysis. *Journal of Materials Chemistry A*, 9(32), 17143–17172.

Wang, Y., Pan, Y., Zhu, L. et al. (2018). Fabrication of 3D heteroatom-doped porous carbons from self-assembly of chelate foams via a solid-state method. *Inorganic Chemistry Frontiers*, 5(3), 656–664.

Xuan, M., Schumacher, C., Bolm, C., Göstl, R., & Herrmann, A. (2022). The mechanochemical synthesis and activation of carbon-rich π-conjugated materials. *Advanced Science*, 9(19), 2105497.

Yamada, Y., Tsung, C. K., Huang, W. et al. (2011). Nanocrystal bilayer for tandem catalysis. *Nature Chemistry*, 3(5), 372–376.

Yan, D., Li, Y., Huo, J., Chen, R., Dai, L., & Wang, S. (2017). Defect chemistry of nonprecious-metal electrocatalysts for oxygen reactions. *Advanced Materials*, 29(48), 1606459.

Yang, B., Huang, Z., Liu, C., Zeng, Z., Fan, X., & Fu, D. (2005). Ti-containing amorphous carbon nanocomposite coatings prepared by means of eight-target arc-assisted middle frequency magnetron sputtering. *Japanese Journal of Applied Physics*, 44(7L), L1022.

Yu, Z., Dang, Q., Liu, C. et al. (2017). Preparation and characterization of poly (maleic acid)-grafted cross-linked chitosan microspheres for Cd (II) adsorption. *Carbohydrate polymers*, 172, 28–39.

Zhang, X., Yang, C., Xue, Z., Zhang, C., Qin, J., & Liu, R. (2020). Spatial separation of charge carriers via heterogeneous structural defects in graphitic carbon nitride for photocatalytic hydrogen evolution. *ACS Applied Nano Materials*, 3(5), 4428–4436.

Zhao, D., Dong, C. L., Wang, B. et al. (2019). Synergy of dopants and defects in graphitic carbon nitride with exceptionally modulated band structures for efficient photocatalytic oxygen evolution. *Advanced Materials*, 31(43), 1903545.

Zhou, Z., Rainforth, W. M., Gass, M. H., Bleloch, A., Ehiassarian, A. P., & Hovsepian, P. E. (2011). C/CrC nanocomposite coating deposited by magnetron sputtering at high ion irradiation conditions. *Journal of Applied Physics*, 110(7), 073301.

6

Design and Properties Evaluation of Carbon-based Nanophotocatalysts

Rohit Kandulna[1]* *and Ramesh Oraon*[2]

[1] *Department of Physics, IIIT Ranchi, Ranchi, Jharkhand, India*
[2] *Department of Chemistry, Central University of Jharkhand, Ranchi, Jharkhand, India*

* Corresponding author: rohitkandulna@iiitranchi.ac.in

6.1 Introduction

One of the most advantageous catalysis techniques is photocatalysis because of its expanding potential and ability to be combined with renewable solar energy. Many uses exist, including the generation of hydrogen from wastewater and the cleansing and decontamination of gaseous and water effluents. Other uses include the creation of biosensors, auto-cleaning surfaces, and novel paths for chemical synthesis. The foundation of all of them is made up of photoactive semiconductor nanoparticles, which can photo-generate charge carriers to catalyze processes.

With a gradual pace of advancement, photo-catalysis has become a global player in some markets. Examples of decontamination and disinfection products include paints and cements, as well as small-capacity reactors. Most of the studies focused on material improvements, primarily mentioning increasing the spectrum's useful range for deterioration and prolonging the photo-generated charge separation time. The major goal of studying these materials was to determine how well dyes and recently discovered compounds degraded, even if some references dealt with gases and others developed novel production processes. Catalysts composed of nanoparticles or other nanostructures, including nanofoam, are referred to as nanocatalysts. A photocatalyst is a substance that speeds up a chemical reaction on exposure to light without getting impacted or affected by the process. A few nanometers in size in at least one dimension, nanophotocatalysts are tiny semiconductor particles. Numerous studies have shown that the characteristics of nanophotocatalysts are distinct from those of bulk materials.

Photocatalysts based on Carbon Scaffolds: Environmental and Industrial Applications, First Edition. Edited by Ramesh Oraon et al.
Published 2026 by Wiley-VCH GmbH.

According to Enesca and L. Andronic's (Enesca & Andronic, 2020) mini-review, enhancements to the structures of photocatalysts, may result in improved energy efficiency and process sustainability. Furthermore, in order to conduct meaningful comparative studies that could lead to the widespread use of photocatalysis, they asserted that the photoactivity results urgently needed to be standardized. In an effort to enhance charge separation and light absorption, heterostructure-based catalysts have drawn the interest of contemporary research projects. Zhang et al. investigated how the photoactivity of titanium oxide (TiO_2) nanofibers produced by electrospinning was affected by various crystal phase compositions (Zhang et al., 2019). Effective electron-hole separation is facilitated at anatase-rutile junctions by the creation of type II band alignment. A. A. Yaqoob et al. published a review work (Yang et al., 2019) of zinc oxide (ZnO) nanocomposites based on graphene oxide (GO). From a critical perspective, this research examined the prospective applications that take into account stability and recyclability in addition to the synthesis processes and photoactivity mechanism. The ZnO type II heterostructure based on GO exhibits favorable properties in charge separation and absorption of visible light.

Additionally, Yang et al. (2019) created nanotubes with strong methyl orange degrading activity. The synthesis process used in this instance, though, was distinct: an alkaline hydrothermal treatment was applied to a titanium (Ti) foil.

Plasmonic photocatalysts, which combine semiconductor and noble metal nanoparticles, offer an additional means of accomplishing visible light-assisted processes. Wei et al. (2019) looked at the molecular characteristics of the TiO_2-based photocatalyst and how they affected the control of electron transport to metal particles (Pt, Ag, and Au). The efficiency of a metal in visible light is dependent on its properties, oxidation state, and photo-absorption capacity (plasmon resonance). Silver has proven to be the best choice for antibacterial applications, even if there isn't a single noble metal that works better in every situation when it comes to contaminant degradation research. Furthermore, the photocatalysts were applied as thin films using a dip-coating technique to glass slides. Significant savings are achieved when the catalyst is removed from the reaction media thanks to the support of photoactive materials.

The breakdown of pharmaceutically active substances, such antibiotics, is a major worry since they are released into the environment and cannot be broken down in traditional wastewater treatment facilities. This challenge has demonstrated the potential of photo-catalysis, and a number of the research in this domain using probe molecule (Enesca & Andronic, 2020; Belet et al., 2019; Patel et al., 2020; Zhang et al., 2020).

Zhang et al. (2020) combined g-C_3N_4 and cadmium sulfide (CdS) with a 2-aminoterephthalate (UiO-66-NH_2) metal–organic framework (MOF) to eliminate tetracycline, while Patel et al. (2020) generated manganese-doped ZnS quantum dots to break down the fluoroquinolone antibiotic norfloxacin. The numerous advantages of MOFs, such as their large number of active sites, elevated surface area, and adaptable porous structure, have garnered increasing interest.

Photo-catalysis is a method that enhances the efficacy of pollution removal only when combined with other physical processes. Pham et al. (2020) created materials that have photochemical and adsorptive qualities in this regard. These materials consist of magnetite (MNP) that has been coated in humic acid (HA). Magnetite enhances photocatalyst recovery from reaction media by acting as a nonactive support due to its magnetic characteristics. This chapter details various modifications that scientists have made to carbon-based photocatalysts in an attempt to improve their catalytic efficiency.

6.2 Carbon Scaffold-based Nanophotocatalysts

Fast charge expenditure during redox reactions, low photo-generated charge carrier recombination, and exceptional light harvesting capabilities are critical for a greater contribution from nanophotocatalysts (Yu et al., 2014). Carbon is suggested as a possible photocatalyst for widespread and industrial use. It is sustainable, affordable, and benign to the environment. The exceptional electrochemical and physicochemical properties of carbon make it a popular material for a wide range of applications, from optical to electronic and solar systems (Kirubasankar et al., 2019; Ngaw et al., 2017; Jintoku et al., 2019; Shi et al., 2019). *Graphene, graphitic carbon nitride (GCN), conducting polymers, and carbon nanotubes (CNTs)* have all been the focus of extensive research in this field (Di et al., 2017; Xia et al., 2007; Cao et al., 2007).

A remarkable nanomaterial, "graphene" exhibits exceptional optical, electrical, and thermal properties (Patel et al., 2019). This renders graphene, subsequent to its revolutionary findings by *Geim and Novoselov* (Geim et al., 2007) a fitting substance for a plethora of purposes, encompassing optical, electronic and biological apparatuses. Kamat et al.'s first discovery of a graphene-based photocatalyst (Lightcap et al., 2010) and Liang et al.'s follow-up study (Liang et al., 2011) validated the advantages of graphene. Graphene is a promising conducting activity that offers a new approach to visible light photocatalytic behavior (Wang et al., 2019; Kandy & Gaikar, 2019a; Liu et al., 2015; Cho et al., 2017; Mateo et al., 2018).

One common 1D carbonaceous substance with a large surface area is CNTs. Graphite sheets are merely rolled into high aspect ratio cylinders to form CNTs (Liang et al., 2017). Still, few research comparing CNT to graphene and GCN as photocatalytic reduction's conducting supports exist. The use of CNTs in a range of applications has expanded due to the development of a more cost-effective synthesis technique (Xu et al., 2019; Li et al., 2018).

The most popular and widely utilized carbon support in this field is GCN due to its exceptional ability to function as a metal-free visible light photocatalyst. GCN (bandgap ~ 2.7 eV) has been engineered and combined with various materials to generate heterostructures that enhance photocatalytic activity (Ouyang et al., 2019; Li et al., 2022). Other types of carbon, such as dopants, naturally occurring materials with regulated shapes, CNT, carbon nanosheets, and carbon nanoparticles, have also been studied by researchers for their potential to exhibit photocatalytic properties (Jiang et al., 2018; Pan et al., 2017). A notable improvement in product yield resulted from the synergistic interaction between the photocatalyst and carbon material. The broad π–π transition structure in the carbon scaffold-based nanocomposite improves the circulation of photo-excited electrons obtained from the photocatalyst and promotes photocatalyst dispersion because of its large surface area (Yu et al., 2014). As a result, it has been suggested that mixing any photocatalyst with carbon scaffold-based nanocomposite is a practical strategy to attain the aims of increased photocatalytic reduction. The electrical characteristics of the aforementioned carbonaceous materials vary significantly, giving them unique performances. Although the materials' 2D layered topologies are similar, graphene is a superior conductor and GCN is a semiconductor (Faraji et al., 2019; Su et al., 2019). Nonetheless, because of their distinct surface characteristics, they work well as conducting supports for research on photocatalytic applications as Fabrication, Performance and Prospects.

6.2.1 Enhanced Nanophotocatalysts Behavior by Carbon-based Materials

The main benefits of carbon-based materials are electron-hole separation, surface area improvement, CO_2 adsorption, and visible light response generation. These are covered in the section that follows.

6.2.1.1 Increase in Surface Area

The increasing usage of nanomaterials as photocatalysts is facilitated by their highly advantageous surface characteristics (Xu et al., 2019; Yang et al., 2018). Elevated surface area of carbon-based materials as CNTs, graphene, and graphene oxide (GO), serve as effective substrates for the widespread distribution of inactivated photocatalysts (Wang et al., 2014). Furthermore, by acting as capping agents to control the transformation of photocatalysts into nanocrystal particles, carbon scaffold-based nanocomposites can increase the particles' surface area (Sun et al., 2018). CdS nanorods were produced recently from a previous work using a template consisting of nanoporous graphene covered on a porous anodic alumina support. Compared to CdS alone, these CdS nanorods developed on nanoporous reduced graphene oxide (RGO) show 4.8 times higher photo-catalytic activity (Kandy & Gaikar, 2019a). After loading on RGO supports, a TiO_2 nanoparticle's specific area increases, developing a large number of reaction sites that can greatly speed up photocatalytic activity on their surface, as shown by Liu et al. (2016). Since hollow and mesoporous materials offer more surface area and active sites for better reactant molecule transit and adsorption, photo-catalysis activity is increased when using them. Mesoporous RGO was found to provide an effective photocatalyst support for copper (II) oxide (CuO) nanoparticles, leading to increased CO_2 photocatalytic reduction as described by Gusain et al. (2016). However, restacking brought on by π–π interactions reduces the accessible surface area of graphene sheets, minimizing the immobilized photocatalyst. The application of crumpled graphene is a simple solution to this problem. Nie et al. (2016) developed crumpled RGO–amine–TiO_2 as a photocatalyst to enable the photocatalytic destruction of carbon-based gases and parallel adsorption. The enormous surface area and electrical conductivity of crumpled and nanoporous graphene make it an excellent conducting support that enhances photocatalytic activity. It is crucial to take into account the interfacial bonding of the RGO/TiO_2 nanocomposite in order to further attain a stronger synergetic effect. Shehzad et al. (2018b) used a method in their work to enhance the interfacial interaction between RGO and TiO_2. To achieve high interfacial bonding, GO and TiO_2 were partially decreased before vacuum heat treatment. Higher surface area nanocomposites showed faster rates of CH_4 and CO production.

GO is commonly utilized as a base material to blend with in the synthesis of RGO/photocatalyst. GO's enormous surface area and abundance of oxygen functions make it an ideal substrate for immobilizing photoactive homogeneous metal complexes. GO inherently exhibits insulating properties due to its broad bandgap.

Modulating the proportion of sp^2 to sp^3 atoms results in the conversion of the material into a semiconductor. The presence of oxygen functions in GO shows to the conversion of sp^2 C-atoms to sp^3 C-atoms, resulting in the formation of a two-dimensional sp^2-sp^3 matrix on the facial of GO. The sp^2 C-atom operates as the conduction band (CB), allowing electrons to move freely, while the sp^3 C-atom traps electrons and functions as the valence band (VB), creating a large number of semiconductor areas on the GO surface. Thermal and photo-assisted methods can be used to decrease the amount of (Eda et al., 2009) GO. The photocatalytic activity of GO after photo-reduction to RGO is analyzed. GO is discovered to have better activity throughout the CO_2 reduction process because of its increased electronic conductivity. Despite the efforts of numerous researchers, achieving efficient contact between photocatalysts and RGO or GO in nanocomposite frameworks remains a significant difficulty. TiO_2 nanotube nanocomposite heterostructures wrapped in GO/RGO layers were created by Rambabu et al. (2019). The development of connecting bridges between nanotubes boosts the photocatalytic activity of TiO_2 nanocrystals.

CNT encourages improved photocatalytic performance of the immobilized photocatalyst, just like RGO and GO do. Xia et al. (2017) investigated the multi-walled carbon nanotube's (MWCNT) effect in photocatalytic CO_2 reduction as a support for the TiO_2 photocatalyst. The various ways for creating the nanocomposite affect the selectivity of the photocatalytic yields. The primary byproducts of photocatalytic CO_2 reduction from MWCNT-based TiO_2 produced by sol–gel and hydrothermal techniques are ethanol (C_2H_5OH) and formic acid (HCOOH). A higher photostability and improved yield are achieved by the sol–gel nanocomposite due to less TiO_2 nanoparticles aggregation on MWCNT support.

6.2.1.2 CO_2 Adsorption

Utilizing materials based on carbon improves CO_2 molecule adsorption and activation. Materials with a p-conjugated structure on their surface, such as graphitic carbon nitride (GCN), polyaniline (PANI), CNTs, RGO, and others, facilitate π–π interactions with CO_2 molecules. The Ag/RGO/CdS nanocomposites made by Zhu et al. (2018) demonstrate improved CO_2 to CO photocatalytic conversion. Silver (Ag) and the RGO cooperated as electron acceptors for CO_2 activation and adsorption, increasing CO_2 conversion. Despite extensive research efforts to create a visible light active photocatalyst, low CO_2 adsorption prevents effective CO_2 conversion. It was discovered that adding CO_2 capture materials by creating holes in the system would help overcome this difficulty. Increased product production requires the CO_2

reactant molecules to move from their adsorption to catalytic sites. In the past, pre-deposited nanoporous RGO on the pore substrate of an alumina based anodic face was used to deposit CdS nanorods. The porous structure provides locations for CO_2 adsorption, which subsequently aids in the high CO_2 photoreduction process. Wang et al. (2019) have improved our understanding of the use of graphene, TiO_2, and porous-hyper cross-linked polymer as a photocatalyst for the manufacture of solar fuel. Since graphene has a short diffusion length, it was functionalized to form a macro-porous cross polymer matrix on the substrate. This helps to collect CO_2 and allows CO_2 to flow around the TiO_2 photocatalyst. According to Cho et al. (2017b), the amine-functionalized graphene/CdS hybrid nanocomposite has a 20-fold higher photocatalytic conversion to CH_4 than RGO/CdS. The reason for this is increased amine functionalization and CO_2 adsorption.

The multifunctional advancement effects of ternary nanocomposites, comprising graphene, a noble metal, and a main photocatalyst, were reported to produce enhanced photocatalytic activity. An anisotropic Bi_2O_3–graphene–ZnO nanocomposite was synthesized by Ali et al. (2018). The ternary nanocomposite's GO surface had an equal distribution of the ZnO and Bi_2O_3 photocatalysts. This nanocomposite shows enhanced CO_2 adsorption on its graphene surface along with a robust photocatalytic conversion of CO_2 to CH_3OH.

6.2.1.3 Induce Enhanced Visible Light Response

Graphene, MWCNT, GCN, fullerenes, and other carbonaceous compounds are added to boost photocatalytic activity in visible light. The numerous advantages of titanium dioxide (TiO_2), such as its low toxicity, exceptional photo-stability, potent oxidizing ability, and reasonable cost, make it a popular photocatalyst for usage in commercial settings. Within the ultraviolet (UV) band, which makes up just 3–5% of the sun spectrum, titanium dioxide (TiO_2) exhibits photocatalytic characteristics. Its application is therefore limited to the visible zone.

As such, it cannot be used as a workable, commercially practical photocatalyst that reacts to visible light. Its functionality in the visible spectrum has been improved through extensive efforts to blend it with carbon-based materials. A stable visible spectrum absorption band with increased absorption intensity was obtained by including graphene into the photocatalyst. The photocatalyst RGO-TiO_2 was developed by Tan et al. (2013b) and is light-activated. The nanocomposites showed a red-shift, or shift toward longer wavelengths, as compared to the UV–vis spectra of pure TiO_2. This alteration implies that the nanocomposites are photoactive when exposed to visible spectrum. The nanocomposite showed excellent CO_2 to CH_4 photocatalytic conversion.

In the presence of visible and ultraviolet A (UVA) light, Olowoyo et al. assessed the CNT/TiO_2 nanocomposite's photocatalytic potential (Olowoyo et al., 2018). In order to get a high photocatalytic yield, it is necessary for CNTs and titanium dioxide (TiO_2) to interact in a suitable manner. Simultaneously, electrons are added to the CB of TiO_2, while photons are absorbed at the contacts between CNT and TiO_2. The MWCNT/TiO_2, a similar nanocomposite, was further doped with Silver (Ag) by Gui et al. (2015), demonstrating plasmonic photo-catalysis mediated by visible light.

Adding visible-light responsive metal ions, such Pt, Au, and Ag, can create impurity states that enhance visible light absorption with a pronounced surface plasmon resonance [SPR] absorbance. The process of photocatalytic reduction is improved by the combined action of Ag and MWCNT. A CVD method was employed by Ong et al. (2013) to develop CNT on Ni/TiO_2 nanoparticles. Compared to TiO_2, CNT@Ni/TiO_2 showed a notable redshift of the absorption edge due to increased optical absorption in the visible band. Using CNT, expanded graphite (EG), graphite powder (GP), and silver bromide (AgBr) as the photocatalyst, Fang et al. (2014) verified the enhanced visible light active performance with CNT-based nanocomposites. All of the carbon-based AgBr nanocomposites were shown to absorb visible light based on their UV–vis spectrum. Comparing AgBr/CNT and AgBr/GP to conventional carbon supports, their enhanced conductivity resulted in a significant photocatalytic CO_2 reduction. Fu et al. (2018) examined the visible light activity of graphene and CNTs as conducting supports together with phosphine oxide (CoP) as a photocatalyst and the photocatalytic reduction of CO_2 using Ruthenium (Ru) as a photo-sensitizer. Graphene and CNTs demonstrated good photocatalytic conversion of CO_2 due to the strong linkages between the CoP and carbon support.

Several modifications have already been made to reduce its inherent restrictions and increase visible light activity, including heteroatom doping. Visible light can activate phosphorus-doped GCN (bandgap of 2.7 eV), a promising photocatalyst due to its long charge carrier lifetime and efficient absorption of visible light. P-doped GCN with an amino-rich surface, a sizable surface area, and a meso-porous nanotube structure was created by Liu et al. (2018). A narrow bandgap is produced by phosphorus doping in the interstitial region by shifting the locations of the VB and CB, which subsequently aids in the absorption of visible light. Rolling GCN sheets into a nanotube shape reduced the van der Waals forces between the interlayer, resulting in the interstitial phosphorous doping. A larger photoreduction of CO_2 to CO and CH_4 was made possible by the unique structure of this photocatalyst. Using N-doped TiO_2, Zhou et al. (2014) suggested a simple process for the synthesis of GCN. By combining the strengths of these two easily accessible and moderately priced photocatalysts – the low stability of GCN and the poor visible light activity of TiO_2 – they create a very stable nanocomposite that is active to visible light. The visible light absorption of the composite was higher than the TiO_2 photocatalyst's usual UV light absorption due to the inclusion of GCN and nitrogen (N). GCN is perfect for absorbing CO_2 using visible light because of the surface doping of phosphorus (P) and nitrogen (N).

Conducting polymers are added to semiconductor nanostructures to increase their visible light activity. By enhancing charge separation, broadening the visible light absorption range, and postponing photo-corrosion, this is accomplished (Lee & Chang, 2019). The CO_2 reduction process poses considerable challenges, which has led to an increasing usage of conducting polymers in other photocatalytically related processes like water splitting and dye degradation. The advantages of conducting polymers must be considered when developing a *visible light active photocatalyst* for CO_2 reduction.

6.2.1.4 Electron-hole Separation

Electrons are excited from the VB to the CB when a photocatalyst is exposed to light. As charge carriers for photocatalytic activities, this results in electrons in the CB and holes in the VB. Nevertheless, within 10 ns of air stimulation, these photogenerated charge carriers recombine, preventing free electrons from being available for various reduction processes (Serpone et al., 1995). Photocatalytic reactions can be facilitated by any substance that has the ability to scavenge these photo-excited electrons from the photocatalyst. Carbon-based materials are known to be very good conducting supports that can capture and recirculate electrons produced by photolysis.

Graphene's outstanding electrical properties and simplicity of fabrication have won it prominence as an inexpensive conducting support. Electrons can be swiftly collected and transferred over graphene as it has a lower Fermi level than most photocatalyst CBs. The VB of the photocatalyst contains free photo-generated holes that aid in the oxidation of water molecules into protons and oxygen. Later on in the photocatalytic CO_2 reduction processes, the free photo-generated electrons existing along the graphene surface are utilized. In nanocomposites, the photocatalyst subsequently functions only as a constant charge generator. The rate of reaction is further accelerated by the wide spatial gaps that exist between the electrons and holes. The photocatalyst and graphene form hetero-junctions that stop electrons from flowing backward from graphene to the photocatalyst. Thus, multi-electron photocatalytic reduction processes are encouraged by graphene that is rich in electrons (An & Yu, 2011). Yu et al. (2014) investigated the function of RGO as an alternative to more expensive noble metals. The visible light range was used to study photocatalytic activity, and it was shown that using cadmium sulfide (CdS) either alone or in conjunction with a CdS–platinum (Pt) nanocomposite enhanced the rate of Methane (CH_4) synthesis. Since RGO effectively separated the charge carriers produced by photolysis and improved the adsorption and activation of carbon molecules, it increased the photocatalytic rate of CH_4 generation. Combining 2D RGO with a 2D photocatalyst results in a high interface contact that inhibits electron-hole recombination. 2D graphene improves electron-hole transport across their heterojunction interface by decreasing the diffusion channel and the time it takes for an electron produced by photolysis to reach the acceptor surface. The 2D/2D heterojunctions exhibit remarkable face-to-face contact, in contrast to the line-to-face and point-to-face contacts present in 1D/2D and 0D/2D heterojunctions, respectively (Ong, 2017). Using protonated GCN and RGO, Ong et al. (2015) created 2D/2D heterostructures. To reduce electron-hole recombination, promote efficient interfacial contact, and speed up electron transit, enough RGO was added. Additionally, light trapping is made easier by the hollow shape, boosting photocatalytic efficacy by multiplexing incident light in the photocatalyst zone. Titanium oxide (TiO_2) nanosheets and RGO were assembled alternately layer by layer into hollow sphere shapes composed of poly (methyl methacrylate) (PMMA) by Tu et al. (2012). The thinness of this photocatalyst and the tightly packed TiO_2 with RGO nanosheets resulted in a ninefold increase in activity. The charge carriers' participation in photo-reduction and faster mobility were encouraged by this configuration. By mixing a 3D graphene aerogel with a micro-porous TiO_2 and molybdenum disulfide (MoS_2) nanocomposite, CO_2

has been reduced photo-catalytically (Jung et al., 2018). Photo-excited electrons have lower diffusion lengths due to the shorter conducting route length of the three-dimensional graphene structure. Consequently, compared to its conventional 2D form, the 3D graphene structure exhibits substantially better photocatalytic activity.

CNTs may receive and transport electrons when used in the right photocatalyst; their Fermi level is about 0.2 V. By employing discrete multiple concentric cylindrical configurations, the CNT-doped photocatalyst inhibits the rate at which photogenerated charge carriers recombine upon the removal of electrons (Fang et al., 2014). The CNT serves as a conductive support for the electrons that are driven from the VB to the CB of TiO_2 by visible light radiation. Recombination rate is delayed as a result of heterostructure creation at the CNT–TiO_2 interface, which helps to improve interfacial charge transfer processes (Ong et al., 2013; Fang et al., 2014). The influence of CNT length on photocatalytic processes was examined by Asi et al., who found that longer CNTs had better photocatalytic activity than shorter ones because of their strong electron transport capabilities (Asi et al., 2013). However, when comparing it to graphene and GCN, there is a scarcity of research articles that utilize CNT as a conducting support for carbon scaffold-based nanophotocatalysts.

The exceptional ability of conducting polymers with rich pi-conjugation systems to scavenge photo-generated electrons and enhance photocatalytic performance makes them attractive materials for supporting photocatalysts (Li et al., 2015). Polythiophene's low photoluminescence, low bandgap, and enhanced charge carrier transfer demonstrates its superior photocatalytic activity when compared to other conducting polymers such as polyaniline (PANI) and polypyrrole (PPY). It serves as an excellent conducting support for the polymer carbon-based bismuth tungstate (Bi_2WO_6) photocatalyst (Dai et al., 2015). The dissociation of photogenerated charge carriers is accelerated by the presence of carbon in polyaniline-containing photocatalysts, as Liu et al. (2015) also found.

In order to decrease the distance over which charge carriers can diffuse from the bulk to the surface, this novel structure includes an inner chamber. Furthermore, it effectively mitigates the issue of competitive recombination of charge carriers that is commonly encountered by typical photocatalysts generated from metals. N-doped TiO_2 and the natural chlorophyll present in spirulina are combined to create the novel photocatalytic nanocomposite. It exhibits improved photocatalytic stability and significant C^{2+} product selectivity. When compared to undoped TiO_2, the composite showed a greater overall light conversion (Wang et al., 2018). 1D carbon nanofibers' low charge transfer resistance allows the photocatalyst to function more effectively after taking in and re-circulating photo-generated charge carriers inside of it (Pan et al., 2017). The doped carbon sites inhibit charge recombination and act as a superb acceptor of photo-generated electrons. According to Shown et al. (2018), interstitial C in the SnS_2 lattice creates a micro-strain, which improves the photo-physical properties of the C-doped SnS_2 photocatalyst. The conductive carbon in these hybrid nanostructures facilitates faster charge transport, reducing the amount of time it takes for active sites to diffuse from the photocatalyst's interior to the surface.

It would be more feasible to utilize the intriguing qualities of carbon-based material supports by cross-linking carbon nanostructures to create an integrated structure. These linked structures show great potential as conductive supports for photocatalysts due to their remarkable features, such as their highly conductive framework that facilitates fast electron uptake and transport.

6.2.2 Challenges Involved Using Carbon-based Materials

The identification of the source of products, photo-stability, and blockage of visible light are the key issues encountered when employing carbon-based materials for carbon cycling.

6.2.2.1 Identification of the Source of Products

The idea that carbon-based materials could speed up the process of converting CO_2 into useful carbon compounds is really intriguing. However, it is not evident whether the compounds generated by Methanol (CH_3OH), CH_4, etc., are the outcome of the reduction of carbon molecules by photocatalysis. High surface area photocatalysts can naturally absorb organic pollutants from the atmosphere, supplying carbon for the reduction process. To get rid of such impurities, photocatalysts might be annealed before use (Grigioni et al., 2017). However, this is not the case if the photocatalytic system contains carbon-based components. Research utilizing isotopic labeling is often used to confirm the origin of products (Fang & Wang, 2018). Zhu et al., (2017) used C-13 in combination with gas chromatography–mass spectrometry (GCMS) to perform isotope tagging of CO_2. The product that was obtained had the origin code C-13 on it. Additional evidence that the products are the outcome of CO_2 reduction by photo-catalysis comes from the absence of species having a C-12 origin. By replacing CO_2 gas with an inert gas, such as N_2, blank experiments can be conducted without isotopic labeling. Carbon-based support material has not undergone photo dissociation if there is no product (Kandy & Gaikar, 2019a).

6.2.2.2 Photo-stability

High photocatalytic activity can be achieved by integrating carbon-based compounds due to their numerous beneficial benefits. However, a crucial consideration for any practical application is the photocatalyst's reusability. In a prior study, the rate of CH_3OH generation on the second day was significantly reduced when CdS was utilized alone as the photocatalyst on peracetic acid (PAA) support. This might result from CdS photo-corroding to $CdSO_4$ (cadmium sulfate) (Kandy & Gaikar, 2019b). RGO was added to the PAA substrate to improve CdS stability and photocatalytic activity. There was no discernible drop in the nanocomposite's photocatalytic activity after three days of use. A new heteroleptic ruthenium (II) semiconducting metal complex adsorbed on a graphene oxide platform was investigated for stability in order to achieve photocatalytic CO_2 reduction (Kumar et al., 2015).

6.2.2.3 Blockage of Visible Light

These photocatalysts have better photocatalytic reduction because of the synergistic positive effect of hybrid systems. Nonetheless, high carbon content would lead to the photocatalyst aggregating, blocking the passage of visible radiation and reflecting as recombination hubs, delaying photoactivity (Yu et al., 2014). When RGO, GCN, or a carbon layer are loaded optimally, light can pass through and the nanocomposite shows its maximum photocatalytic activity (Wang et al., 2017). Pan et al. (2017) found that in order to keep the Indium (III) oxide (In_2O_3) photocatalyst from obstructing light absorption, a thin coating of carbon must be placed over it. Therefore, as illustrated in Figure 6.1, it is imperative to optimize the carbon concentration so as not to obstruct the photoactive material's integral photocatalytic activity in the carbon/TiO_2 photocatalyst's nanocomposite structure.

The photocatalytic stability is impacted by the catalyst that evaporates off of sheets. It was found that the nanocomposite retained its photocatalytic activity without any metal or ligand leaching following three more rounds of the photocatalytic process. Olowoyo et al. showed how CNT can improve photocatalytic stability (Olowoyo et al., 2018). The durability and reusability of the nanocomposite are demonstrated by the gradual increase in the CH_3OH production rate. Both GCN alone and its nanocomposite show good chemical and thermal stability, which makes them extremely stable visible light active photocatalysts. Metal nanoparticles scattered across the GCN matrix are further stabilized by the addition of nitrogen vacancies (Tu et al., 2017). This class of very stable carbon-modified photocatalysts starts a viable pathway for solar energy to fuel conversion. The remarkable photo-stability exhibited by carbon-based photocatalysts suggests that their composition and structure remain stable even after repeated use.

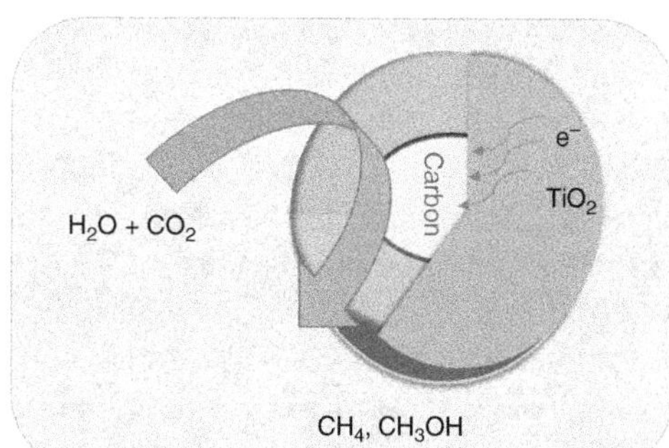

Figure 6.1 Diagram showing how the carbon content affects the TiO_2 photocatalyst's ability to enhance CO_2 photocatalytic reduction.

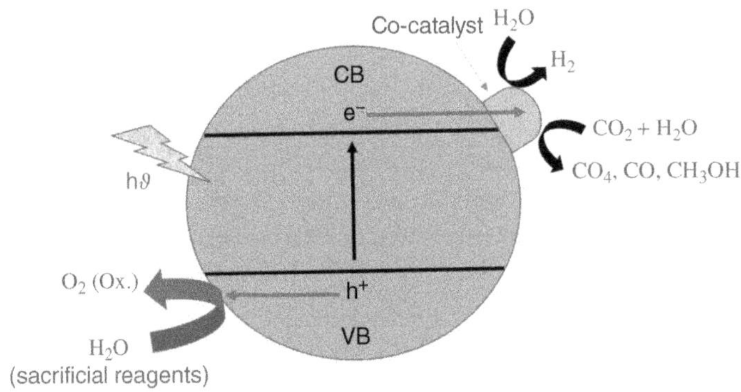

Figure 6.2 Diagrammatic representation of the cocatalyst's role in reducing H_2O or CO_2 to H_2 or CO and organic molecules when H_2O is present. *Source*: Fan et al. (2013) / Americal Chemical Society.

6.3 Semiconductor-based Carbon Scaffold-based Nanophotocatalysts

Using photoelectrochemical water splitting on TiO_2 electrodes exposed to UV light, Fujishima and Honda observed the origin of H_2 for the first time in 1972 (Fujishima & Honda, 1972). Inoue et al. (1979) investigated the photoelectrocatalytic reduction of CO_2 to organic compounds, including formic acid, methanol, and formaldehyde, in a range of semiconductors, including TiO_2, ZnO, CdS, silicon carbide (SiC), and tungsten (VI) oxide (WO_3), by exposing aqueous suspensions saturated with CO_2 to xenon (Xe) lamp radiation (Inoue et al., 1979). The nanophotocatalytic domain has been the topic of much research since then; in recent decades, this research has accelerated significantly due to the world's deteriorating environmental circumstances and the rapidly developing demand for alternate energy sources. Developments in nanotechnology have sped up the search for novel photocatalysts by creating nanomaterials with a range of forms and morphologies (Rajeshwar et al., 2001).

As shown in Figure 6.2, semiconductor-based photo-catalysis can be used to reduce CO_2 with H_2O and split H_2O into H_2 and O_2. These three main processes are basically engaged in this process. The process of electron-hole pair generation occurs when a semiconductor absorbs photons. It also happens when electrons and holes produced by photons go to surfaces or combine again. Lastly, two categories of processes take place on surface sites: (i) electrons reducing H_2O or CO_2 (acceptor) into H_2 or CO and organic compounds (eqns. (2)–(7)) and (ii) O_2 is produced when holes oxidize H_2O (donor) (eq1). The semiconductor's bandgap determines how much photon energy is needed for photo-excitation.

$$H_2O + 2h^+ \longrightarrow ½\,O_2 + 2H^+ \tag{1}$$

$$2H^+ + 2e^- \longrightarrow H_2 \tag{2}$$

$$CO_2 + 2H^+ + 2e^- \longrightarrow CO + H_2O \tag{3}$$

$$CO_2 + 2H^+ + 2e^- \longrightarrow HCOOH \tag{4}$$

$$CO_2 + 4H^+ + 4e^- \longrightarrow HCHO + H_2O \tag{5}$$

$$CO_2 + 6H^+ + 6e^- \longrightarrow CH_3OH + H_2O \tag{6}$$

$$CO_2 + 8H^+ + 8e^- \longrightarrow CH_4 + 2H_2O \tag{7}$$

Semiconductors with a shorter bandgap, such as CdS (2.4 eV), may operate under visible light, while semiconductors with a wider bandgap, such as TiO_2 (3.2 eV), can only use UV radiation, which accounts for 4% of solar energy. Another important factor that must coincide with the redox potentials of the pertinent chemical species is the band-edge position, as shown in Figure 6.3. In the case of H_2 development and CO_2 reduction, the valance-band edge should be more positive than the redox potential of O_2/H_2O, while the conduction-band edge should be higher (more negative) than the redox potential of H_2/H_2O or CH_4/CO_2. Hence, designing semiconductor photocatalysts' energy bands to allow them to utilize visible light is one of the primary issues in photo-catalysis. A further significant challenge in photo-catalysis is increasing efficiency. Photosynthesis-induced electron and hole recombination is thought to be the primary cause of photo-catalysis's low efficiency. Two competing mechanisms are identified: charge separation/migration and recombination (Fan et al., 2013). Chemical interactions with adsorbed species take longer than photogenerated electron-hole pair recombination, which typically takes place on the scale of 10^{-9} s. The most difficult photocatalytic reactions, such as the reduction of CO_2 and the creation of H_2, require accelerating the separation and migration of the photo-excited electrons and holes in order to avoid recombination.

A number of tactics have been put out to help with the charge migration and separation (Liu et al., 2010). Using nanocomposites as photocatalysts instead of single semiconductors is one of the most feasible and efficient methods to increase photocatalytic performance. In semiconductor-based nanocomposites, the charge-migrating

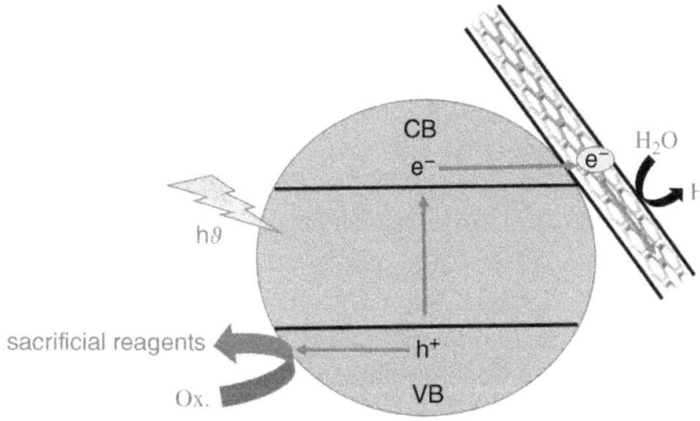

Figure 6.3 Diagram illustrating the primary role of CNT in semiconductor–CNT nanocomposites as an acceptor and electron-transfer channel. *Source*: Fan et al. (2013) / Americal Chemical Society.

component may be a metal oxide or sulfide, carbon, a transition or noble metal, or another semiconductor with the proper band-edge locations. Usually, the semiconductor absorbs radiation. Maintaining separation between reduction and oxidation events can also increase photocatalytic efficiency. Furthermore, the second and third elements in the nanocomposites may operate as cocatalysts by offering active centers for reactant activation or the ensuing reduction/oxidation processes.

6.3.1 Semiconductor–carbon Composites

6.3.1.1 Carbon Nanotubes

Photocatalysis research has shown a great deal of interest in semiconductor–carbon nanotube (CNT) nanocomposites due to robust characteristics of CNTs (Leary & Westwood, 2011). The semiconductor catalyst in the nanocomposite can release photogenerated electrons for absorption thanks to the substantial electron-storage capacity of CNTs. At the same time, CNTs' long-range p electronic conjugation facilitates electron transmission. Consequently, they could boost the photocatalytic activity for the generation of H_2, as seen in Figure 6.3, by preventing the photo-generated electrons and holes from recombining. Second, CNTs' large specific surface area (>150 $m^2\ g^{-1}$) may promote reactant adsorption, hence boosting photocatalytic activity. This is especially significant for the degradation of contaminants by photo-catalysis. Lastly, by functioning as photo-sensitizers, CNTs may increase the wavelength of excitation. The ability of CNTs to absorb visible light and promote the transfer of photo-induced electrons into the conduction band of UV light-activated semiconductors like TiO_2 is the reason why TiO_2–CNT composites show improved photo-efficiency in phenol degradation under visible light (Wang et al., 2005). Additionally, the functional groups on the CNT surfaces may have an effect on the photocatalytic activities of the nanocomposites. The majority of research on semiconductor–CNT nanocomposite photocatalysts has been done thus far with an emphasis on environmental uses, such as the photo-degradation of organic molecules (Tang et al., 2011; Ou et al., 2006).

Ou et al. (2006) created a CNT and Ni-doped TiO_2 nanocomposite using a modified chemical vapor deposition technique. When exposed to visible light, this nanocomposite may speed up the process of producing H_2 from a methanol and water solution. The Ni doped TiO_2 without CNT produced very little H_2 when exposed to visible light, whereas the nanocomposite with 4.4% CNT showed an H_2 evolution rate of 38.1 $\mu mol\, h^{-1}$ from a CH_3OH/H_2O (1/50) solution. According to Wang et al. (2005), CNTs' function as photo-sensitizers clarified their boosting effect.

Water-thermal synthesis of CdS–CNT nanocomposites with varying CNT concentrations by Peng et al. (2011) allowed CdS nanoparticles to grow directly on the functionalized CNT surfaces. The CdS nanoparticles, which range in size from 12 to 28 nm, were found to be in intimate contact with the CNT outer surfaces, according to the characterizations. When exposed to visible light, the CdS–CNT nanocomposites with the right amounts of CNT showed faster rates of H_2 evolution from an aqueous solution containing sodium sulfide (Na_2S) and sodium sulfite (Na_2SO_3) as sacrificial reagents than did CdS alone. The 10% CdS–CNT nanocomposite had a greater rate of H_2 evolution (174.2 μmol h^{-1}) than CdS (97.9 μmol h^{-1}). Apart from its principal purpose of preventing photo-generated charge recombination, the additional responsibilities of the CdS–CNT system for the production of H_2 were also examined. The addition of CNTs obviously improved the durability of the catalyst; after >11 hours of radiation for CdS, the generation of H_2 exhibited a declining trend, but the nanocomposite remained steady after 15 hours of reaction. After 15 hours of reaction, the photoreaction solution for CdS and 10 wt% CdS–CNT revealed Cd^{2+} concentrations of 10.6 and 0 μgmL^{-1}, respectively, indicating that CNTs could successfully stop CdS from photo-corroding.

Furthermore, the photocatalytic evolution of H_2 from the aqueous solution containing the sacrificial reagents Na_2S and Na_2SO_3 under visible light irradiation was examined for the ternary composites of CdS–CNT–M (where M is a transition metal) (Kim & Park, 2011). First, the transition metal was chemically reduced using Sodium borohydride ($NaBH_4$) to load it onto CNTs. Next, the CNTs loaded with transition metal were distributed throughout the Cadmium acetate ($Cd(CH_3COO)_2$) aqueous solution. This was done by adding Na_2S, filtering, washing, and drying the mixture. Among the CdS-CNT-M catalysts, the CdS-CNT-Pt catalyst exhibited the quickest rate of H_2 evolution, which was roughly an order of magnitude faster than that of the CdS-CNT-Pt catalyst. Additionally, the importance of CNT pretreatment was demonstrated; specifically, acid pretreatment was found to improve Pt loading onto CNTs and accelerate photocatalytic activity. The CdS particles, which were 460 nm, and the Pt nanoparticles, which were 10–30 nm in size, were found on CNTs and appeared to stick to the CNT fairly strongly.

6.3.1.2 Graphene

Two-dimensional sp^2-hybridized carbon nanosheets, or graphene, have a number of unique properties, including as excellent mechanical strength, high specific surface area (2600 $m^2\ g^{-1}$), and exceptional charge carrier mobility (200 000 $cm^2\ V^{-1}\ s^{-1}$) (Allen et al., 2009; Novoselov, 2011). One of the most intriguing materials with a vast array of potential uses is graphene (Guo & Dong, 2011; Bai & Shen, 2012).

Throughout the last three to four years, graphene has also been used for photocatalysis (Zhang et al., 2012a; Zhang et al., 2012b). Semiconductor–graphene nanocomposite photocatalysts are usually made by combining semiconductors and graphene. The photocatalytic performance can be enhanced for three possible reasons by incorporating graphene: (1) increasing reactant adsorption because of graphene's large specific surface area; (2) enhanced charge separation and transfer as a result of graphene's high charge carrier mobility; and (3) broadening the range of light absorption. Strong connections between graphene and semiconductors can be formed in graphene-based nanocomposites thanks to graphene's two-dimensional open structure and the presence of oxygen-containing functional groups on graphene oxide (GO), one of the most often used graphene precursors. Several methods for creating graphene-based nanocomposites have been developed (Xiang et al.,

2012a; Machado & Serp, 2012). Graphene-based nanocomposites can be easily prepared by adding semiconductors or their precursors to an aqueous solution of GO, then reducing the solution or growing it in one pot.

TiO_2–graphene composites were the subject of ground-breaking research by Kamat et al. (Williams et al., 2008; Lightcap et al., 2010). They explained that GO may be reduced to graphene by photogenerated electrons from UV-irradiated TiO_2. In order to change Ag ions into Ag nanoparticles, photo-generated electrons might also be transmitted over graphene at a distance from the TiO_2. These investigations unequivocally show how well graphene sheets transfer electrons. According to Li and colleagues, TiO_2 (P25)–graphene nanocomposites exhibited more photodegradation of methylene blue than P25 (Zhang et al., 2009). P25-graphene may even be effective in photo-degrading methylene in visible radiation. Graphene-based nanocomposite photocatalysts have been the focus of numerous organizations since these pioneering investigations (Zhang et al., 2012b; Zhang et al., 2012c). GO and tetrabutyl titanate were utilized as precursors by Cui and colleagues to produce TiO_2-graphene nanocomposites by the sol–gel technique. The graphene sheets in the nanocomposites were coated with spherical anatase TiO_2 nanoparticles, which ranged in size from 10 to 15 nm (Zhang et al., 2010a). The aqueous solution containing the sacrificial reagents Na_2S and Na_2SO_3 may experience acceleration in the development of H_2 due to the presence of nanocomposites under both UV and visible radiation. TiO_2-graphene produced by a hydrothermal technique is effective for H_2 evolution, according to a follow-up study conducted by the same research team (Zhang et al., 2012b).

An interesting comparison is being made between the catalytic properties of TiO_2–graphene and TiO_2–CNT nanocomposites. According to a previous study by Xu and colleagues, P25-GR and P25-CNT performed as well when it came to photodegrading volatile aromatic pollutants (Zhang et al., 2010b). P25/GR (1/0.2) or P25/CNT (1/0.3) was the ideal mass ratio, according to the results, which demonstrated that the P25–GR nanocomposite generated H_2 at a substantially higher pace than the P25–CNT nanocomposite (Fan et al., 2011). Structure–performance study suggests that a greater interaction between P25 nanoparticles and graphene sheets is the cause of the P25–GR–hydrothermal's enhanced photocatalytic activity for H_2 production. P25 and graphene have a larger surface area than graphene, as seen in Figure 6.4, which could potentially speed up the transfer of photogenerated electrons from P25 to graphene and more effectively prevent charge carrier recombination.

The photocatalytic selective oxidation of alcohols requires enhanced interfacial contact between TiO_2 and graphene, according to a follow-up work by Xu and colleagues (Zhang et al., 2011). This was presented as graphene's primary benefit over CNTs in the construction of effective carbon-based nanocomposite photocatalysts (Zhang et al., 2011). Furthermore, it was demonstrated that the photocatalytic evaluation of TiO_2–graphene nanocomposites for alcohol oxidation could be enhanced by combining the techniques of decreasing graphene defects and enhancing the interfacial contact between TiO_2 and graphene (Zhang et al., 2012a).

Yu and co-workers used GO and anatase nanosheets with exposed (100) facets that had been prepared as precursors in a microwave-assisted hydrothermal process

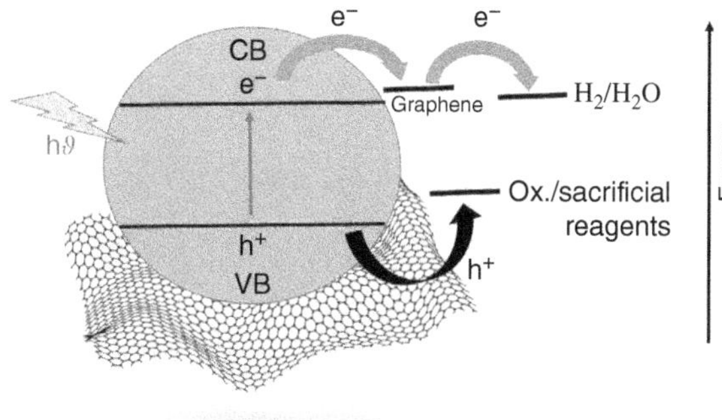

Figure 6.4 The function of graphene in semiconductor-graphene composites for H_2 photocatalysis. *Source*: Fan et al. (2013).

to further improve the contact between the two materials by creating a composite of both TiO_2 and graphene nanosheets (Xiang et al., 2011b). The graphene sheets in the nanocomposite were evenly dispersed face-to-face, according to the TEM characterizations. The average side dimension of the TiO_2 nanosheets was between 50 and 80 nm, and their thickness was between 6 and 8 nm. This close contact made it possible for the electrons to migrate from TiO_2 to graphene more quickly, which greatly enhanced the photocatalytic behavior under UV radiation A 3.1% apparent quantum efficiency was observed for this nanocomposite at 365 nm. The TiO_2-graphene, which was made up of graphene sheets and P25 nanoparticles, demonstrated a significantly lower rate of H_2 evolution under the same reaction conditions as this nanocomposite, confirming the importance of the interfacial contact.

A few researchers have examined the effects of coexisting graphene and a noble metal cocatalyst on the photocatalytic characteristics of TiO_2-based nanocomposites. In comparison to Pt/TiO_2 or TiO_2–GO, the Pt/TiO_2–GO composite generated more H_2 from methanol solution when exposed to UV radiation, according to Park et al. (2011). A TiO_2–MoS_2–graphene composite was created by Yu et al. (Xiang et al., 2012b) utilizing a two-step hydrothermal procedure to enable the UV-induced evolution of H_2 from an ethanol aqueous solution. The MoS_2–graphene composite was first synthesized by hydrothermally reacting methanamide (H_2CSNH_2) and sodium molybdate (Na_2MoO_4) in an aqueous graphene oxide (GO) solution to create smaller MoS_2 nanosheets scattered throughout graphene sheets. Tetrabutyl titanate (Ti $(OC_4H_9)_4$) and the MoS_2–graphene composite underwent hydrothermal treatment in an ethanol/water solution to produce TiO_2 nanoparticles on the composite with sizes ranging from 7 to 10 nm. Compared to the binary nanocomposites of TiO_2–MoS_2 and TiO_2–graphene, the ternary nanocomposite with the proper composition – 99.5% TiO_2 and 0.5% MoS–graphene, which is composed of 95% MoS_2 and 5% graphene – had a noticeably higher rate of H_2 evolution. At 365 nm, it also offered an apparent quantum efficiency of 9.7%. It is well known that MoS_2 cocatalyzes the development of H_2 (Zong et al., 2008). The inclusion of MoS_2 in the TiO_2–graphene composite may accelerate the synthesis of H_2 as it facilitates the adsorption and activation of H_2O.

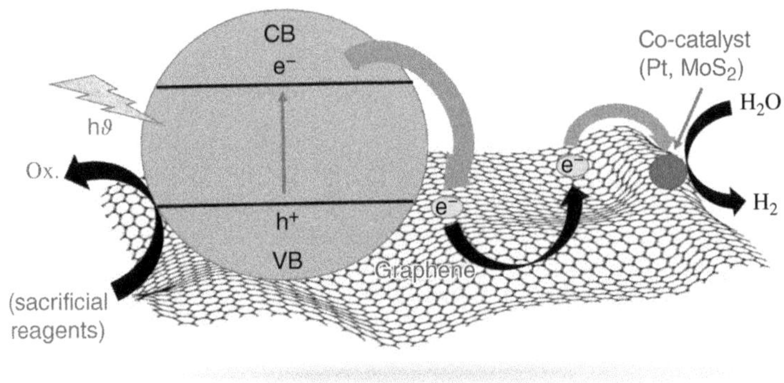

Figure 6.5 Diagrammatic illustration of the synergistic effect for the photocatalytic development of H_2 in ternary composite photocatalysts based on graphene. *Source*: Fan et al. (2013).

Figure 6.5 could be used to interpret the synergistic impact for such a ternary composite photocatalyst based on graphene. The electrons produced by photolysis are moved from the semiconductor's conduction band to the tightly connected graphene. Electron and hole recombination is delayed due to their high mobility on the graphene sheets. In order to function as active sites and activate and reduce H_2O to H_2, the MoS_2 nanosheets affixed to graphene can take electrons.

Graphene has also been used to enhance the performance of other semiconductors with lower bandgaps (such as CdS) for visible light-driven H_2 generation. The remarkable mobility of the electrons in graphene sheets speeds up the separation of electrons and holes. Graphene sheets in comparable topologies may also accept photogenerated electrons from semiconductor conduction bands. Li et al. used a solvothermal approach to manufacture the CdS–graphene composites, which had different compositions. They initially mixed GO and $Cd(AC)_2$ in dimethyl sulfoxide (DMSO) before autoclaving the mixture for 12 hours at 180 °C (Li et al., 2011). Cubic CdS nanoparticles with an average size of 3 nm were discovered during the analysis of the graphene sheets of the nanocomposites. The rate of H_2 production was increased to $1.12\,\mu mol\,h^{-1}$ by adding an aqueous solution containing lactic acid as a sacrificial reagent to a composite with an ideal graphene content of 1 and 0.5 wt% Pt cocatalysts. When exposed to light, this was around five times as much as the catalyst without graphene. At a wavelength of 420 nm, the apparent quantum efficiency over CdS–graphene was 22.5%, with graphene accounting for 1% of the total.

The photocatalytic production of H_2 from an aqueous solution under visible light exposure was investigated using a composite composed of CdS and N-doped graphene (N-graphene) and the sacrificial reagents Na_2S and Na_2SO_3 (Jia et al., 2011). At the ideal concentration of N-graphene (2 wt%), the rate of H_2 generation over the CdS–N-graphene was $210\,\mu mol\,h^{-1}$, which was about five times greater than CdS ($40\,\mu mol\,h^{-1}$). It has been found that the CdS–N-graphene nanocomposite is more stable than Pt/CdS, which seems to be beneficial for photo-catalysis as well. Theoretically, N-graphene could function as a barrier to prevent CdS from photocorroding when exposed to visible light.

Under visible light irradiation, the photocatalytic performances of the CdS–graphene (CdS–GR) and CdS–CNT nanocomposites for H_2 evolution were compared by introducing an aqueous solution of Na_2S to a mixed aqueous solution of Cd $(AC)_2$ and GO. A hydrothermal treatment at 453 K for 40 hours was the next phase (Ye et al., 2012). The characterizations showed that the CdS crystallites, which were dispersed across the graphene sheets or CNT surfaces in the composite materials, had an average size of 35 nm. The crystallites showed both hexagonal and cubic phases.

The mass ratio of CNT/CdS or GR/CdS served as the basis for the photocatalytic assessment of the two composites. For both systems, the CNT/CdS or GR/CdS ratios were ideal. The optimal GR/CdS and CNT/CdS ratios for H_2 production were 0.01/1 and 0.05/1, respectively. In comparison to CdS alone (14.5 $\mu mol\,h^{-1}$), the rate of H_2 generation over the CdS–GR composite (70 $\mu mol\,h^{-1}$) and the CdS–CNT composite (52 $\mu mol\,h^{-1}$) was higher at the optimal mass ratios. The mechanical combination of graphene and CdS showed activity equivalent to that of CdS alone at a mass ratio of 0.01/1. The CdS–GR composite had better photocatalytic capabilities than the CdS-CNT due to its wider contact surfaces, which resembled the TiO_2-GR system previously discussed. More graphene or CNTs could improve the interfacial contact, but this could hinder the CdS's ability to absorb light. In order to create a nanocomposite of CdS and modified graphene (CdS–mGR) and produce H_2 from an aqueous solution of Na_2S and Na_2SO_3, a Pt-loaded CdS under visible light irradiation was contrasted with a *one-pot solution* approach using sulfonated graphene (Lv et al., 2012). In comparison to a 2 wt% Pt/CdS loading, it was discovered that a 2 wt% mGR loading over the CdS–mGR increased the rate of H_2 production.

Combining graphene with a metal-free visible light-driven graphitic carbon nitride (g-C_3N_4) photocatalyst can significantly boost photocatalytic activity (Xiang et al., 2011a). When exposed to visible light, the rate of H_2 evolution from an aqueous solution of methanol may increase from 147 to 451 $\mu mol\,h^{-1}$ with 1 wt% addition of graphene to Pt-loaded g-C_3N_4.

The bandgap energy was dropped from 4.2 to 2.3 eV and the absorption band was relocated from 290 to 550 nm as a result of the nitrogen doping of $Sr_2Ta_2O_7$ to generate $Sr_2Ta_2O_{7x}N_x$ containing Ta—N bond (Mukherji et al., 2011). By adding Pt-loaded graphene, the photocatalytic capabilities of the $Sr_2Ta_2O_{7x}N_x$ under visible radiation could be markedly improved. The Pt/$Sr_2Ta_2O_{7x}N_x$ composite demonstrated a quantum efficiency of 4.26%, whereas a 5 wt% graphene $Sr_2Ta_2O_{7x}N_x$ composite had a quantum efficiency of 6.42% and an H_2 evolution rate of 293 $\mu mol\,h^{-1}$. The Pt nanoparticles on graphene substrate, which are involved in H_2 development, receive photo-generated electrons from $Sr_2Ta_2O_{7x}N_x$.

Semiconductor-graphene nanocomposites have been used in very few researches for photocatalytic CO_2 reduction. Two types of TiO_2-graphene nanocomposites were studied by Liang et al. for the photocatalytic reduction of carbon (Liang et al., 2011).

In the majority of photocatalytic studies, this nanocomposite – which also contained reduced GO (RGO) sheets and TiO_2 nanoparticles – was used as the precursor. As an alternative, graphite was directly exfoliated in N, N-dimethylformamide

(DMF) solvent using ultrasonic energy to create solvent-exfoliated graphene (SEG). The P25–SEG, which had a 0.27 wt% SEG content, produced CH_4 at a rate that was 4.5 times faster than P25 when exposed to UV light for photocatalytic CO_2 reduction. Even though the absolute value was less than that achieved under UV radiation, the rate of CH_4 synthesis under visible light irradiation was 7.2 times higher for the P25–SEG with 0.55 wt% SEG than for P25 alone. However, under UV light, P25–RGO composites showed essentially no change in their photocatalytic activity, whereas under visible radiation, the P25–RGO hybrid nanocomposite showed a maximum enhancement of 2.3 times. Defect density appears to be the primary distinction between SEG and RGO; SEG has fewer faults than RGO. Thus, photocatalysis of nanocomposites benefits from the less faulty SEG. This is consistent with the plan put forward by Xu and colleagues to prepare an effective graphene-based photocatalyst by reducing graphene defects and enhancing graphene-semiconductor interfacial contact.

$Ti_{0.91}O_2$-graphene hollow spheres were built layer by layer onto poly(methyl methacrylate) (PMMA) beads (Zhang et al., 2011). The PMMA templates was eliminated and GO was changed into graphene at the same time using microwave irradiation. CO_2 was decreased with H_2O under UV radiation using the hollow-sphere nanocomposite. CO and CH_4 were generated over the $Ti_{0.91}O_2$-graphene hollow spheres at 8.91 and 1.14 $\mu mol\,g^{-1}h^{-1}$, respectively. The fact that blank trials had neither CO nor CH_4 demonstrated that the input CO_2 was the carbon source. A rate of 1.41 $\mu mol\,g^{-1}h^{-1}$ of CH_4 was generated by the $Ti_{0.91}O_2$ hollow sphere in the absence of CO. The rate of CO_2 conversion to both CO and CH_4 increased fivefold over the $Ti_{0.91}O_2$-graphene in comparison to the pristine state as a result. This shows that graphene significantly boosts the photocatalytic activity of the $Ti_{0.91}O_2$ hollow spheres (Zhang et al., 2012a). The face-to-face orientation of the graphene and semiconductor sheets in the nanocomposite enables fast passage of photo-generated electrons from $Ti_{0.91}O_2$ into graphene, therefore separating the electron-hole pairs. It was suggested that the electrons transported to graphene be quickly distributed over a larger area of the material in order to avoid electron accumulation and lower the local electron density. This made it favorable for the two electrons of CO_2 to be reduced to CO instead of the eight electrons needed for the reduction of CO_2 to CH_4 (Tu et al., 2012).

6.4 Conclusion

In conclusion, numerous carbon-based nanophotocatalysts are described, along with their tailored materials for their diverse uses, particularly in photocatalytic reduction/conversion of CO_2 and H_2 production. In graphene, CNTs, and graphene nanonetwork (GCN), as well as in its more basic forms such as nanofibers, nanosheets, and nanoparticles, the chapter explores how carbon can drive more powerful photocatalytic reactions. There has been discussion on the fundamental enhancing process for photocatalytic CO_2 reduction. New, sophisticated photocatalysts may result from this, offering hope for resolving global problems pertaining to energy and the environment. The mechanism that alters the production of the intended products in a specific way throughout the process has to be explained in more detail. A field of study that offers fresh opportunities in the near future has been established by several attempts to boost photocatalytic activity. The addition of carbon to these nanocomposites can boost their surface area, CO_2 adsorption, absorption of visible light, and the quick separation and transfer of photogenerated electrons, all of which can increase their photocatalytic activity. It is necessary to quickly separate and transmit photogenerated electrons in order to facilitate photocatalytic processes. When photocatalysts and carbon are combined, a wide range of multi-component heterostructures can be created that improve charge separation and transportation, increase the lifetime of charge carriers, and increase their photocatalytic activity. Since carbon-based compounds exist in their most stable form, it is anticipated that they will eventually become efficient and stable photocatalysts. Carbon-based photocatalysts for CO_2 reduction have advanced significantly, allowing for the reasonable design of a carbon-based photocatalyst with high activity and stable CO_2-to-solar-fuel conversion. Unlike their usual, less noticeable UV activity, graphene supports coated with TiO_2 show apparent light activity. By employing crumpled and nanoporous graphene as a conducting base, the photocatalyst's high surface area and electrical conductivity are enhanced by meticulous engineering. CNTs feature distinct multiple concentric cylindrical shapes, which slows the rate at which photo-generated charge carriers recombine after collecting electrons from the catalyst, similar to graphene. GCN is acknowledged for its inherent merits as a metal-free visible light active photocatalyst. Various structural changes to GCN have been examined in order to produce efficient heterojunctions that slow down recombination events, thereby facilitating CO_2 reduction via photocatalysis. The development of carbon-based materials with remarkable electron transport properties has advanced significantly, according to the statement, and these materials – in particular, carbon nanostructures like RGO, CNT, and GCN – are now appropriate for use as next-generation functional materials in photocatalytic applications.

Moreover, utilizing nanocomposites as photocatalysts has been demonstrated in some research to significantly speed up the rate of CO_2 conversion or H_2 evolution when single-component semiconductors are used instead. Spatial separation of reduction and oxidation processes, better separation and migration of photogenerated electrons and holes, and a higher probability of charge recombination can all increase photocatalytic activity. In the description and characterization of semiconductor–carbon nanocomposites (graphene and CNT), the main goal has been to pinpoint the crucial elements affecting the photocatalytic behaviors of the nanocomposite catalysts for the two challenging processes. The photocatalytic activity of semiconductors can be significantly increased by adding carbon nanotubes or graphene sheets. Photogenerated electrons may be drawn to these carbon compounds due to their significant charge carrier mobilities, which could enhance charge separation. A critical element of catalysts made of semiconductor–carbon nanocomposite materials is the interfacial interaction between graphene sheets and semiconductor particles. Additionally, potential nanocomposites with specific morphologies or nanostructures have been made possible by the characteristic 1D and

2D architectures of these carbon materials. Graphene sheets are layered with semiconductor nanosheets firmly, face-to-face, to form a highly active photocatalyst for H_2 evolution. Interesting synergistic systems with high photocatalytic activity can be created by depositing a metal co-catalyst on graphene sheets at a distance from the semiconductor particles.

Carbon nanostructures are functionalized, and then they are cross linked with functional cross linkers to create an intricately linked, highly ordered structure. By taking use of these fascinating features of carbon-based materials, it may be possible to build photocatalysts that are more efficient and highly selective.

Acknowledgments

The authors would like to extend their heartfelt gratitude to IIIT Ranchi and IIT-ISM Dhanbad for their invaluable administrative assistance.

Funding and Competing Interests

This research did not receive any specific grant from funding agencies in the public, commercial, or non-profitable sector.

Conflict of Interest:

The authors declare that they have no conflict of interest.

References

Ali, A., Biswas, M. R. U. D., & Oh, W. (2018). Novel and simple process for the photocatalytic reduction of CO_2 with ternary Bi_2O_3–graphene–ZnO nanocomposite. *Journal of Materials Science Materials in Electronics*, 29(12), 10222–10233. https://doi.org/10.1007/s10854-018-9073-5.

Allen, M. J., Tung, V. C., & Kaner, R. B. (2009). Honeycomb carbon: A review of graphene. *Chemical Reviews*, 110(1), 132–145. https://doi.org/10.1021/cr900070d.

An, X. & Yu, J. C. (2011). Graphene-based photocatalytic composites. *RSC Advances*, 1(8), 1426. https://doi.org/10.1039/c1ra00382h.

Asi, M. A., Zhu, L., He, C. et al. (2013). Visible-light-harvesting reduction of CO_2 to chemical fuels with plasmonic Ag@AgBr/CNT nanocomposites. *Catalysis Today*, 216, 268–275. https://doi.org/10.1016/j.cattod.2013.05.021.

Bai, S. & Shen, X. (2012). Graphene–inorganic nanocomposites. *RSC Advances*, 2(1), 64–98. https://doi.org/10.1039/c1ra00260k.

Belet, A., Wolfs, C., Mahy, J. G. et al. (2019). Sol-Gel syntheses of photocatalysts for the removal of pharmaceutical products in water. *Nanomaterials*, 9(1), 126. https://doi.org/10.3390/nano9010126.

Cao, L., Sahu, S., Anilkumar, P. et al. (2011). Carbon nanoparticles as Visible-Light photocatalysts for efficient CO_2 conversion and beyond. *Journal of the American Chemical Society*, 133(13), 4754–4757. https://doi.org/10.1021/ja200804h.

Cho, K. M., Kim, K. H., Park, K. et al. (2017). Amine-functionalized graphene/CDS composite for photocatalytic reduction of CO_2. *ACS Catalysis*, 7(10), 7064–7069. https://doi.org/10.1021/acscatal.7b01908.

Cho, K. M., Kim, K. H., Park, K. et al. (2017b). Amine-functionalized graphene/CDS composite for photocatalytic reduction of CO_2. *ACS Catalysis*, 7(10), 7064–7069. https://doi.org/10.1021/acscatal.7b01908.

Dai, W., Xu, H., Yu, J. et al. (2015). Photocatalytic reduction of CO_2 into methanol and ethanol over conducting polymers modified Bi_2WO_6 microspheres under visible light. *Applied Surface Science*, 356, 173–180. https://doi.org/10.1016/j.apsusc.2015.08.059.

Di, T., Zhu, B., Cheng, B., Yu, J., & Xu, J. (2017). A direct Z-scheme g-C_3N_4/SnS_2 photocatalyst with superior visible-light CO_2 reduction performance. *Journal of Catalysis*, 352, 532–541. https://doi.org/10.1016/j.jcat.2017.06.006.

Eda, G., Mattevi, C., Yamaguchi, H., Kim, H., & Chhowalla, M. (2009). Insulator to semimetal transition in graphene oxide. *The Journal of Physical Chemistry C*, 113(35), 15768–15771. https://doi.org/10.1021/jp9051402.

Enesca, A. & Andronic, L. (2020). The influence of photoactive heterostructures on the photocatalytic removal of dyes and pharmaceutical active compounds: A mini-review. *Nanomaterials*, 10(9), 1766. https://doi.org/10.3390/nano10091766.

Fan, W., Lai, Q., Zhang, Q., & Wang, Y. (2011). Nanocomposites of TIO_2 and reduced graphene oxide as efficient photocatalysts for hydrogen evolution. *The Journal of Physical Chemistry C*, 115(21), 10694–10701. https://doi.org/10.1021/jp2008804.

Fan, W., Zhang, Q., & Wang, Y. (2013). Semiconductor-based nanocomposites for photocatalytic H_2 production and CO_2 conversion. *Physical Chemistry Chemical Physics*, 15(8), 2632. https://doi.org/10.1039/C2CP43524A.

Fang, Y. & Wang, X. (2018). Photocatalytic CO_2 conversion by polymeric carbon nitrides. *Chemical Communications*, 54(45), 5674–5687. https://doi.org/10.1039/c8cc02046a.

Fang, Z., Li, S., Gong, Y. et al. (2014). Comparison of catalytic activity of carbon-based AgBr nanocomposites for conversion of CO_2 under visible light. *Journal of Saudi Chemical Society*, 18(4), 299–307. https://doi.org/10.1016/j.jscs.2013.08.003.

Faraji, M., Yousefi, M., Yousefzadeh, S. et al. (2019). Two-dimensional materials in semiconductor photoelectrocatalytic systems for water splitting. *Energy & Environmental Science*, 12(1), 59–95. https://doi.org/10.1039/c8ee00886h.

Fu, Z., Xu, R., Moore, J. T. et al. (2018). Highly efficient photocatalytic system constructed from CoP/Carbon nanotubes or graphene for visible-light-driven CO_2 reduction. *Chemistry - A European Journal*, 24(17), 4273–4278. https://doi.org/10.1002/chem.201800335.

Fujishima, A. & Honda, K. (1972). Electrochemical photolysis of water at a semiconductor electrode. *Nature*, 238(5358), 37–38. https://doi.org/10.1038/238037a0.

Geim, A. K. & Novoselov, K. S. (2007). The rise of graphene. *Nature Materials*, 6(3), 183–191. https://doi.org/10.1038/nmat1849.

Grigioni, I., Dozzi, M. V., Bernareggi, M., Chiarello, G. L., & Selli, E. (2017). Photocatalytic CO_2 reduction vs. H_2 production: The effects of surface carbon-containing impurities on the performance of TiO_2 -based photocatalysts. *Catalysis Today*, 281, 214–220. https://doi.org/10.1016/j.cattod.2016.05.040.

Gui, M. M., Wong, W. M. P., Chai, S., & Mohamed, A. R. (2015). One-pot synthesis of Ag-MWCNT@TiO_2 core–shell nanocomposites for photocatalytic reduction of CO_2 with water under visible light irradiation. *Chemical Engineering Journal*, 278, 272–278. https://doi.org/10.1016/j.cej.2014.09.022.

Guo, S. & Dong, S. (2011). Graphene nanosheet: Synthesis, molecular engineering, thin film, hybrids, and energy and analytical applications. *Chemical Society Reviews*, 40(5), 2644. https://doi.org/10.1039/c0cs00079e.

Gusain, R., Kumar, P., Sharma, O. P., Jain, S. L., & Khatri, O. P. (2016). Reduced graphene oxide–CuO nanocomposites for photocatalytic conversion of CO_2 into methanol under visible light irradiation. *Applied Catalysis B Environment and Energy*, 181, 352–362. https://doi.org/10.1016/j.apcatb.2015.08.012.

Inoue, T., Fujishima, A., Konishi, S., & Honda, K. (1979). Photoelectrocatalytic reduction of carbon dioxide in aqueous suspensions of semiconductor powders. *Nature*, 277(5698), 637–638. https://doi.org/10.1038/277637a0.

Jia, L., Wang, D., Huang, Y., Xu, A., & Yu, H. (2011). Highly durable N-doped graphene/CdS nanocomposites with enhanced photocatalytic hydrogen evolution from water under visible light irradiation. *The Journal of Physical Chemistry C*, 115(23), 11466–11473. https://doi.org/10.1021/jp2023617.

Jiang, Z., Sun, H., Wang, T. et al. (2018). Nature-based catalyst for visible-light-driven photocatalytic CO_2 reduction. *Energy & Environmental Science*, 11(9), 2382–2389. https://doi.org/10.1039/c8ee01781f.

Jintoku, H., Matsuzawa, Y., & Yoshida, M. (2018). Switching the optical and electrical properties of carbon nanotube hybrid films using a photoresponsive dispersant as a dopant. *RSC Advances*, 8(20), 11186–11190. https://doi.org/10.1039/c8ra01447g.

Jung, H., Cho, K. M., Kim, K. H. et al. (2018). Highly efficient and stable CO_2 reduction photocatalyst with a hierarchical structure of mesoporous TiO_2 on 3D graphene with few-layered MoS_2. *ACS Sustainable Chemistry & Engineering*, 6(5), 5718–5724. https://doi.org/10.1021/acssuschemeng.8b00002.

Kandy, M. M. & Gaikar, V. G. (2019a). Continuous photocatalytic reduction of CO_2 using nanoporous reduced graphene oxide (RGO)/Cadmium sulfide (CDS) as catalyst on porous anodic alumina (PAA)/Aluminum support. *Journal of Nanoscience and Nanotechnology*, 19(8), 5323–5331. https://doi.org/10.1166/jnn.2019.16817.

Kandy, M. M. & Gaikar, V. G. (2019b). Enhanced photocatalytic reduction of CO_2 using CdS/Mn_2O_3 nanocomposite photocatalysts on porous anodic alumina support with solar concentrators. *Renewable Energy*, 139, 915–923. https://doi.org/10.1016/j.renene.2019.03.002.

Kim, Y. K. & Park, H. (2011). Light-harvesting multi-walled carbon nanotubes and CdS hybrids: Application to photocatalytic hydrogen production from water. *Energy & Environmental Science*, 4(3), 685–694. https://doi.org/10.1039/c0ee00330a.

Kirubasankar, B., Vijayan, S., & Angaiah, S. (2019). Sonochemical synthesis of a 2D–2D MoSe 2/graphene nanohybrid electrode material for asymmetric supercapacitors. *Sustainable Energy Fuels*, 3, 467–477.

Kumar, P., Bansiwal, A., Labhsetwar, N., & Jain, S. L. (2015). Visible light assisted photocatalytic reduction of CO_2 using a graphene oxide supported heteroleptic ruthenium complex. *Green Chemistry*, 17(3), 1605–1609. https://doi.org/10.1039/c4gc01400f.

Leary, R. & Westwood, A. (2011). Carbonaceous nanomaterials for the enhancement of TiO_2 photocatalysis. *Carbon*, 49(3), 741–772. https://doi.org/10.1016/j.carbon.2010.10.010.

Lee, S. & Chang, C. (2019). Recent developments about conductive polymer based composite photocatalysts. *Polymers*, 11(2), 206. https://doi.org/10.3390/polym11020206.

Li, J., Xiao, Q., Li, L., Shen, J., & Hu, D. (2015). Novel ternary composites: Preparation, performance and application of $ZnFe_2O_4/TiO_2$/polyaniline. *Applied Surface Science*, 331, 108–114. https://doi.org/10.1016/j.apsusc.2015.01.001.

Li, Q., Guo, B., Yu, J. et al. (2011). Highly efficient visible-light-driven photocatalytic hydrogen production of CDS-cluster-decorated graphene nanosheets. *Journal of the American Chemical Society*, 133(28), 10878–10884. https://doi.org/10.1021/ja2025454.

Li, Z., Li, H., Wang, S., Yang, F., & Zhou, W. (2022). Mesoporous black $TiO_2/MoS_2/Cu_2S$ hierarchical tandem heterojunctions toward optimized photothermal-photocatalytic fuel production. *Chemical Engineering Journal*, 427, 131830. https://doi.org/10.1016/j.cej.2021.131830.

Li, Z., Yuan, D., Wu, H., Li, W., & Gu, D. (2018). A novel route to synthesize carbon spheres and carbon nanotubes from carbon dioxide in a molten carbonate electrolyzer. *Inorganic Chemistry Frontiers*, 5(1), 208–216. https://doi.org/10.1039/c7qi00479f.

Liang, L., Xie, W., Fang, S. et al. (2017). High-efficiency dispersion and sorting of single-walled carbon nanotubes via non-covalent interactions. *Journal of Materials Chemistry C*, 5(44), 11339–11368. https://doi.org/10.1039/c7tc04390b.

Liang, Y. T., Vijayan, B. K., Gray, K. A., & Hersam, M. C. (2011). Minimizing graphene defects enhances Titania Nanocomposite-Based photocatalytic reduction of CO_2 for improved solar fuel production. *Nano Letters*, 11(7), 2865–2870. https://doi.org/10.1021/nl2012906.

Lightcap, I. V., Kosel, T. H., & Kamat, P. V. (2010). Anchoring semiconductor and metal nanoparticles on a two-dimensional catalyst mat. Storing and shuttling electrons with reduced graphene oxide. *Nano Letters*, 10(2), 577–583. https://doi.org/10.1021/nl9035109.

Liu, B., Ye, L., Wang, R. et al. (2018). Phosphorus-doped graphitic carbon nitride nanotubes with amino-rich surface for efficient CO_2 capture, Enhanced photocatalytic activity, and product selectivity. *ACS Applied Materials & Interfaces*, 10(4), 4001–4009. https://doi.org/10.1021/acsami.7b17503.

Liu, G., Wang, L., Yang, H. G., Cheng, H., & Lu, G. Q. (2010). Titania-based photocatalysts—crystal growth, doping and heterostructuring. *Journal of Materials Chemistry*, 20(5), 831–843. https://doi.org/10.1039/b909930a.

Liu, G., Xie, S., Zhang, Q., Tian, Z., & Wang, Y. (2015). Carbon dioxide-enhanced photosynthesis of methane and hydrogen from carbon dioxide and water over Pt-promoted polyaniline–TiO_2 nanocomposites. *Chemical Communications*, 51(71), 13654–13657. https://doi.org/10.1039/c5cc05113d.

Liu, J., Niu, Y., He, X., Qi, J. & Li, X. (2016). Photocatalytic reduction of CO_2 using TiO_2-graphene nanocomposites. *Journal of Nanomaterials*, 6012896.

Lv, X. J., Fu, W. F., Chang, H. X. et al. (2012). Hydrogen evolution from water using semiconductor nanoparticle/graphene composite photocatalysts without noble metals. *Journal of Materials Chemistry*, 22(4), 1539–1546. https://doi.org/10.1039/c1jm14502a.

Machado, B. F. & Serp, P. (2012). Graphene-based materials for catalysis. *Catalysis Science & Technology*, 2(1), 54–75. https://doi.org/10.1039/c1cy00361e.

Mateo, D., Asiri, A. M., Albero, J., & García, H. (2018). The mechanism of photocatalytic CO_2 reduction by graphene-supported Cu_2O probed by sacrificial electron donors. *Photochemical & Photobiological Sciences*, 17(6), 829–834. https://doi.org/10.1039/c7pp00442g.

Mukherji, A., Seger, B., Lu, G. Q., & Wang, L. (2011). Nitrogen doped $Sr_2Ta_2O_7$ coupled with graphene sheets as photocatalysts for increased photocatalytic hydrogen production. *ACS Nano*, 5(5), 3483–3492. https://doi.org/10.1021/nn102469e.

Ngaw, C. K., Zhao, C., Wang, V. B. et al. (2017). A graphene/carbon nanotube biofilm based solar-microbial fuel device for enhanced hydrogen generation. *Sustainable Energy & Fuels*, 1(1), 191–198. https://doi.org/10.1039/c6se00018e.

Nie, Y., Wang, W., Jiang, Y., Fortner, J., & Biswas, P. (2016). Crumpled reduced graphene oxide–amine–titanium dioxide nanocomposites for simultaneous carbon dioxide adsorption and photoreduction. *Catalysis Science & Technology*, 6(16), 6187–6196. https://doi.org/10.1039/c6cy00828c.

Novoselov, K. S. (2011). Graphene: Materials in the flatland (Nobel Lecture). *Angewandte Chemie International Edition*, 50(31), 6986–7002. https://doi.org/10.1002/anie.201101502.

Olowoyo, J. O., Kumar, M., Jain, S. L., Babalola, J. O., Vorontsov, A. V., & Kumar, U. (2018). Insights into reinforced photocatalytic activity of the CNT–TiO_2 nanocomposite for CO_2 reduction and water splitting. *The Journal of Physical Chemistry C*, 123(1), 367–378. https://doi.org/10.1021/acs.jpcc.8b07894.

Ong, W. (2017). 2D/2D graphitic carbon nitride (g-C_3N_4) heterojunction nanocomposites for photocatalysis: Why does face-to-face interface matter? *Frontiers in Materials*, 4, 11. https://doi.org/10.3389/fmats.2017.00011.

Ong, W., Gui, M. M., Chai, S., & Mohamed, A. R. (2013). Direct growth of carbon nanotubes on Ni/TiO_2 as next generation catalysts for photoreduction of CO_2 to methane by water under visible light irradiation. *RSC Advances*, 3(14), 4505. https://doi.org/10.1039/c3ra00030c.

Ong, W., Tan, L., Chai, S., Yong, S., & Mohamed, A. R. (2015). Surface charge modification via protonation of graphitic carbon nitride (g-C_3N_4) for electrostatic self-assembly construction of 2D/2D reduced graphene oxide (rGO)/g-C_3N_4 nanostructures toward enhanced photocatalytic reduction of carbon dioxide to methane. *Nano Energy*, 13, 757–770. https://doi.org/10.1016/j.nanoen.2015.03.014.

Ou, Y., Lin, J., Fang, S., & Liao, D. (2006). MWNT–TiO_2:Ni composite catalyst: A new class of catalyst for photocatalytic H_2 evolution from water under visible light illumination. *Chemical Physics Letters*, 429 (1–3), 199–203. https://doi.org/10.1016/j.cplett.2006.08.024.

Ouyang, T., Ye, Y., Wu, C., Xiao, K., & Liu, Z. (2019). Heterostructures composed of N-Doped carbon nanotubes encapsulating cobalt and B-MO_2C nanoparticles as bifunctional electrodes for water splitting. *Angewandte Chemie International Edition*, 58(15), 4923–4928. https://doi.org/10.1002/anie.201814262.

Pan, Y., You, Y., Xin, S. et al. (2017). Photocatalytic CO_2 reduction by carbon-coated indium-oxide nanobelts. *Journal of the American Chemical Society*, 139(11), 4123–4129. https://doi.org/10.1021/jacs.7b00266.

Park, Y., Kang, S., & Choi, W. (2011). Exfoliated and reorganized graphite oxide on titania nanoparticles as an auxiliary co-catalyst for photocatalytic solar conversion. *Physical Chemistry Chemical Physics*, 13(20), 9425. https://doi.org/10.1039/c1cp20697d.

Patel, J., Singh, A. K., & Carabineiro, S. a. C. (2020). Assessing the photocatalytic degradation of fluoroquinolone norfloxacin by MN:ZNS Quantum dots: Kinetic study, degradation pathway and influencing factors. *Nanomaterials*, 10(5), 964. https://doi.org/10.3390/nano10050964.

Patel, K. D., Singh, R. K., & Kim, H. (2019). Carbon-based nanomaterials as an emerging platform for theranostics. *Materials Horizons*, 6(3), 434–469. https://doi.org/10.1039/c8mh00966j.

Peng, T., Zeng, P., Ke, D., Liu, X., & Zhang, X. (2011). Hydrothermal preparation of multiwalled carbon nanotubes (MWCNTs)/CdS nanocomposite and its efficient photocatalytic hydrogen production under visible light irradiation. *Energy & Fuels*, 25(5), 2203–2210. https://doi.org/10.1021/ef200369z.

Pham, P., Rashid, M., Cai, Y., Yoshinaga, M., Dionysiou, D. D., & O'Shea, K. (2020). Removal of As(III) from water using the adsorptive and photocatalytic properties of humic acid-coated magnetite nanoparticles. *Nanomaterials*, 10(8), 1604. https://doi.org/10.3390/nano10081604.

Rajeshwar, K., De Tacconi, N. R., & Chenthamarakshan, C. R. (2001). Semiconductor-based composite materials: Preparation, properties, and performance. *Chemistry of Materials*, 13(9), 2765–2782. https://doi.org/10.1021/cm010254z.

Rambabu, Y., Kumar, U., Singhal, N. et al. (2019). Photocatalytic reduction of carbon dioxide using graphene oxide wrapped TiO_2 nanotubes. *Applied Surface Science*, 485, 48–55. https://doi.org/10.1016/j.apsusc.2019.04.041.

Serpone, N., Lawless, D., & Khairutdinov, R. (1995). Size effects on the photophysical properties of colloidal anatase TiO_2 particles: Size quantization versus direct transitions in this indirect semiconductor? *The Journal of Physical Chemistry*, 99(45), 16646–16654. https://doi.org/10.1021/j100045a026.

Shehzad, N., Tahir, M., Johari, K., Murugesan, T., & Hussain, M. (2018b). Improved interfacial bonding of graphene-TiO_2 with enhanced photocatalytic reduction of

CO_2 into solar fuel. *Journal of Environmental Chemical Engineering*, 6(6), 6947–6957. https://doi.org/10.1016/j.jece.2018.10.065.

Shi, H., Long, S., Hou, J. et al. (2019). Defects promote ultrafast charge separation in graphitic carbon nitride for enhanced visible-light-driven CO_2 reduction activity. *Chemistry - A European Journal*, 25(19), 5028–5035. https://doi.org/10.1002/chem.201805923.

Shown, I., Samireddi, S., Chang, Y. C. et al. (2018). Carbon-doped SnS_2 nanostructure as a high-efficiency solar fuel catalyst under visible light. *Nature Communications*, 9(1), 169. https://doi.org/10.1038/s41467-017-02547-4.

Su, Z., Zhang, B., Shi, J. et al. (2019). An NH_2-MIL-125 (Ti)/Pt/g-C_3N_4 catalyst promoting visible-light photocatalytic H_2 production. *Sustainable Energy & Fuels*, 3(5), 1233–1238. https://doi.org/10.1039/c9se00118b.

Sun, S., Yu, X., Yang, Q., Yang, Z., & Liang, S. (2019). Mesocrystals for photocatalysis: A comprehensive review on synthesis engineering and functional modifications. *Nanoscale Advances*, 1(1), 34–63. https://doi.org/10.1039/c8na00196k.

Tan, L., Ong, W., Chai, S., & Mohamed, A. R. (2013b). Reduced graphene oxide-TiO_2 nanocomposite as a promising visible-light-active photocatalyst for the conversion of carbon dioxide. *Nanoscale Research Letters*, 8(1). https://doi.org/10.1186/1556-276x-8-465.

Tang, Z., Li, F., Zhang, Y., Fu, X., & Xu, Y. (2011). Composites of titanate nanotube and carbon nanotube as photocatalyst with high mineralization ratio for gas-phase degradation of volatile aromatic pollutant. *The Journal of Physical Chemistry C*, 115(16), 7880–7886. https://doi.org/10.1021/jp1115838.

Tu, W., Xu, Y., Wang, J. et al. (2017). Investigating the role of tunable nitrogen vacancies in graphitic carbon nitride nanosheets for efficient visible-light-driven H_2 evolution and CO_2 reduction. *ACS Sustainable Chemistry & Engineering*, 5(8), 7260–7268. https://doi.org/10.1021/acssuschemeng.7b01477.

Tu, W., Zhou, Y., Liu, Q. et al. (2012). Robust Hollow spheres consisting of alternating titania nanosheets and graphene nanosheets with high photocatalytic activity for CO_2 conversion into renewable fuels. *Advanced Functional Materials*, 22(6), 1215–1221. https://doi.org/10.1002/adfm.201102566.

Wang, J., Huang, H., Xu, Z., Kou, J., & Lu, C. (2014). The potential of carbon-based materials for photocatalytic application. *Current Organic Chemistry*, 18(10), 1346–1364. https://doi.org/10.2174/13852728196661404242140222.

Wang, P., Yin, G., Bi, Q. et al. (2018). Efficient photocatalytic reduction of CO_2 using Carbon-Doped amorphous titanium oxide. *ChemCatChem*, 10(17), 3854–3861. https://doi.org/10.1002/cctc.201800476.

Wang, S., Xu, M., Peng, T. et al. (2019). Porous hypercrosslinked polymer-TiO_2-graphene composite photocatalysts for visible-light-driven CO_2 conversion. *Nature Communications*, 10(1), 676. https://doi.org/10.1038/s41467-019-08651-x.

Wang, W., Serp, P., Kalck, P., & Faria, J. L. (2005). Visible light photodegradation of phenol on MWNT-TiO_2 composite catalysts prepared by a modified sol–gel method. *Journal of Molecular Catalysis a Chemical*, 235 (1–2), 194–199. https://doi.org/10.1016/j.molcata.2005.02.027.

Wang, W., Xu, D., Cheng, B., Yu, J., & Jiang, C. (2017). Hybrid carbon@TiO_2 hollow spheres with enhanced photocatalytic CO_2 reduction activity. *Journal of Materials Chemistry A*, 5(10), 5020–5029. https://doi.org/10.1039/c6ta11121a.

Wei, Z., Endo-Kimura, M., Wang, K., Colbeau-Justin, C., & Kowalska, E. (2019). Influence of semiconductor morphology on photocatalytic activity of plasmonic photocatalysts: Titanate nanowires and octahedral anatase nanoparticles. *Nanomaterials*, 9(10), 1447. https://doi.org/10.3390/nano9101447.

Williams, G., Seger, B., & Kamat, P. V. (2008). TIO_2-Graphene nanocomposites. UV-Assisted photocatalytic reduction of graphene oxide. *ACS Nano*, 2(7), 1487–1491. https://doi.org/10.1021/nn800251f.

Xia, P., Zhu, B., Cheng, B., Yu, J., & Xu, J. (2017). 2D/2D G-C_3N_4/MNO_2 nanocomposite as a direct Z-Scheme photocatalyst for enhanced photocatalytic activity. *ACS Sustainable Chemistry & Engineering*, 6(1), 965–973. https://doi.org/10.1021/acssuschemeng.7b03289.

Xia, X., Jia, Z., Yu, Y., Liang, Y., Wang, Z., & Ma, L. (2007). Preparation of multi-walled carbon nanotube supported TiO_2 and its photocatalytic activity in the reduction of CO_2 with H_2O. *Carbon*, 45(4), 717–721. https://doi.org/10.1016/j.carbon.2006.11.028.

Xiang, Q., Yu, J. & Jaroniec, M. (2011a). Preparation and enhanced visible-light photocatalytic H_2-production activity of graphene/C_3N_4 composites. *Journal of Physical Chemistry C*, 115, 7355–7363.

Xiang, Q., Yu, J., & Jaroniec, M. (2011b). Enhanced photocatalytic H_2-production activity of graphene-modified titania nanosheets. *Nanoscale*, 3(9), 3670. https://doi.org/10.1039/c1nr10610d.

Xiang, Q., Yu, J., & Jaroniec, M. (2012a). Graphene-based semiconductor photocatalysts. *Chemical Society Reviews*, 41(2), 782–796. https://doi.org/10.1039/c1cs15172j.

Xiang, Q., Yu, J., & Jaroniec, M. (2012b). Synergetic effect of MOS_2 and graphene as cocatalysts for enhanced photocatalytic H_2 production activity of TIO2 nanoparticles. *Journal of the American Chemical Society*, 134(15), 6575–6578. https://doi.org/10.1021/ja302846n.

Xu, C., Anusuyadevi, P. R., Aymonier, C., Luque, R., & Marre, S. (2019). Nanostructured materials for photocatalysis. *Chemical Society Reviews*, 48(14), 3868–3902. https://doi.org/10.1039/c9cs00102f.

Xu, Q., Li, W., Ding, L., Yang, W., Xiao, H., & Ong, W. (2019). Function-driven engineering of 1D carbon nanotubes and 0D carbon dots: Mechanism, properties and applications. *Nanoscale*, 11(4), 1475–1504. https://doi.org/10.1039/c8nr08738e.

Yang, J., Du, J., Li, X. et al. (2019). Highly hydrophilic TIO_2 nanotubes network by alkaline hydrothermal method for photocatalysis degradation of methyl orange. *Nanomaterials*, 9(4), 526. https://doi.org/10.3390/nano9040526.

Yang, Y., Ajmal, S., Zheng, X., & Zhang, L. (2018). Efficient nanomaterials for harvesting clean fuels from electrochemical and photoelectrochemical CO_2 reduction. *Sustainable Energy & Fuels*, 2(3), 510–537. https://doi.org/10.1039/c7se00371d.

Ye, A., Fan, W., Zhang, Q., Deng, W., & Wang, Y. (2012). CdS–graphene and CdS–CNT nanocomposites as visible-light photocatalysts for hydrogen evolution and organic dye degradation. *Catalysis Science & Technology*, 2(5), 969. https://doi.org/10.1039/c2cy20027a.

Yu, J., Jin, J., Cheng, B., & Jaroniec, M. (2014). A noble metal-free reduced graphene oxide–CdS nanorod composite for the enhanced visible-light photocatalytic reduction of CO_2 to solar fuel. *Journal of Materials Chemistry A*, 2(10), 3407. https://doi.org/10.1039/c3ta14493c.

Zhang, H., Li, J., He, X., & Liu, B. (2020). Preparation of a g-C_3N_4/UiO-66-NH_2/CdS photocatalyst with enhanced visible light photocatalytic activity for tetracycline degradation. *Nanomaterials*, 10(9), 1824. https://doi.org/10.3390/nano10091824.

Zhang, H., Lv, X., Li, Y., Wang, Y., & Li, J. (2009). P25-graphene composite as a high performance photocatalyst. *ACS Nano*, 4(1), 380–386. https://doi.org/10.1021/nn901221k.

Zhang, H., Yu, M., & Qin, X. (2019). Photocatalytic activity of TiO_2 nanofibers: The surface crystalline phase matters. *Nanomaterials*, 9(4), 535. https://doi.org/10.3390/nano9040535.

Zhang, N., Zhang, Y., & Xu, Y. (2012c). Recent progress on graphene-based photocatalysts: Current status and future perspectives. *Nanoscale*, 4(19), 5792. https://doi.org/10.1039/c2nr31480k.

Zhang, N., Zhang, Y., Pan, X., Yang, M., & Xu, Y. (2012a). Constructing ternary CdS–graphene–TiO_2 hybrids on the flatland of graphene oxide with enhanced visible-light photoactivity for selective transformation. *The Journal of Physical Chemistry C*, 116(34), 18023–18031. https://doi.org/10.1021/jp303503c.

Zhang, X. Y., Li, H. P., Cui, X. L., & Lin, Y. (2010a). Graphene/TiO_2 nanocomposites: synthesis, characterization and application in hydrogen evolution from water photocatalytic splitting. *Journal of Materials Chemistry*, 20(14), 2801. https://doi.org/10.1039/b917240h.

Zhang, X., Sun, Y., Cui, X., & Jiang, Z. (2012b). A green and facile synthesis of TiO_2/graphene nanocomposites and their photocatalytic activity for hydrogen evolution. *International Journal of Hydrogen Energy*, 37(1), 811–815. https://doi.org/10.1016/j.ijhydene.2011.04.053.

Zhang, Y., Tang, Z., Fu, X., & Xu, Y. (2010b). TiO_2–graphene nanocomposites for gas-phase photocatalytic degradation of volatile aromatic pollutant: Is TiO_2–Graphene truly different from other TiO_2–carbon composite materials? *ACS Nano*, 4(12), 7303–7314. https://doi.org/10.1021/nn1024219.

Zhang, Y., Tang, Z., Fu, X., & Xu, Y. (2011). Engineering the unique 2D mat of graphene to achieve graphene-TiO_2 nanocomposite for photocatalytic selective transformation: What advantage does graphene have over its forebear carbon nanotube? *ACS Nano*, 5(9), 7426–7435. https://doi.org/10.1021/nn202519j.

Zhang, Y., Zhang, N., Tang, Z., & Xu, Y. (2012a). Improving the photocatalytic performance of graphene–TiO2 nanocomposites via a combined strategy of decreasing defects of graphene and increasing interfacial contact. *Physical Chemistry Chemical Physics*, 14(25), 9167. https://doi.org/10.1039/c2cp41318c.

Zhang, Y., Zhang, N., Tang, Z., & Xu, Y. (2012b). Graphene transforms wide band gap ZNS to a visible light photocatalyst. The new role of graphene as a macromolecular photosensitizer. *ACS Nano*, 6(11), 9777–9789. https://doi.org/10.1021/nn304154s.

Zhou, S., Liu, Y., Li, J. et al. (2014). Facile in situ synthesis of graphitic carbon nitride (g-C_3N_4)-N-TiO_2 heterojunction as an efficient photocatalyst for the selective photoreduction of CO_2 to CO. *Applied Catalysis B Environment and Energy*, 158–159, 20–29. https://doi.org/10.1016/j.apcatb.2014.03.037.
Zhu, Y., Gao, C., Bai, S. et al. (2017). Hydriding Pd cocatalysts: An approach to giant enhancement on photocatalytic CO_2 reduction into CH_4. *Nano Research*, 10(10), 3396–3406. https://doi.org/10.1007/s12274-017-1552-0.
Zhu, Z., Han, Y., Chen, C., Ding, Z., Long, J., & Hou, Y. (2018). Reduced graphene oxide-cadmium sulfide nanorods decorated with silver nanoparticles for efficient photocatalytic reduction carbon dioxide under visible light. *ChemCatChem*, 10(7), 1627–1634. https://doi.org/10.1002/cctc.201701573.
Zong, X., Yan, H., Wu, G. et al. (2008). Enhancement of photocatalytic H_2 evolution on CdS by loading MoS_2 as cocatalyst under visible light irradiation. *Journal of the American Chemical Society*, 130(23), 7176–7177. https://doi.org/10.1021/ja8007825.

7

Carbon Scaffold-based Nanophotocatalyst in Water Splitting and Fuel Production: Mechanism, Functionalization, Challenges, and Prospects

*Md. Iftikar Hussain[1], Gyandeep Saikia[1], Dipjyoti Dutta[2], Ranjeeta Singh[1], Gunaprabha Kurmi[1], and Rekha Rani Dutta[1]**

[1]*Department of Chemistry, School of Basic Sciences, The Assam Kaziranga University, Jorhat, Assam, India*
[2]*Material Sciences and Technology Division (MSTD), CSIR-North East Institute of Science and Technology (NEIST), Jorhat, Assam, India*

*Corresponding author: rekha@kazirangauniversity.in

7.1 Introduction

As the world population continues to grow, the energy demand has increased tremendously. However, the conventional methods of energy production, such as burning coal in thermal power plants, are no longer sufficient to meet the ever-increasing energy demands. Therefore, an alternative and eco-friendly source of energy is highly required to address the current energy crisis. Among the various alternatives, hydrogen gas produced through a water-splitting reaction is a promising solution due to its green-energy nature (Vesborg et al., 2015). However, the current methods of producing H_2 using transition metal oxide and nonmetal oxide photocatalysts via water splitting are inefficient (Al-Naggar et al., 2023). Low light absorption and poor charge separation are the main issues reported in this regard (Zheng et al., 2023). In this context, carbon-based materials gained considerable attention due to their unique properties, including high conductivity, chemical stability, and abundance. Carbon-based nanomaterials, in particular, offer a versatile platform for photocatalytic applications, owing to their tunable properties and potential for integration with other functional materials. Carbon scaffold-based nanophotocatalysts represent a promising class of materials for water splitting and fuel production. These nanomaterials combine the intrinsic properties of carbon

Photocatalysts based on Carbon Scaffolds: Environmental and Industrial Applications, First Edition.
Edited by Ramesh Oraon et al.
Published 2026 by Wiley-VCH GmbH.

scaffolds with the photocatalytic activity of semiconductor nanoparticles, resulting in synergistic effects that enhance overall performance.

In this chapter, we will provide an in-depth analysis of the mechanism of carbon scaffold-based nanophotocatalysts in the water-splitting reaction. This process involves a series of complex steps, starting with the absorption of light by the photocatalyst. The absorbed light energy generates charge carriers, and the ions then move to the catalytic surface, where they actively take part in redox reactions, leading to the separation of water into oxygen and hydrogen. We will also delve into the functionalization of the catalysts, which involves the introduction of specific functional groups or dopants. This process can optimize the photocatalytic activity of the catalysts and improve their stability under different conditions. Moreover, we will discuss the challenges associated with the practical implementation of these photocatalysts. These challenges include ensuring the high efficiency and stability of the photocatalysts while keeping them cost-effective. In the last, we will discuss the future prospects of these photocatalysts and their potential applications in various fields. The systems mostly discussed in this chapter will be of heteroatom-doped carbon scaffolding hybrid materials, graphene-based carbon scaffolding systems and nano-functionalized single-walled carbon nanotubes and carbon dots-based scaffolding systems. All the systems are well known for their semiconductor properties, and hence, this property can be utilized in a water-splitting reaction to produce green hydrogen.

7.2 Mechanism Behind Photocatalytic Water-splitting Reaction

The mechanism of the carbon scaffold-based nanophotocatalyst toward the water-splitting process consists of three main phases. First, electron-hole pairs are created by using a semiconductor nanophotocatalyst, which is necessary for light absorption. Surface migration and charge separation are included in the second stage. The charge separation step undergoes a redox process in which water molecules break down into H_2 and O_2 through the production of hole and electrons (Amy & Linsebigler, 1995; Mao, 2007). To put it briefly, semiconductor photocatalysts–which are able to absorb light and excite their electrons from the valence band (VB) to the conduction band (CB)–are used in the process of photocatalytic water-splitting. Consequently, electron-hole (e/h) pairs are created, which can react with water to split it into its constituent elements–hydrogen and oxygen–into smaller pieces. Photons with energies up to or beyond the energy bandgap (E_g) that separates the VB and CB can be absorbed by semiconductors. It forms positive-charged holes (h^+) in VB when it absorbs a photon, which promotes electrons from VB to CB. Electron promotion to CB results in a negative charge. By oxidation, this e/h pair is used to split water into hydrogen and oxygen (Mohsin et al., 2023; Moniz et al., 2015). The separation of hydrogen and oxygen from water is an endothermic process. Since water can only be dissociated by radiation with wavelengths shorter than 190 nm and is transparent to visible light (Gesellschaft, 1910), the mechanism operates in the presence of ultraviolet (UV) radiation (Honda, 1972). Under UV irradiation, holes formed in the valence band VB excite electrons in the conduction band (CB). To interact with certain reagents, these charge carriers split and travel to the semiconductor surface (Li et al., 2015); Tong et al. (2012). A bandgap greater than 1.23 eV is required in a semiconductor to show photocatalytic activity (Hisatomi et al., 2014; Xie et al., 2013).

One of the most promising developments in the development of clean and renewable fuel alternatives to fossil fuels is the synthesis of hydrogen from solar energy using photocatalytic active materials. Many strategies have been developed to move photocatalyst activity into the visible range while preserving stability and efficiency in order to exploit solar energy more effectively (Subbaraman et al., 2011). Large bandgaps, high hydrogen overpotentials, and fast electron-hole pair recombination are some of the disadvantages of the original titanium metal oxide photocatalyst, which have since been addressed by doping and carbon coupling, among other techniques (Jafari et al., 2016). Because of its large bandgap of 3.0–3.2 eV, which makes up just 5% of the solar spectrum Pulido Melián et al. (2013), TiO_2 can only be activated by UV radiation. Proper modification is required to permit visible-light harvesting while preventing recombination of photogenerated electron-hole pairs. Traditional TiO_2 compounds are doped with a wide variety of metals and nonmetals, especially anionic doping using dopants B, C, N, F, S, and Cl. Extended light absorption into the visible-light range is exhibited by the unique Br- and Cl-doped TiO_2, whilst nonmetal dopants such as carbon scaffold were found to be the primary factor in decreasing the bandgap, leading to materials with enhanced solar light-induced water splitting activity (Tang et al., 2011; Piskunov et al., 2015; Luo et al., 2004; Wang et al., 2012a).

7.2.1 Mechanism Behind PEC Splitting of Water

Another exciting technique that has the potential to operate safely, economically, and cleanly is solar PEC (photoelectrochemical) hydrogen generation shown in (Figure 7.1). A PEC semiconductor device with the same properties may split water into hydrogen and oxygen by converting photon energy into electrochemical energy when it is submerged in an aqueous electrolyte and exposed to sunlight (Dismukes, 1997). Thus, chemical bonding transforms variable solar energy into a more storable form of energy. The upstream outcome of the photocatalytic water-splitting is an increase in Gibbs free energy of 237 kJ/mol.

Sunlight is the best source of energy for both photocatalytic and PEC water splitting activation (Chandrasekaran et al., 2016). Moreover, redox reagents or an external bias voltage drive the behavior of semiconducting materials with smaller bandgaps or lower flat-band potentials, and electrons and holes can recombine with one another without participating in chemical reactions (Gratezel, 2001). The matching cocatalysts and buffer layers are added to the photocatalysts and photoelectrodes to help in charge separation and surface reactions. Alternatively, using multiphoton processes, the water-splitting processes can be initiated independently by connecting two or more tiny bandgap semiconductors. Although surface oxidation processes, ion absorption, surface imperfections, and exposed

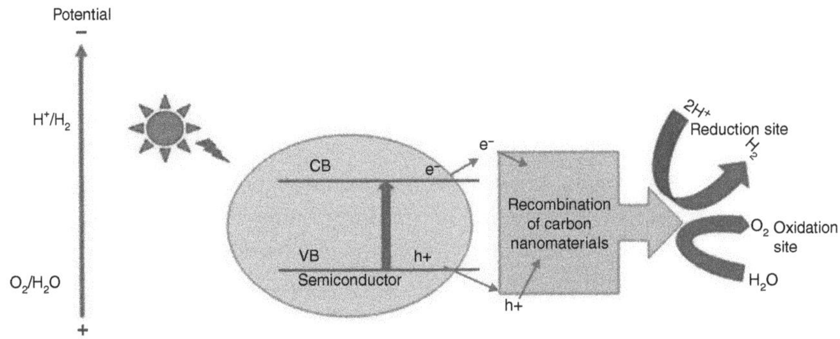

Figure 7.1 Schematic diagram for PEC mechanism of water splitting. *Source*: Adapted from Moniz et al. (2015).

crystallographic orientation are all known to have a significant impact on the flat-band potentials, these factors are rarely considered when developing and accessing photochemical water-splitting catalysts (Li & Wu, 2015). The final production of hydrogen or oxygen from the semiconductor photocatalyst system is impacted by each of these stages. Furthermore, efficient charge separation and rapid charge carrier transportation that avoid bulk/surface charge recombination are essential for photocatalytic water-splitting. Using aqueous solutions containing the sacrificial reagents can minimize the backward reaction of H_2 and O_2 to water on the photocatalyst surface (Chen et al., Semiconductor-based Photocatalytic Hydrogen Generation, 2010; Li & Wu, 2015). Therefore, redox reactions on the surface of the photocatalyst particles and charge separation within the particles must take place during the lifetimes of photoexcited carriers. Developing low-cost, high-performing photocatalytic systems is essential for converting light into energy. For water splitting to be successful, charge separation in the photocatalyst particles and redox reactions on their surface must consequently take place within the lifetimes of photoexcited carriers.

Thus, to make the water-splitting reaction more successful, carbon scaffold-based hybrid materials or namely nanophotocatalysts, have emerged as potential applicants in continuous research. Here are a few examples of nanophotocatalysts that have been used to produce green hydrogen as fuel: two-dimensional graphene, one-dimensional carbon nanotubes, nanofibers, zero dimensional carbon dots, and heteroatom-doped carbon scaffolding systems.

7.3 Functionalization of Carbon Scaffold-based Nanophotocatalyst

Functionalization of carbon scaffold-based water-splitting and fuel production involves modifying the surface of the carbon material to enhance its catalytic activity and selectivity in the water-splitting reaction. Some of the functionalization techniques are summarized below.

7.3.1 Doping with Heteroatom

Heteroatoms, such as nitrogen, sulfur, and phosphorus, effectively modify the electronic structure of carbon scaffold and create active sites for catalysis (Li et al., 2016). By incorporating these elements into the carbon lattice, researchers can enhance the material's conductivity and improve its efficiency in driving the water-splitting reaction (Wang et al., 2011). The different methods used for doping nitrogen onto carbon lattice are given below.

7.3.1.1 Chemical Vapor Deposition (CVD) Approach

The synthesis of various carbon nanomaterials, like graphene (Reina et al., 2009), carbon nanotubes (Zhang et al., 2003), carbon nanofibers (Che et al., 1998), and N-doped carbon nanotubes (Maldonado et al., 2006), is commonly achieved using the CVD method. During nitrogen doping of graphene via the CVD method, copper (Cu) or nickel (Ni) typically serves as the metal catalyst substrate. A carbon source combined with a nitrogen-containing gas is introduced at high temperatures. This environment causes the precursor materials to dissociate and then recombine, resulting in the formation of nitrogen-doped graphene (N-graphene) on the catalyst surface through a precipitation process (Wei et al., 2009; Luo et al., 2011; Qu et al., 2010; Di et al., 2008).

7.3.1.2 Segregation Growth Approach

The segregation growth approach applied to carbon scaffold-based nanophotocatalysts involves mixing precursor materials containing carbon and other desired elements or compounds, such as metal ions or metal oxide precursors, and subjecting them to high temperatures in a controlled environment. During heating, one component segregates to the surface of the carbon scaffold, forming a gradient in composition shown in (Figure 7.4). This results in the deposition and growth of metal nanoparticles or metal oxide coatings onto or into the carbon scaffold. The resulting nanomaterials demonstrate improved photocatalytic activity for fuel production by water splitting, attributed to the synergistic effects between the carbon scaffold and the functional components (Ismail et al., 2014). An example of this method is the formation of nitrogen-doped graphene (Wang et al., 2012), where nitrogen-containing boron layers and carbon-containing nickel layers are sequentially deposited on the SiO_2/Si substrate by electron beam evaporation, then during the vacuum annealing process, the boron atoms are trapped by nickel, and the carbon atoms will segregate out onto the nickel surface and combine to form N-graphene (Wang et al., 2012b; Zhang et al., 2011) shown in Figure 7.2.

7.3.1.3 Solvothermal Approach

The functionalization of carbon scaffold-based nanophotocatalysts through the solvothermal approach involves dissolving precursor materials in a solvent within a sealed reaction vessel, which is heated to elevated temperatures and pressures. Under these conditions, nucleation and growth of nanocrystals occur, with functional groups or nanoparticles being incorporated onto or into the carbon scaffold. This process enhances the photocatalytic activity of the material for water splitting

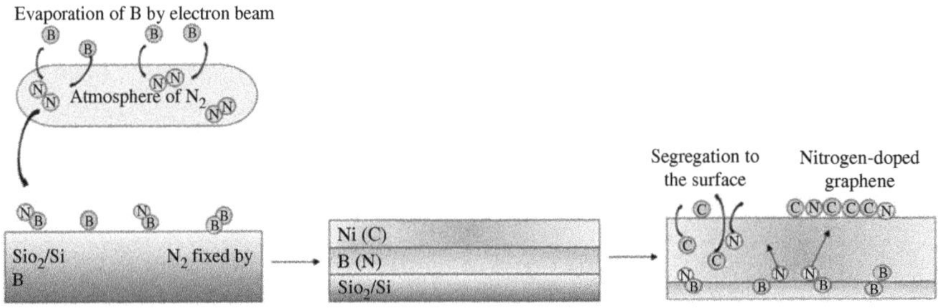

Figure 7.2 Schematic illustration of the segregation technique for growing nitrogen-doped graphene. *Source*: Zhang et al. (2011) / John Wiley & Sons.

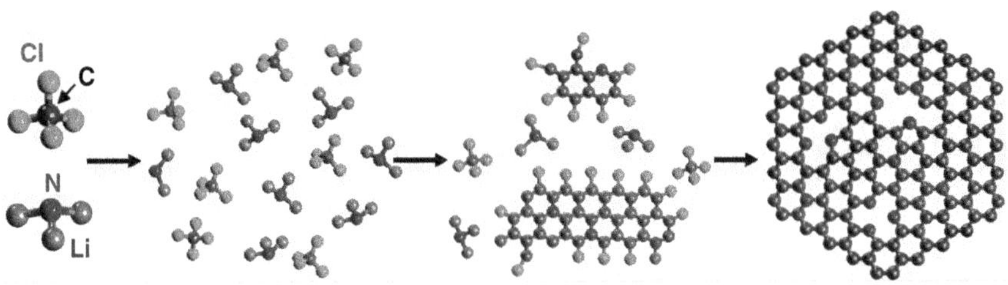

Figure 7.3 Solvothermal method of synthesis of N-graphene. *Source*: Deng et al. (2011) / American Chemical Society.

and fuel production. One example is CCl_4; N-graphene with different nitrogen contents was obtained by mixing lithium nitride (Li_3N) with tetrachloromethane (CCl_4) at ~300 °C (Figure 7.3) or cyanuric chloride ($N_3C_3Cl_3$) with Li_3N (Deng et al., 2011).

7.3.2 Nanostructuring

Nanostructuring refers to the process of reducing the size of carbon materials to the nanoscale. This approach offers several advantages, like a larger surface area, improved mass transport, and enhanced electron transfer kinetics (Valiev, 2002). Due to the reduced size, nanomaterials are typically harder than their bulk counterparts, as the nano-scale size limits the particles' ability to bend, resulting in increased hardness. However, this increased hardness can alter other physical properties, such as malleability, brittleness, ductility, and toughness. Moreover, reducing particle size to below 10 nm can induce quantum confinement effects, leading to an increased bandgap and better charge separation (Valiev, 2002; Guo et al., 2013). These nanomaterials can also be synthesized into various morphologies, including tubes, rods, flowers, and spheres, each exhibiting distinct properties (Gleiter, 2000; Arora & Sharma, 2014).

In water splitting and fuel production, nanostructured carbon materials have demonstrated significant potential as catalysts for both the hydrogen evolution reaction (HER) and the oxygen evolution reaction (OER). For example, nanostructured carbon materials, such as CNTs decorated with metal nanoparticles or nitrogen-doped graphene, exhibit enhanced catalytic activity and stability compared to their bulk counterparts. These materials promote the adsorption and dissociation of water molecules, facilitate charge transfer during electrochemical reactions, and mitigate the formation of reaction intermediates that can poison the catalyst surface. Some common methods of nanostructuring carbon scaffold synthesis are summarized below:

7.3.2.1 Arc Discharge

In this innovative technique, an electric arc is meticulously generated between two electrodes within an inert environment, resulting in the precise deposition of nanomaterials onto the electrode surface shown in (Figure 7.4). This method, which is widely used for producing carbon nanotubes and fullerenes, is one of the oldest and most effective techniques for high-quality carbon nanotube production (Arora & Sharma, 2014).

7.3.2.2 Sonochemical

During this process, an ultrasound wave causes bubbles to form in the sample. These bubbles then oscillate, collecting ultrasonic energy and growing in size. Eventually, the excessive growth in size causes the bubbles to collapse explosively, releasing stored energy that promotes the formation of nanomaterials (Figure 7.5). Jeong et al. synthesize SWCNTs (single-walled carbon nanotubes) using a sonochemical technique. For the synthesis, they immersed silica powder in a ferrocene-xylene mixture solution. They ultrasonicated the mixture for 20 minutes at room temperature and atmospheric pressure, resulting in high-purity SWCNTs on the surface of silica powder (Jeong et al., 2004).

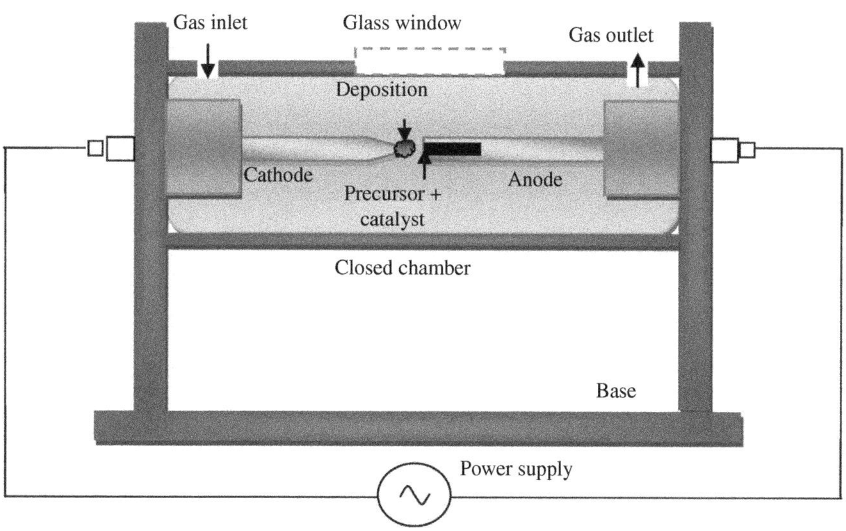

Figure 7.4 Arc discharge method for synthesis of carbon nanotubes. *Source*: Arora & Sharma (2014) / with permission of Elsevier.

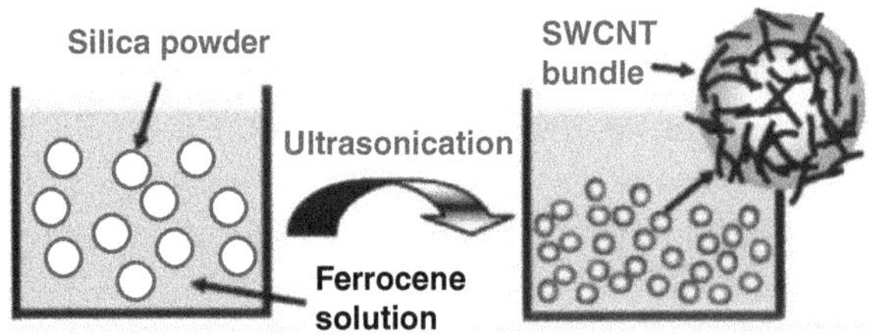

Figure 7.5 Synthesis of SWCNTs by sonochemical method. *Source*: Jeong et al. (2004) / American Chemical Society.

7.3.2.3 Exfoliation

Exfoliation is primarily used to fabricate "ultrathin 2D nanomaterials." The task can be accomplished using one of two methods: mechanical exfoliation, which gives a low yield, and chemical exfoliation, which oxidizes the substrate and leads to the weakening of bonding of the substrate layer and leads to the formation of nanomaterials (Coroş et al., 2016).

7.3.2.4 Hydrothermal

Herein, the precursor is heated in an autoclave, using water as the solvent. It is a relatively easy and simple process and therefore widely used for the synthesis of carbon-based nanomaterials (Kondalkar et al., 2014; Wang et al., 2017).

7.3.2.5 Microwave-assisted Synthesis

This method also uses heat to synthesize nanomaterials. It employs microwave to heat the reaction mixture, and it is the fastest way for nanostructuring, and the yield of this process is also good (Li et al., 2016).

7.3.2.6 Ball Milling

This is a physical method used for nanostructuring. In this method, the samples are obtained in powder form and subsequently undergo mechanical treatment to achieve a finer, more homogeneous nanoparticle structure (Dhand et al., 2015).

7.3.3 Metal Deposition

Metal deposition involves immobilizing metal nanoparticles onto carbon scaffolds to create hybrid catalysts with improved catalytic activity and stability. This method is widely used in water splitting and other electrochemical reactions, leveraging the synergistic effects between metal catalysts and carbon materials. For instance, Liu et al. (2020) prepared carbon metal oxide/hydroxide composites using graphene solutions, where charge-stabilized graphene layers acted directly as a reducing agent (Vallés et al., 2008; Hof & Pénicaud, 2018). This approach allowed the random distribution of small, size-calibrated nanoparticles onto the carbon frameworks, while simultaneously controlling the grafting ratio and size distribution (Hof et al.,

2017; Hof et al., 2019). The resulting nanocomposites displayed superior HER performance compared to commercial Pt/C and exhibited comparable activity to commercial RuO_2 for the OER (Liu et al., 2020).

7.4 Challenges

Although carbon scaffold-based nanophotocatalysts have so many advantages in the context of water splitting and fuel production, they also face some challenges that hinder their practical application to some extent. Some of the challenges are summarized in the following paragraphs.

7.4.1 Limited Light Absorption

Carbon-based materials like graphene, carbon nanotubes, and carbon dots exhibit limited light absorption in the visible range due to their wide bandgap and low extinction coefficients. For example, when we consider graphene, which consists of a single layer of carbon atoms arranged in a hexagonal pattern exhibits strong light absorption in the UV region but it has limited absorption in the visible range (Bae et al., 2010). Similarly, carbon nanotubes possess unique optical properties determined by their chirality, diameter, and length, leading to variable absorption characteristics across the electromagnetic spectrum (Jorio et al., 2008).

7.4.2 Efficient Charge Separation

Carbon scaffold-based materials often face challenges in minimizing electron-hole recombination and maximizing charge utilization at defect sites or interfaces within the material structure (Zhang et al., 2016).

7.4.3 Selectivity and Stability

Carbon scaffold-based nanophotocatalysts should exhibit high selectivity toward the desired reaction and should also be stable under harsh reaction conditions. However, it is not possible to achieve both selectivity and stability at the same time for carbon scaffold-based photocatalysts. For example, carbon scaffold-based nanophotocatalysts functionalized with specific organic ligands or metal nanoparticles exhibit enhanced selectivity toward certain reactions, but their stability under photocatalytic conditions requires careful consideration (Zhang et al., 2017).

7.4.4 Scalability and Reproducibility

Synthesizing carbon scaffold-based nanophotocatalysts on a laboratory scale is not so difficult, but synthesizing them in industrial-scale applications introduces some challenges in maintaining reproducibility, scalability, and cost-effectiveness. The controlling parameters, such as particle size distribution, morphology, and surface functionalization, become increasingly complex at larger scales (Tian et al., 2019).

7.5 Overall Summary and Future Prospects

In the current emergent fields of material chemistry, researchers are continuously searching for new and advanced materials for solar water-splitting reactions and green fuel production. Since the working principle of water splitting is associated with visible-light absorption, charge separation as well as interfacial photochemical reactions, henceforth more improvements are required in this regard. Although traditional metal oxides and other inorganic semiconductor-based photocatalysts have been playing an important role in this advancement still some limitations are here regarding environmental sustainability, stability, toxicity, and cost-effectiveness. Therefore, it is potentially important to find some alternatives to overcome these challenges. Carbon scaffolds, carbon-based nanomaterials, or carbon-based hybrid nanomaterials have emerged as a potential applicant in the field of solar water splitting and green hydrogen production in recent development owing to their specific hole scavenger/electron-donating behavior with enhanced visible light absorption as well as better charge separation characteristics. Carbon is supported with different groups and heteroatoms through conjugation and doping to form the hybrid or scaffolding material which can be employed as a nanophotocatalyst in the water-splitting reactions to produce hydrogen as a green alternative to conventional fuel. This chapter includes a discussion of hybrid materials like heteroatom doped graphene-based carbon scaffolding, single-walled carbon nanotubes, and carbon dot-based scaffolding nanophotocatalyst materials, their functionalization behavior, challenges, and mechanisms toward water splitting as well as green hydrogen production. Though these carbon scaffolding materials have a high impact in solar water splitting and fuel production, still a few number of literatures are available about their photocatalytic efficiency and detailed structure–property correlations. Further research is also needed in the detailed theoretical study of structure–property optimization and intrinsic semiconductor properties of these materials. Although carbon scaffolding nanophotocatalysts are unable to fulfill the current global energy demand, in near future, unstoppable development on enhanced efficiency, environmental sustainability, and cost efficiency will surely make these materials as one of the most effectual alternatives for renewable energy assets.

References

Al-Naggar, A. H., Shinde, N. M., Kim, J. S., & Mane, R. S. (2023). Water splitting performance of metal and non-metal-doped transition metal oxide electrocatalysts. *Coordination Chemistry Reviews*, 474, 214864.

Amy, L. & Linsebigler, G. L. (1995). Photocatalysis on TiO_2 surfaces: Principles, mechanisms, and selected results. *Chemical Reviews*, 95(3), 735–758.
Arora, N. & Sharma, N. N. (2014). Arc discharge synthesis of carbon nanotubes: Comprehensive review. *Diamond and Related Materials*, 50, 135–150.
Bae, S., Kim, H., Lee, Y. et al. (2010). Roll-to-roll production of 30-inch graphene films for transparent electrodes. *Nature Nanotechnology*, 5(8), 574–578.
Chandrasekaran, S., Chung, J. S., Kim, E. J., & Hur, S. H. (2016). Advanced nano-structured materials for photocatalytic water splitting. *Journal of Electrochemical Science and Engineering*, 7(1), 1–12.
Che, G., Lakshmi, B. B., Martin, C. R., Fisher, E. R., & Ruoff, R. S. (1998). Chemical vapor deposition based synthesis of carbon nanotubes and nanofibers using a template method. *Chemistry of Materials*, 10(1), 260–267.
Chen, H., Linfeng, H., Chen, M., & Yan, Y. (2014). Nickel-cobalt layered double hydroxide nanosheets for high-performance supercapacitor electrode materials. *Advanced Functional Materials*, 934–942.
Chen, X., Shen, S., Guo, L., & Mao, S. S. (2010). Semiconductor-based photocatalytic hydrogen generation. *Chemical Reviews*, 110(11), 6503–6570.
Coroş, M., Pogăcean, F., Roşu, M. C. et al. (2016). Simple and cost-effective synthesis of graphene by electrochemical exfoliation of graphite rods. *RSC Advances*, 6(4), 2651–2661.
Deng, D., Pan, X., Yu, L. et al. (2011). Toward N-doped graphene via solvothermal synthesis. *Chemistry of Materials*, 23(5), 1188–1193.
Dhand, C., Dwivedi, N., Loh, X. J. et al. (2015). Methods and strategies for the synthesis of diverse nanoparticles and their applications: A comprehensive overview. *RSC Advances*, 5(127), 105003–105037.
Di, C. A., Wei, D., Yu, G., Liu, Y., Guo, Y., & Zhu, D. (2008). Patterned graphene as source/drain electrodes for bottom contact organic field-effect transistors. *Advanced Materials*, 20(17), 3289–3293.
Dismukes, W. R. (1997). Synthetic water-oxidation catalysts for artificial photosynthetic water oxidation. *Chemical Reviews*, 97(1), 1–24.
Gesellschaft, B. d. (1910). Studien über photochemische Gleichgewichte. IV. Das Lichtgleichgewicht Knallgas-Wasserdampf. *Chemistry Europe*, 43(1), 880–884.
Gleiter, H. (2000). Nanostructured materials: Basic concepts and microstructure. *Acta Materialia*, 48(1), 1–29.
Gratezel, M. (2001). Photoelectrochemical cells. *Nature*, 414(6861), 338–344.
Guo, D., Xie, G., & Luo, J. (2013). Mechanical properties of nanoparticles: Basics and applications. *Journal of Physics D: Applied Physics*, 47(1), 013001.
Hisatomi, T., Kubota, J., & Domen, K. (2014). Recent advances in semiconductors for photocatalytic and photoelectrochemical water splitting. *Chemical Society Reviews*, 43(22), 7520–7535.
Hof, F. & Pénicaud, A. (2018). Graphenide solutions: A chemical platform for nanoparticle–nanocarbon composites. *Chemistry–A European Journal*, 24(61), 16246–16250.

Hof, F., Boni, A., Valenti, G., Huang, K., Paolucci, F., & Pénicaud, A. (2017). From food waste to efficient bifunctional nonprecious electrocatalyst. *Chemistry–A European Journal*, 23(61), 15283–15288.

Hof, F., Liu, M., Valenti, G., Picheau, E., Paolucci, F., & Pénicaud, A. (2019). Size control of nanographene supported iron oxide nanoparticles enhances their electrocatalytic performance for the oxygen reduction and oxygen evolution reactions. *The Journal of Physical Chemistry C*, 123(34), 20774–20780.

Honda, A. F. (1972). Electrochemical photolysis of water at a semiconductor electrode. *Nature*, 238, 37–38.

Ismail, A. A. & Bahnemann, D. W. (2014). Photochemical splitting of water for hydrogen production by photocatalysis: A review. *Solar Energy Materials and Solar Cells*, 128, 85–101.

Jafari, T., Moharreri, E., Amin, A. S., Miao, R., Song, W., & Suib, S. L. (2016). Photocatalytic water splitting—The untamed dream: A review of recent advances. *Molecules*, 21(7), 900.

Jeong, S. H., Ko, J. H., Park, J. B., & Park, W. (2004). A sonochemical route to single-walled carbon nanotubes under ambient conditions. *Journal of the American Chemical Society*, 126(49), 15982–15983.

Jorio, A., Dresselhaus, G., & Dresselhaus, M. S. (2008). Carbon nanotube characterization with Raman spectroscopy. *Topics in Applied Physics*, 111, 59–81.

Kondalkar, V. V., Kharade, R. R., Mali, S. S. et al. (2014). Nanobrick-like WO_3 thin films: Hydrothermal synthesis and electrochromic application. *Superlattices and Microstructures*, 73, 290–295.

Li, H., Shao, F. Q., Zou, S. Y. et al. (2016). Microwave-assisted synthesis of N, P-doped carbon dots for fluorescent cell imaging. *Microchimica Acta*, 183, 821–826.

Li, J. & Wu, N. (2015). Semiconductor-based photocatalysts and photoelectrochemical cells for solar fuel generation: A review. *Catalysis Science and Technology*, 5(3), 1360–1384.

Li, X., Yu, J., Low, J., Fang, Y., Xiao, J., & Chen, X. (2015). Engineering heterogeneous semiconductors for solar water splitting. *Journal of Materials Chemistry A*, 3(6), 2485–2534.

Liu, M., Hof, F., Moro, M., Valenti, G., Paolucci, F., & Pénicaud, A. (2020). Carbon supported noble metal nanoparticles as efficient catalysts for electrochemical water splitting. *Nanoscale*, 12(39), 20165–20170.

Luo, H., Takata, T., Lee, Y., & Zhao, J. (2004). Photocatalytic activity enhancing for titanium dioxide by co-doping with bromine and chlorine. *Chemistry of Materials*, 16(5), 846–849.

Luo, Z., Lim, S., Tian, Z. et al. (2011). Pyridinic N doped graphene: synthesis, electronic structure, and electrocatalytic property. *Journal of Materials Chemistry*, 21(22), 8038–8044.

Maldonado, S., Morin, S., & Stevenson, K. J. (2006). Structure, composition, and chemical reactivity of carbon nanotubes by selective nitrogen doping. *Carbon*, 44(8), 1429–1437.

Mao, X. C. (2007). Titanium dioxide nanomaterials: Synthesis, properties, modifications, and applications. *Chemical Reviews*, 107(7), 2891–2959.

Mohsin, M., Ishaq, T., Bhatti, I. A. et al. (2023). Semiconductor nanomaterial photocatalysts for water-splitting hydrogen production: The holy grail of converting solar energy to fuel. *Nanomaterials*, 13(3), 546.

Moniz, S. J. A., Shevlin, S., Martin, D., Xiao Guo, Z., & Tang, J. (2015). Visible-light driven heterojunction photocatalysts for water splitting – a critical review. *Energy and Environmental Science*, 8(3), 731–759.

Piskunov, S., Lisovski, O., Begens, J. et al. (2015). C-, N-, S-, and Fe-Doped TiO_2 and $SrTiO_3$ nanotubes for visible-light-driven photocatalytic water splitting: Prediction from first principles. *The Journal of Physical Chemistry C*, 119(32), 18686–18696.

Pulido Melián, E., González-Díaz, O. M., Ortega Méndez, A. et al. (2013). Efficient and affordable hydrogen production by water photo-splitting using TiO_2-based photocatalysts. *International Journal of Hydrogen Energy*, 38(5), 2144–2155.

Qu, L., Liu, Y., Baek, J. B., & Dai, L. (2010). Nitrogen-doped graphene as efficient metal-free electrocatalyst for oxygen reduction in fuel cells. *ACS Nano*, 4(3), 1321–1326.

Reina, A., Jia, X., Ho, J. et al. (2009). Large area, few-layer graphene films on arbitrary substrates by chemical vapor deposition. *Nano Letters*, 9(1), 30–35.

Subbaraman, R., Tripkovic, D., Strmcnik, D. et al. (2011). Enhancing hydrogen evolution activity in water splitting by tailoring Li^+-$Ni(OH)_2$-Pt interfaces. *Science*, 334(6060), 1256–1260.

Tang, J., Cowan, A. J., Durrant, J. R., & Klug, D. (2011). Mechanism of O_2 production from water splitting: Nature of charge carriers in nitrogen doped nanocrystalline TiO_2 films and factors limiting O_2 production. *The Journal of Physical Chemistry C*, 115(7), 3143–3150.

Tian, J., Zhao, Y., Zeng, Z. et al. (2019). Challenges and opportunities of carbon dots in environment and health. *Frontiers of Environmental Science & Engineering*, 13(2), 19.

Tong, H., Ouyang, S., Bi, Y., Umezawa, N., Oshikiri, M., & Ye, J. (2012). Nano-photocatalytic materials: Possibilities and challenges. *Advanced Materials*, 24(2), 229–251.

Valiev, R. (2002). Nanomaterial advantage. *Nature*, 419(6910), 887–889.

Vallés, C., Drummond, C., Saadaoui, H. et al. (2008). Solutions of negatively charged graphene sheets and ribbons. *Journal of the American Chemical Society*, 130(47), 15802–15804.

Vesborg, P. C. K., Seger, B., & Chorkendorff, I. (2015). Recent development in hydrogen evolution catalyst and their practical implementation. *The Journal of Physical Chemistry Letters*, 6(6), 951–957.

Wang, H., Dong, S.-J., Chang, Y., & Faria, J. L. (2012a). Enhancing the photocatalytic properties of TiO_2 by coupling with carbon nanotubes and supporting gold. *Journal of Hazardous Materials*, 235-236, 230–236.

Wang, H., Maiyalagan, T., & Wang, X. (2012b). Review on recent progress in nitrogen-doped graphene: Synthesis, characterization, and its potential applications. *ACS Catalysis*, 2(5), 781–794.

Wang, X., Li, X., Zhang, L., Yoon, Y., & Weber, P. K. (2011). Nitrogen-doped graphene nanostructures. *Advanced Materials*, 23(17), 2015–2040.

Wang, X., Zheng, Y., Yuan, J. et al. (2017). Porous NiCo diselenide nanosheets arrayed on carbon cloth as promising advanced catalyst used in water splitting. *Electrochimica Acta*, 225, 503–513.

Wei, D., Liu, Y., Wang, Y., Zhang, H., Huang, L., & Yu, G. (2009). Synthesis of N-doped graphene by chemical vapor deposition and its electrical properties. *Nano Letters*, 9(5), 1752–1758.

Xie, G., Zhang, K., Guo, B., Liu, Q., Fang, L., & Gong, JR. (2013). Graphene-based materials for hydrogen generation from light-driven water splitting. *Advanced Materials*, 25(28), 3820–3839.

Zhang, C., Fu, L., Liu, N., Liu, M., Wang, Y., & Liu, Z. (2011). Synthesis of nitrogen-doped graphene using embedded carbon and nitrogen sources. *Advanced Materials*, 23(8), 1020–1024.

Zhang, J., Yu, J., Zhang, Y., & Li, Y. (2016). Defect-rich ultrathin $ZnIn_2S_4$ nanosheets for efficient visible-light-driven photocatalytic hydrogen production. *Applied Catalysis B: Environmental*, 180, 530–537.

Zhang, J., Zou, H., Qing, Q. et al. (2003). Effect of chemical oxidation on the structure of single-walled carbon nanotubes. *The Journal of Physical Chemistry B*, 107(16), 3712–3718.

Zhang, N., Yang, M., Liu, S. et al. (2017). Carbon-based metal-free nanocatalysts for electrocatalysis beyond hydrogen evolution reactions. *Chemical Society Reviews*, 46(10), 2826–2846.

Zheng, D., Xue, Y., Wang, J., Varbanov, P. S., Klemeš, J. J., & Yin, C. (2023). Nanocatalysts in photocatalytic water splitting for green hydrogen generation: Challenges and opportunities. *Journal of Cleaner Production*, 414, 137700.

8

Industry Economics for Carbon Scaffold-based Nano Photocatalysts: Past, Present, and the Prospects

*Rajaji Sagar Lal[1], Tanmaya Biswal[1], Kumar Abhisek[2], Helen Treasa Mathew[2], and Ramesh Oraon[1]**

[1] *Department of Chemistry, Central University of Jharkhand, Ranchi, India*
[2] *Department of Metallurgical and Materials Engineering, Central University of Jharkhand, Ranchi, India*

*Corresponding author: ramesh.oraon@cuj.ac.in

8.1 Introduction

Carbon scaffold-based nano photocatalysts are a fast-evolving field of research providing a roadmap for sustainable growth of the global economy. Carbon-based composites have the potential to be highly effective photocatalysts because of their improved photocatalytic activity due to the coupling effect resulting from the integration of carbon species (Bhanderi et al., 2024). The process through which photocatalysts improve the photocatalytic reaction rate is termed photocatalysis. After each round of interactions, the catalyst's excited state regenerates itself and forms an intermediate by interacting with reaction partners. In general, photocatalytic reactions are classified into three categories. 1) Homogenous photocatalysis, 2) Heterogeneous photocatalysis, and 3) Plasmonic antenna photocatalysis. In homogenous photocatalysis, the phase of the photocatalyst and reactants is equal. One of the best examples of homogenous photocatalysis is the self-cleaning property of the atmosphere. With the help of this process, the atmosphere gets rid of large organic molecules (He et al., 2021). When both the photocatalyst and the reactant are in different phases, the process is called heterogenous photocatalysis. Dehydration, detoxification through gaseous elimination, decontamination of wastewater, $^{18}O_2$–$^{16}O_2$, and deuterium alkane isotope exchange are some of the most common examples of heterogeneous photocatalysis reactions (Linsebigler et al., 1995). Whereas, in a plasmonic antenna reactor-based photocatalysis process,

Photocatalysts based on Carbon Scaffolds: Environmental and Industrial Applications, First Edition. Edited by Ramesh Oraon et al.
Published 2026 by Wiley-VCH GmbH.

the photocatalyst is associated with an antenna (cocatalyst) attached to it, which boosts the efficacy of the photocatalyst by amplifying the light adsorption capacity of the photocatalyst. For example, the conversion of hydrogen sulfide (H_2S) to hydrogen (H_2) in the presence of a SiO_2 catalyst can be enhanced by embedding "Au" into the SiO_2 catalyst as a light absorber. This process is considered an alternative to the Claus process (Lou et al., 2022).

On exploring the applications of the photocatalysts, it was found that these materials can potentially alter the rate of a larger number of light-driven reactions, providing a sustainable solution for resolving the environmental pollutants. For example, when exposed to light, TiO_2 can break down volatile organic compounds (VOCs), dyes, and organic compounds. So, it can be used in air purification, wastewater treatment, and the disposal of hazardous chemicals from industrial wastes (Al-Nuaim et al., 2023). Similarly, TiO_2-coated materials can be employed as excellent photocatalysts to fabricate self-cleaning surfaces that can break down organic materials like filth, dirt, and oil (Koe et al., 2020). Being photoactive, photocatalysts are used in dye-sensitized solar cells (DSSCs) to enhance energy conversion efficiency (Tontapha et al., 2021). Furthermore, these materials play an important role in the green production of H_2 and O_2 from water in the presence of light (Ma et al., 2023). Under illumination, these materials inhibit the growth of the microbes on the surface upon which they are coated (Rilda et al., 2022). This helps in maintaining hygiene; these are utilized in public areas, hospitals, and also in household items (Kim et al., 2018).

Transition metal oxides (TMOs) are the most common class of photocatalysts widely studied by the scientific community (Krishnan et al., 2024). In response to light, electron-hole pairs are produced by them, which then react with water, hydroxyl groups, and oxygen to form reactive oxygen species (ROS) such as superoxide radical anions (O^{2-}) and hydroxyl radicals (OH) (Illarionov et al., 2020). TiO_2, ZnO, Mn_3O_4, Co_3O_4, $MnCo_2O_4$, and others are some of the widely known photocatalysts that participate in oxidation reactions to degrade harmful and odor-causing compounds to make the environment clean and safe (Krishnan et al., 2024; Hassaan et al., 2023; Karthikeyan & Mariappan, 2023; Sowmiya & Vijayalakshmi, 2024).

Transition metal dichalcogenides (TMDs) are a wonder class of photocatalytic materials that can potentially replace TMOs soon. They possess superior photocatalytic behavior owing to their good electrical conductivity, a huge number of active sites, high surface area, and narrow bandgap (Yang et al., 2023). However, the inertness of basal planes and sluggish dynamics of charge carriers are some major setbacks in their application (Li et al., 2018; Voiry et al., 2013). MoS_2, MSe_2, WS_2, $NiSe_2$, VSe_2, VS_2, and others are a few examples of TMD-based photocatalysts (Yang et al., 2023).

Because of their excellent electrochemical and physicochemical properties, carbon-based photocatalysts have drawn considerable interest from researchers (Kandy, 2020). Carbon scaffold photocatalysts are materials where carbon structures such as graphene, graphitic carbon nitride (g-C_3N_4), or carbon nanotubes (CNTs) act as support matrices in photocatalytic reactions. Carbon materials contain a large number of active sites due to large surface areas for photocatalysts (Kandy, 2020).

Some materials can absorb in the visible light region by enhancing the range of photocatalysts (Djurišić et al., 2020). Carbon materials like CNTs and graphene can decrease recombination to enhance charge transport capacity (Yang et al., 2017). One frequently employed common kind of support material is graphene, in which a single sheet of carbon atoms is arranged in a hexagonal lattice. It has a large surface area and excellent electrical conductivity, which ultimately increases the photocatalytic activity in the visible light region. To enhance the efficiency of reactions, graphene can scavenge photogenerated electrons from the photocatalyst surface (Sajna et al., 2022). A single-layered two-dimensional carbon support material is called g-C_3N_4. For visible light absorption, it acts as a photocatalyst (Purabgola et al., 2022). Similarly, CNTs are excellent carbon-based photocatalysts because of their strong mechanical strength, vast surface area, and great electrical conductivity. By enhancing charge transport and separation, CNTs can improve photocatalytic activity (Khan et al., 2017).

Based on a thorough literature survey, we have elucidated the development and application of carbon-based photocatalysts that facilitate light-driven reactions. Also, an economic assessment was carried out to understand the practical/industrial applicability of such a catalyst to drive the path for sustainable economic development.

8.2 Photocatalytic Activity of Different Carbon Scaffold-based Photocatalysts

8.2.1 Carbon Nanotubes

In 1991, "Sumio Iijima" was the first to report on the development of CNTs using a simple arc evaporation method (Iijima, 1991). CNTs belong to the subfamily of fullerenes, which are carbon allotropes (Ōsawa, 2015). They are regarded as an allotrope of carbon in which atoms of carbon have undergone sp^2 hybridization. The cylindrical nanostructure of CNTs contains C-atoms arranged in a hexagonal lattice with an inner diameter of 3–30 nm (Aqel et al., 2012). When unlayered graphene is rolled up into cylindrical form, it leads to the development of CNTs (Gupta et al., 2019; Eatemadi et al., 2014).

CNTs are suitable catalyst supports due to their large specific surface area, excellent thermal stability, and outstanding adsorption capability (Tao et al., 2024). Because of their high electrical conductivity, incredible mechanical strength, and remarkable electrocatalytic ability, CNTs are widely utilized in supercapacitor electrodes for energy storage. Functionalized CNTs are well-suited for the synthesis of composite materials. For example, because of the availability of complex pore architecture and reactive sites, they are used for sequestering carbon dioxide from indoor and outdoor environments (Tao et al., 2024). Furthermore, CNTs are also used in environmental technology to treat pollutants, clean up the environment, and create eco-friendly products (Kumar Jagadeesan et al., 2020). In addition to this, CNTs are also useful in medicinal applications like biosensing, cancer treatment, inducing hyperthermia, antimicrobial therapy, and tissue engineering due to their special combination of characteristics (Zafar et al., 2022).

8.2.1.1 Classification of CNT

Based on the structural view of the layer or wall, CNTs are of three different kinds:

A. **Single-walled CNTs (SWCNTs):** SWCNTs are made of a single graphene sheet, which is rolled into a cylindrical shape. The diameter of these nanotubes is usually between 0.5 and 2 nm (Kumar Jagadeesan et al., 2020; Hasnain & Nayak, 2019). These tubes possess unusual features such as great tensile strength and flexibility, outstanding thermal conductivity, and the ability to be either metallic or semiconducting depending on their electrical chirality. SWCNTs can be used in sensors, nanoelectronics, and as conductive additions in composite materials because of their unique properties (Hasnain & Nayak, 2019).
B. **Multi-walled CNTs (MWCNTs):** In contrast, MWCNTs consist of several concentric graphene cylinders that form a tube-in-tube structure. The diameter of MWCNTs varies from a few nanometers to over 100 nanometers (Gupta et al., 2019; Kumar Jagadeesan et al., 2020). Although characteristics can change depending on the number of walls and how they are arranged, MWCNTs are renowned for their exceptionally high mechanical strength and endurance, excellent thermal stability, and usually metallic electrical properties. They serve as conductive fillers, structural applications, and composite materials (Kumar Jagadeesan et al., 2020).
C. **Double-walled CNTs (DWCNTs):** DWCNTs combine the characteristics of SWCNTs and MWCNTs, consisting of two concentric graphene cylinders. They possess improved physicochemical behavior as compared to SWCNTs. Because of this, DWCNTs are suitable for applications that require great strength and stability (Hasnain & Nayak, 2019) (Figure 8.1).

8.2.2 Synthesis

Several synthetic pathways were used to synthesize CNTs, including arc discharge, laser ablation, chemical vapor deposition (CVD), flame synthesis, and electrochemical synthesis. Table 8.1 presents a brief discussion of the preparation of CNTs, their advantages, and limitations.

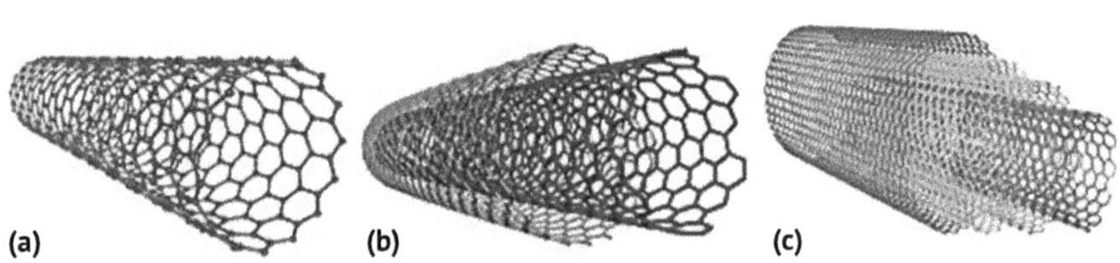

Figure 8.1 Types of CNTs: (a) SWCNT, (b) DWCNT, and (c) MWCNT. *Source*: Sobamowo et al. (2021)/with permission of Elsevier.

Table 8.1 Summary of common synthesis methods for carbon nanotubes (CNTs), along with their advantages and limitations.

Method	Synthetic procedure	Advantages	Limitations
Arc discharge	• In this method, carbon atoms from the anode are sublimized and condensed at high temperatures with the help of plasma arc under an inert atmosphere, Janas (2020).	• Yields both single-walled and multi-walled CNTs. High-quality CNTs, Yahyazadeh et al. (2024). • The simplest method for the fabrication of CNTs, Saifuddin et al. (2013).	• Requires purification due to mixed carbonaceous materials. • This method is expensive as it operates under high temperatures, low pressure, and costly inert gases, Jeon et al. (2011).
Laser Ablation	• With the help of a high-power laser, a graphite target was evaporated in an inert atmosphere to obtain high-quality CNTs, Mubarak et al. (2014).	• Precise control over SWCNT diameter and length. High-purity SWCNTs, Das et al. (2016).	• Expensive due to laser cost and energy consumption, Saifuddin et al. (2013).
CVD	• This method involves breaking down hydrocarbon gases in the presence of a metal catalyst at high temperatures, Ma et al. (2016).	• Customizable CNT characteristics. • A scalable method for large-scale production of CNTs, Pierson (1999).	• Requires careful temperature and gas flow control. Purification may be needed if catalyst particles remain. • The CNT developed is prone to defects, Saifuddin et al. (2013).
PECVD (Plasma-enhanced CVD)	• Similar to CVD but uses plasma to accelerate gas breakdown. Yields vertically aligned CNTs, Kumar (2011).	• Lower synthesis temperatures. • Vertically aligned CNTs for field emission applications, Jiang et al. (2006).	• Requires specialized plasma equipment. Potential harm to CNTs, Kumar Jagadeesan et al. (2020).

(Continued)

Table 8.1 (Continued)

Method	Synthetic procedure	Advantages	Limitations
High-pressure carbon monoxide process (HiPco)	• Breaks down carbon monoxide at high temperatures and pressures. Yields high-quality SWCNTs, Shoukat and Khan (2022).	• Continuous production of SWCNTs with fewer defects. • High purity, Shoukat and Khan (2022).	• Complex process requiring precise control. • High-pressure equipment is needed, Shoukat and Khan (2022).
Flame synthesis	• CNTs are formed when hydrocarbon fuels burn in a controlled atmosphere. Metal catalysts can be added, Gore and Sane (2011).	• Easy, economical, and scalable. • Suitable for large-scale manufacturing, Hamzah et al. (2017).	• Produces a mixture of CNTs and other carbonaceous elements. Purification required, Gore and Sane (2011).
Electrochemical Synthesis	• In this process, CNTs are deposited onto a substrate from an electrolytic solution through the process of electrolysis, Zhou et al. (2006).	• Customizable CNT properties. • Exact control over the deposition, Douglas et al. (2018).	• Lower CNT yield compared to other methods. Requires specialized equipment, Shoukat and Khan (2021).

8.2.3 Application of CNTs

CNTs are well known for their exceptional optoelectronic activity, excellent mechanical strength, high surface area, and superior electroluminescence nature, which has led to their application in the field of photocatalysis (Natarajan et al., 2017; Rasool et al., 2022). CNTs have been used as photocatalysts to speed up the reaction rate of various light-driven reactions. This section discusses this further with suitable examples.

8.2.3.1 Pollutant Degradation

As a leading photocatalytic material, CNTs are used to remove harmful chemicals from the air and aqueous medium efficiently (Figure 8.2). In comparison to others, CNTs facilitate photocatalytic activity by reducing the recombination rate of photogenerated electron-hole pairs. As a result, they are found to detoxify many wastes, including colors, insecticides, pharmaceutical wastes, etc. Furthermore, because of their large surface area, CNTs also help photocatalysts disperse more effectively and make more active sites available for reactions (Saleh, 2013). To increase their

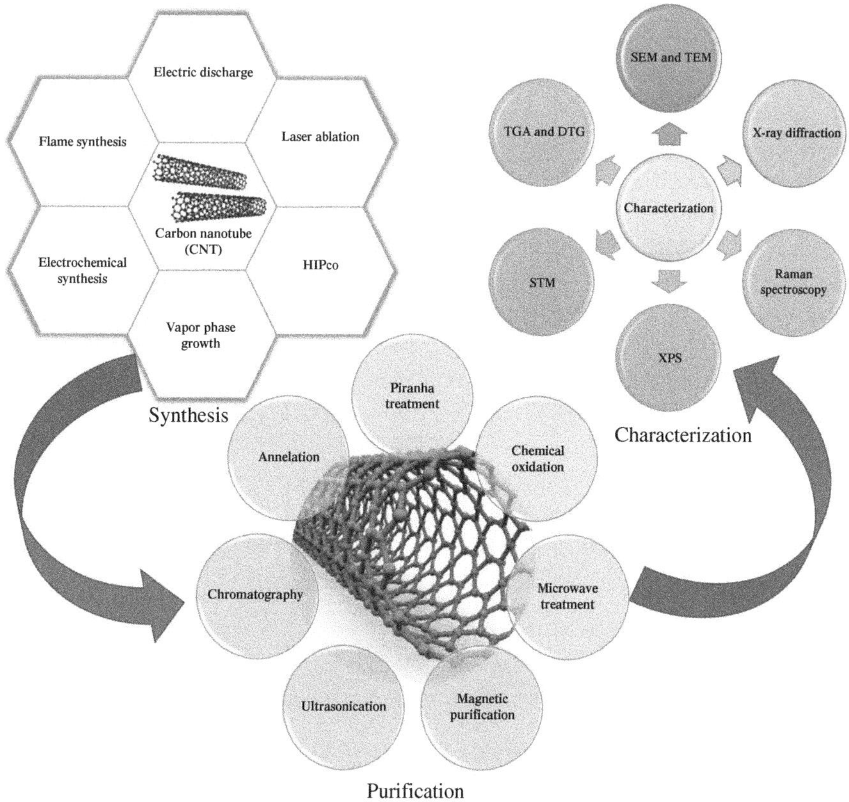

Figure 8.2 Various methods for synthesis, purification, and characterization of CNT.

photocatalytic efficacy, CNTs are frequently mixed with conventional photocatalysts such as tungsten trioxide (WO_3), zinc oxide (ZnO), and titanium dioxide (TiO_2). The photogenerated charge carriers' lifetime was extended, and their charge separation was improved by the CNTs' role as electron acceptors. As a result, the decomposition process becomes faster (Zhang et al., 2022). According to a study, as compared to TiO_2 alone, CNT/TiO_2 composites greatly accelerated the pace at which methylene blue degraded under UV light (Saleh, 2013). A study on CNT/AgS_2 composite shows that 2.4% CNT/AgS_2 shows a higher decomposition rate toward photodegradation of Rhodamine-B dye as compared with the other combinations (Di et al., 2019). Similarly, it was found that the presence of CNT enhances the photocatalytic dye degradation by 8.64 times that of the pure WSe_2 (Wang et al., 2017b).

8.2.3.2 Hydrogen Production

To generate hydrogen, a clean and renewable energy source, photocatalytic water splitting requires CNTs. They increase charge separation and light absorption, which raises photocatalyst efficiency. By acting as electron acceptors, CNTs can help the hydrogen evolution reaction by transferring electrons from the photocatalyst to the water molecules. When this is done, instead of just employing photocatalysts, higher rates of hydrogen production are achieved (Hassaan et al., 2023). In a study, under solar light irradiation, CNT-modified TiO_2 photocatalysts demonstrated a considerable increase in hydrogen production from water splitting (Hassaan et al., 2023). The procedure was more efficient overall because of the CNTs' improved light absorption and provision of a conductive channel for electron transport. A ternary composite of MoS_2/CNTs/CdS shows and 15% improvement in the H_2 production with pure CdS. As per the report, the CNT has played the role of a charge mediator for enhancing the photocatalytic activity (Mou et al., 2024).

8.2.3.3 Carbon Dioxide Reduction

The adoption of fossil fuel-derived energy contributed to rising atmospheric CO_2 emissions, a major contributor to global warming and climate change (Ng et al., 2021). To minimize carbon dioxide (CO_2) emissions while generating useful compounds, CNTs are used in the photocatalytic conversion of CO_2 to valuable hydrocarbons and alcohols. By offering an extensive surface area for CO_2 adsorption and allowing the transmission of photogenerated electrons to CO_2 molecules, CNTs improve the reduction efficiency and photocatalytic activity (Stroyuk et al., 2020). This process is a multielectron transfer process with possible products as listed below.

$$CO_2 + 2H^+ + 2e^- \rightarrow HCOOH,\ E_{redox} = -0.20\ V \tag{8.1}$$

$$CO_2 + 2H^+ + 2e^- \rightarrow CO + H_2O,\ E_{redox} = -0.12\ V \tag{8.2}$$

$$CO_2 + 4H^+ + 4e^- \rightarrow HCHO + H_2O,\ E_{redox} = -0.07\ V \tag{8.3}$$

$$CO_2 + 6H^+ + 6e^- \rightarrow CH_3OH + H_2O,\ E_{redox} = +0.03\ V \tag{8.4}$$

$$CO_2 + 8H^+ + 8e^- \rightarrow CH_4 + 2H_2O,\ E_{redox} = +0.17\ V \tag{8.5}$$

According to a study, under UV light, CNT/TiO_2 composites demonstrated improved photocatalytic conversion of CO_2 to methane (Stroyuk et al., 2020). Methane yields increased due to the CNTs' enhanced CO_2 adsorption and electron transport. Similarly, Lashgari et al. developed a ternary catalyst using Zn, CuO, and CNTs, which helps in the production of some essential chemicals such as CH_3OH, HCHO, and HOOC—COOH from CO_2 (Lashgari et al., 2017). As CO_2 conversion involves multi-electron transfer, hence selectivity of the catalyst plays a critical role in improving the yield of the product. For example, CNTs/NiO and CNTs/NiO/Fe_2O_3 catalysts selectively convert CO_2 to CH_3OH with and conversion rate of 93% (Lashgari & Soodi, 2020).

8.2.3.4 Antibacterial Applications

Using CNTs in photocatalytic systems has significant antimicrobial effects. They can boost the production of ROS, which are efficient at killing bacteria, when paired with photocatalysts. The photocatalyst produces ROS, including superoxide anions and hydroxyl radicals, when exposed to light. CNTs contribute to the effective transport of electrons, which increases ROS generation and improves antibacterial activity (Stroyuk et al., 2020). According to a study, when exposed to UV light, CNT/TiO_2 composites showed better antibacterial activity against E. coli than TiO_2 alone (Stroyuk et al., 2020). The better antibacterial effectiveness was caused by the increased formation of ROS brought about by the presence of CNTs.

8.2.3.5 Self-cleaning Surfaces

Employing CNTs, self-cleaning surfaces have been developed that, when exposed to light, may break down organic pollutants and pathogens, preserving hygienic conditions. CNT-based composites' photocatalytic activity aids in the breakdown of organic materials on the surface, avoiding the buildup of dirt and microorganisms (Subramanian et al., 2013). CNT/TiO_2 coatings on glass surfaces had good self-cleaning qualities when exposed to UV light, according to a study (Subramanian et al., 2013). The coatings preserved the glass's cleanliness and transparency by breaking down organic impurities. Pandiyan et al. observed that SnO_2-doped GO/CNT showed enhanced bactericidal properties with an efficiency of 50% reduction in bacterial growth (Pandiyan et al., 2019). Similarly, Alsafari et al. developed ternary metal oxide/CNT nanocomposite, which improved the antibacterial properties of the ternary composite by several folds. These examples explain the role of CNT in improving the antimicrobial properties of the material (Alsafari et al., 2022).

8.2.3.6 Solar Cells

DSSCs are made more efficient by adding CNTs, which boost charge transport and light absorption. By offering a conductive network which allows it easier for electrons to move from dye molecules to the electrode, CNTs lower recombination losses and raise the solar cells' overall efficiency (Kuvarega et al., 2021). According to a study, CNT-modified DSSCs outperformed traditional DSSCs in terms of power conversion efficiency (Kuvarega et al., 2021). Better performance was achieved by the CNTs through decreased electron recombination and enhanced charge transport.

8.2.4 Challenges of CNT

Numerous issues that the CNT business faces are frequently covered in research articles. These are a few major challenges:

A. **High Production Expenses:** The costly synthesis of superior CNTs restricts their extensive application in industry (Tao et al., 2024; Zafar et al., 2022).
B. **Scalability:** Making CNTs on a big scale without reducing quality is a very difficult task (Tao et al., 2024).
C. **Purity and Defects:** For CNTs to function well in a variety of applications, their purity must be guaranteed, and manufacturing flaws must be kept to a minimum (Zafar et al., 2022).
D. **Toxicity and Environmental Impact:** The possible toxicity of CNTs and their environmental impact raise questions about their use, particularly in biological applications (Song et al., 2018; Murjani et al., 2022).

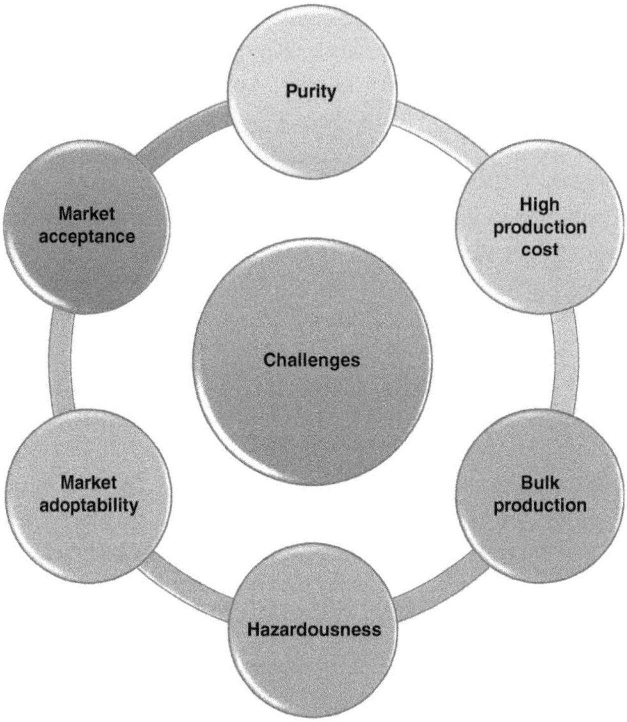

Figure 8.3 Challenges with CNTs.

E. **Integration with Other Materials:** The performance of CNTs in composite materials is impacted by the difficult process of successfully integrating them with other materials, such as metals and polymers (Daneshvar et al., 2021).
F. **Market Acceptance:** Growth in the CNT-based product market can be influenced by public opinion and acceptance. It is imperative to tackle safety and environmental issues in order to get wider acceptance (Daneshvar et al., 2021).

These difficulties show how much more research and development is required in order to get beyond obstacles and realize the full potential of CNTs in a variety of industries (Figure 8.3).

8.2.5 Future Prospectus

CNTs have extremely exciting potential for the future, with possible uses across multiple industries. The following are some important areas where CNTs are anticipated to have a major impact:

A. **Energy Storage and Conversion:** For their vast surface area and strong electrical conductivity, CNTs are being widely explored for application in batteries and supercapacitors. They can improve fuel cell and lithium-ion battery performance, resulting in more effective energy conversion and storage systems (Tao et al., 2024).

B. **Electronics and Semiconductors:** CNTs are perfect for use in transistors, sensors, and other electronic components because of their remarkable electrical capabilities. They might eventually take the place of silicon in semiconductor devices, resulting in quicker and more effective electronics (Tao et al., 2024).
C. **Biomedical Applications:** CNTs are being investigated for tissue engineering, imaging, and drug delivery. Their biocompatibility and capacity to cross cell membranes make them ideal for targeted drug delivery systems and cutting-edge medical diagnostics (Murjani et al., 2022).
D. **Composite Materials:** Composite materials' mechanical qualities can be greatly improved by CNTs, making them lighter and stronger. This is useful in the construction, automotive, and aerospace industries where lightweight, high-strength materials are essential (Yang, 2023).
E. **Environmental Applications:** Owing to their strong adsorption ability and chemical stability, CNTs can be employed in air and water filtration systems. Air and water quality can be enhanced, and pollutants can be eliminated with their assistance (Tao et al., 2024).
F. **Construction Industry:** CNTs can be added to concrete and other building materials during construction to enhance their thermal, mechanical, and strength qualities. Infrastructure that is more resilient and sustainable may result from this (Tao et al., 2024).

Notwithstanding these encouraging uses, issues like high production costs and the requirement for scalable manufacturing procedures remain. These problems are being researched in order to improve the accessibility of CNTs for general applications.

8.3 Graphene

Graphene is a two-dimensional material with outstanding attributes composed of a single layer of carbon atoms organized in a hexagonal lattice. Graphene is an extremely adaptable material due to its exceptional mechanical, electrical, thermal, and optical capabilities (Urade et al., 2023).

With its exceptional qualities and special combination, graphene has the potential to completely transform a wide range of sectors. Despite being incredibly light, its tensile strength of 130 GPa and Young's modulus of 1 TPa demonstrate its amazing mechanical strength, making it almost 200 times stronger than steel. Graphene exhibits superior electron mobility to copper at normal temperature (about 200 000 cm^2/Vs), which makes it a perfect material for electronic uses. Its thermal conductivity is also the highest of any known material, reaching up to 5000 W/mK, which is critical for applications needing effective heat dissipation. Because graphene absorbs only 2.3% of light, it can be used as a transparent conductive film for solar cells, LCDs, and touchscreens. Graphene is virtually transparent. Graphene is incredibly flexible despite its strength; it can be stretched up to 20% over its initial length without breaking (Urade et al., 2023).

Due to its exceptional flexibility and conductivity, graphene is being investigated for potential applications in quantum computing as well as transistors, sensors, and flexible displays. It provides more capacity and quicker charging periods for batteries and supercapacitors, improving their performance. Graphene can be used in tissue engineering, biosensors, and drug delivery systems due to its conductivity and biocompatibility. The strength and endurance of materials like metals and plastics are greatly increased when graphene is added; this is advantageous for the aerospace, automotive, and construction industries (Madurani et al., 2020). Nevertheless, despite graphene's extraordinary qualities, its commercial manufacture is still difficult. Techniques like mechanical exfoliation and CVD are costly and challenging to scale. Research and development on the integration of graphene into current industrial processes is also necessary. For graphene to be widely used, these production and integration obstacles must be overcome (Muley & Ravindra, 2019).

8.3.1 Types of Graphene

Graphene is a single layer of hexagonally organized carbon atoms that can be categorized into many forms according to its functionalization, number of layers, and structure. These are graphene's primary varieties:

Graphene comes in many forms, each with its own properties and uses:

A. **Monolayer Graphene:** One layer of carbon atoms, super strong, super conductive, and super thermal (Urade et al., 2023).
B. **Bilayer and Few-layer Graphene:** Two to ten layers, many of the same properties as monolayer but with some differences (Urade et al., 2023).
C. **Graphene Oxide (GO):** Has oxygen groups, water soluble, used in many applications (Santhiran et al., 2021).
D. **Reduced Graphene Oxide (rGO):** Partially restores graphene's conductivity by removing some oxygen groups (Santhiran et al., 2021).
E. **Graphene Nanoribbons (GNRs):** Thin strips with adjustable bandgaps, good for electronics (Nguyen et al., 2022).
F. **Graphene Quantum Dots (GQDs):** Tiny pieces with special optical and electrical properties, used in optoelectronics and sensing (Nguyen et al., 2022).
G. **Functionalized Graphene:** Chemically modified for better properties or new functions, used in drug delivery and sensors (Muley & Ravindra, 2019).
H. **Epitaxial Graphene:** High-quality graphene grown on substrates like silicon carbide, used in electronics (Muley & Ravindra, 2019).
I. **Graphene Aerogels:** 3D porous structures, good for energy storage and environmental cleanup (Muley & Ravindra, 2019).
J. **Graphene Foams:** Same as aerogels but made differently, used where you need lightweight conductive materials (Muley & Ravindra, 2019).

Graphene is a versatile material in nanotechnology and materials science due to its diverse kinds, which offer a wide range of features and uses. The summary of different types of graphene is shown in Figure 8.4.

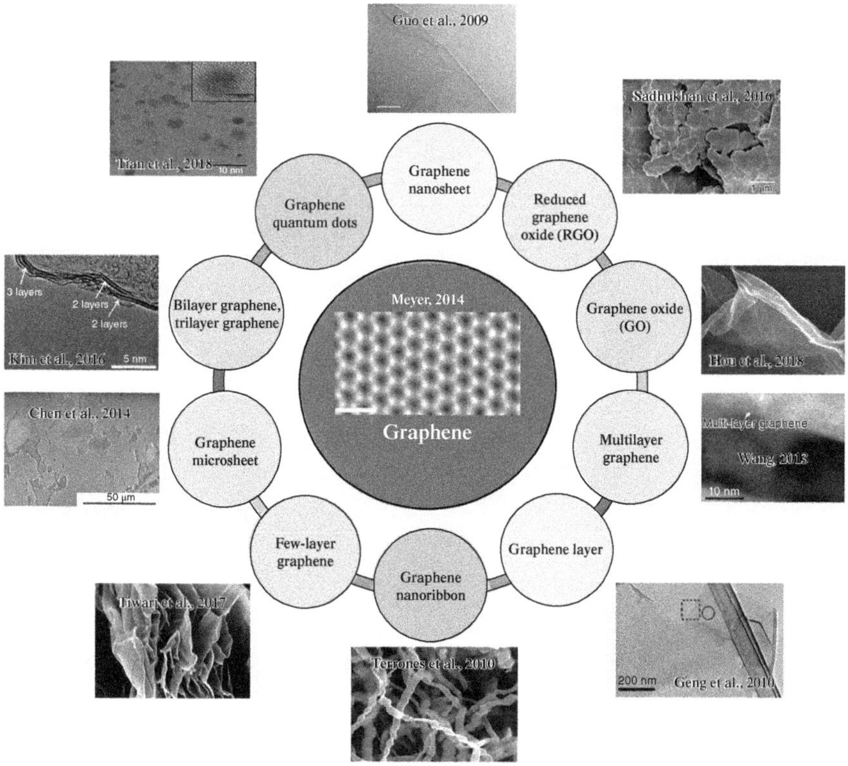

Figure 8.4 Different types of graphene. *Source*: Reproduced with permission from Madurani et al. (2020)/IOP Science/CC BY-NC-ND 4.0.

8.3.2 Synthesis

Graphene, a single layer of carbon atoms organized in a hexagonal lattice, has attracted a lot of interest for its remarkable features. There are two primary methods for synthesizing graphene: top-down and bottom-up techniques. Within these categories, researchers have devised a variety of strategies to maximize the quality, scalability, and cost-effectiveness of graphene manufacturing. Each method has advantages and challenges of its own. A brief discussion of the preparation of graphene, its advantages, and limitations is presented in Table 8.2.

8.3.3 Application of Graphene

8.3.3.1 Photodegradation of Organic Pollutants

Photocatalysts based on graphene have demonstrated exceptional efficacy in reducing organic pollutants found in both air and water. For example, the photodegradation of dyes, drugs, and other organic pollutants has been thoroughly researched in graphene oxide (GO) and reduced graphene oxide (rGO) composites with titanium dioxide (TiO_2). Pollutant adsorption is facilitated by graphene's

Table 8.2 Summary of common synthesis methods for graphene, along with their advantages and limitations.

Method	Description	Advantages	Challenges	References
Top-down methods				
Mechanical exfoliation	Peeling graphite with tape	High quality	Not scalable for industrial use	Urade et al. (2023)
Liquid phase exfoliation	Using ultrasonic energy to disperse graphite in a solvent	Produces large quantities	Variable flake size	Santhiran et al. (2021)
Chemical exfoliation	Oxidizing graphite to create graphene oxide, then reducing it	Scalable, cost-effective	Alters the electrical properties	Santhiran et al. (2021)
Electrochemical exfoliation	Using an electrochemical cell to exfoliate graphite	High quality, fewer defects	Complex setup	Yan et al. (2020)
Bottom-up methods				
CVD	Decomposing carbon-containing gas on a substrate	High-quality, large-area films	High temperatures required	Madurani et al. (2020)
Plasma-enhanced CVD (PECVD)	Accelerating gas breakdown with plasma	Lower synthesis temperature, fewer flaws	Requires plasma setup	Madurani et al. (2020)
Epitaxial growth on silicon carbide	Heating SiC to sublime silicon and leave graphene	High quality, superior electrical properties	Expensive, limited wafer size	Santhiran et al. (2021)
Unzipping of CNTs	Cutting nanotubes to create graphene nanoribbons	Unique electrical properties	Complex process	Liu et al. (2021)

large surface area, and photocatalytic activity is provided by TiO_2. Research findings indicate that the addition of graphene improves charge separation and decreases electron-hole pair recombination, resulting in increased photocatalytic efficiency (Kuvarega et al., 2021).

8.3.3.2 Microbial Inactivation

Water disinfection studies have examined the antibacterial capabilities of graphene-based photocatalysts. Upon exposure to UV and visible light, graphene/TiO_2 composites have demonstrated the ability to effectively inactivate germs and viruses

(Trivedi et al., 2024). ROS are produced, which harm the DNA and cell walls of microorganisms. This is the mechanism. Because of graphene's excellent conductivity, electrons can move between molecules more effectively, increasing ROS generation and disinfection effectiveness. Water disinfection has investigated the antibacterial capabilities of graphene-based photocatalysts. Inactivation of bacteria and viruses by UV and visible light irradiation has been demonstrated for graphene/TiO_2 composites. ROS are produced, which harm the DNA and cell walls of microorganisms (Trivedi et al., 2024).

8.3.3.3 Heavy Metal Reduction

Heavy metal reduction in contaminated water has also been accomplished by the use of graphene-based photocatalysts. For instance, hexavalent chromium (Cr(VI)) has been reduced to its less hazardous trivalent form (Cr(III)) using rGO/ZnO composites (Sajna et al., 2022). Cr(VI) may be adsorbed onto graphene due to its large surface area and functional groups, and ZnO offers the photocatalytic activity needed to reduce Cr(VI). For the removal of hazardous heavy metals from industrial effluent, this application is essential.

8.3.3.4 Water Splitting for Hydrogen Production

The process of "water splitting" produces hydrogen and oxygen from water by photocatalysis. The effective charge separation and transport facilitated by graphene have led to improved hydrogen production in graphene-based photocatalysts, including graphene/CdS and graphene/TiO_2 composites. Because graphene has a large surface area, semiconductor nanoparticles can disperse more easily, increasing the number of active sites in the water-splitting reaction. This application has the potential to produce hydrogen sustainably and serve as a clean energy source (Yunus et al., 2023).

8.3.3.5 Carbon Dioxide Reduction

Another significant use for graphene-based photocatalysts is the conversion of carbon dioxide (CO_2) to useful compounds like methanol and methane. It has been demonstrated that graphene/ZnO and graphene/TiO_2 composites improve CO_2 photocatalytic reduction when exposed to visible light. Because of graphene's high conductivity, photogenerated electrons can more easily reach CO_2 molecules, accelerating their reduction. By tackling energy and environmental issues, this application helps with carbon recycling and energy storage (Kuvarega et al., 2021).

8.3.3.6 Selective Oxidation Reaction

Organic molecules have been selectively oxidized with the help of photocatalysts based on graphene. GQDs, for instance, have been applied to the photocatalytic oxidation of alcohols to produce ketones and aldehydes. GQDs have special optical and electrical characteristics that improve their photocatalytic activity, including as edge effects and quantum confinement. Pharmaceutical and fine chemical synthesis benefits greatly from this use (Mathew et al., 2023).

8.3.3.7 C—C Bond Formation

One of the most important reactions in organic synthesis is the creation of carbon-carbon (C—C) bonds. Photocatalytic C—C coupling processes have made use of graphene-based photocatalysts, such as graphene/ZnO and graphene/TiO_2. Reactant adsorption and activation are facilitated by graphene's large surface area and conductivity, which improves reaction speeds and selectivity. For the synthesis of complex organic compounds and materials, this application is essential (Budiarso et al., 2024).

8.3.3.8 Photocatalytic Disinfection

Potential for disinfecting water and air has been demonstrated using graphene-based photocatalysts. The composites exhibit broad-spectrum antibacterial action against a variety of diseases due to the combination of semiconductors such as ZnO and TiO_2 with graphene's photocatalytic activity. The process includes the production of ROS, which injure the DNA and cell walls of microbes and cause them to become inactive. Limiting the spread of contagious diseases and guaranteeing clean drinking water are two major benefits of this application (Ramos et al., 2024; Upadhyay et al., 2014).

8.3.3.9 Photocatalytic Hydrogen Evolution

One important step in the creation of hydrogen fuel is the hydrogen evolution reaction (HER). The improved HER activity of graphene-based photocatalysts, like graphene/CdS and graphene/TiO_2 composites, can be attributed to the effective charge separation and transport that graphene provides. Because graphene has a large surface area, semiconductor nanoparticles can disperse more easily, increasing the number of active sites for the HER. This application has the potential to produce hydrogen sustainably and serve as a clean energy source (Qin et al., 2022).

8.3.3.10 Photocatalytic CO_2 Reduction

The conversion of CO_2 to useful compounds like methanol and methane is another significant use for graphene-based photocatalysts. It has been demonstrated that graphene/ZnO and graphene/TiO_2 composites improve CO_2 photocatalytic reduction when exposed to visible light. Because of graphene's high conductivity, photogenerated electrons can more easily reach CO_2 molecules, accelerating their reduction. By tackling energy and environmental issues, this application helps with carbon recycling and energy storage (Fang et al., 2023; Yang & Xu, 2016).

8.3.3.11 Photocatalytic Nitrogen Fixation

The process of nitrogen fixation involves changing atmospheric nitrogen (N_2) into ammonia (NH_3), which is an essential ingredient in fertilizers. The photocatalytic nitrogen fixation reaction has made use of graphene-based photocatalysts, such as graphene/ZnO and graphene/TiO_2 composites. Graphene's large surface area and strong conductivity make it easier for N_2 molecules to adsorb and activate, which improves reaction speeds and selectivity. For food production and sustainable agriculture, this use is essential (Kuvarega et al., 2021; Huang et al., 2021).

8.3.4 Challenges of Graphene

A. **Synthesis and Production:**
   a. **Quality Control:** It is still challenging to guarantee the homogeneity and repeatability of graphene sheets, particularly for large-scale manufacturing (Capaz, 2022).
   b. **Scalability:** To achieve increased production rates without sacrificing quality, methods such as CVD must be refined (Capaz, 2022).
   c. **Cost:** One obstacle to widespread adoption may be the high cost of raw materials and production techniques (Zurutuza & Marinelli, 2014).

B. **Environmental and Health Risks:**
   a. **Toxicity:** There is still much to learn about the possible health hazards connected to graphene and its byproducts (Guo et al., 2024).
   b. **Environmental Impact:** The use of hazardous chemicals in the production process and the trash it produces raise questions about the sustainability of the environment (Guo et al., 2024).

C. **Material Properties:**
   a. **Bandgap Engineering:** The absence of a natural bandgap in graphene restricts its application in semiconductors. One of the main challenges is to create a bandgap without sacrificing its other features (Huang et al., 2021).
   b. **Functionalization:** Adding functional groups to graphene can improve its properties for particular uses, but this can complicate the material and compromise its overall performance (Zare et al., 2021).

D. **Commercialization:**
   a. **Market Integration:** The commercial success of graphene depends on the identification of "killer" applications that warrant the expense and intricacy of its manufacture.

E. **Standards and Regulations:** The establishment of a stable market is hampered by the absence of established procedures for production and quality evaluation.

F. **Research and Development:**
   a. **Interdisciplinary Collaboration:** It can be difficult to arrange effective collaboration amongst several disciplines when doing graphene research.
   b. **Finance and Investment:** To break through technological obstacles and develop graphene technologies, a consistent investment in research and development is required.

8.3.5 Future Prospectus

Graphene appears to have a very bright future ahead of it, with major developments and uses anticipated in a number of industries (Bramhaiah & Bhattacharyya, 2022; Dubey et al., 2023). These are a few important prospects:

a. **Electronics and Telecommunications:** Graphene is perfect for next-generation electronics, such as flexible screens, high-speed transistors, and sophisticated sensors, due to its remarkable electrical conductivity and flexibility (Dubey et al., 2016).

b. **Energy Storage:** The use of graphene is set to revolutionize energy storage technologies. Higher energy densities, quicker charging times, and longer lifespans can result from its application in batteries and supercapacitors, which will especially help portable gadgets and electric cars (Mombeshora & Muchuweni, 2023).
c. **Biomedical Applications:** Graphene's special qualities and biocompatibility are opening new possibilities for tissue engineering, biosensors, and drug delivery. It could lead to the creation of cutting-edge medical technologies and therapies (Li et al., 2021).
d. **Composite Materials:** Composites with added graphene have better strength, light weight, and heat conductivity. It is anticipated that these materials will be used in the construction, automotive, and aerospace sectors to enhance durability and performance (Ibrahim et al., 2021).
e. **Environmental Solutions:** Graphene has the potential to be extremely important for environmental protection in applications like air and water filtration. It is efficient at eliminating impurities and pollutants due to its large surface area and reactivity (Sharma et al., 2024).
f. **Market Growth:** From 2024 to 2030, the graphene market is expected to develop at a compound annual growth rate (CAGR) of about 31.3%, according to projections. By 2027, the market size is anticipated to surpass 1.5 billion US$ (Schmaltz et al., 2024).

8.4 Carbon Quantum Dots

Zero-dimensional carbon nanomaterials, or carbon quantum dots (CQDs), are often less than 10 nm. Their distinctive qualities render them extremely beneficial for a wide range of applications. Due to their size-dependent electronic transitions and surface states, CQDs emit strong and controllable fluorescence (Wang & Hu, 2014). The extensive carboxyl group count on the surface of CQDs contributes to their exceptional water solubility (Kong et al., 2024). CQDs are also well-known for having minimal biotoxicity, which qualifies them for use in biomedical applications (Kong et al., 2024; Sayapina et al., 2023). Their fluorescence qualities are sustained over time due to their resistance to photobleaching (Sayapina et al., 2023; Wang et al., 2019). Additionally, CQDs have strong electrical conductivity, which is advantageous for uses in catalysis and electronics. Moreover, they can be employed in biological contexts without harming them thanks to their biocompatibility (Kong et al., 2024; Wang et al., 2019). Because of these qualities, CQDs are very adaptable and helpful in a variety of industries, including biomedicine, optronics, catalysis, and sensing (Wang & Hu, 2014; Kong et al., 2014).

8.4.1 Types of Carbon Quantum Dots

With their distinct qualities and wide range of uses, CQDs are an intriguing family of nanomaterials. These are some typical forms of CQDs, according to research:

A. **Graphene Quantum Dots:** GQDs are generated from graphene and have superior electrical characteristics. Usually, techniques like hydrothermal cutting, chemical exfoliation, and electrochemical oxidation are used to create them. Due to their intense photoluminescence, high surface area, and superior biocompatibility, GQDs are well-suited for use in optoelectronics, bioimaging, and sensors (Wang & Hu, 2014).

B. **Carbon Nanodots (CNDs):** Small carbon nanoparticles smaller than 10 nm are called CNDs. They can be produced from a variety of carbon sources, such as trash that is rich in carbon, biomass, and organic materials. Because of their high fluorescence, excellent solubility in water, and low toxicity, CNDs are perfect for use in environmental sensing, medication delivery, and bioimaging (Kong et al., 2014).

C. **Polymer Dots (PDs):** Another kind of CQD that is made from polymer precursors is called a polymer dot. Their unique ability to control photoluminescence is a result of structural modifications to the polymer. Because of their great brightness and stability, PDs are employed in light-emitting devices, sensors, and bioimaging applications (Magesh et al., 2022).

D. **Doped CQDs:** Doped CQDs are made of carbon and one or more heteroatoms, such as phosphorus, sulfur, or nitrogen. Through doping, CQDs' optical and electrical characteristics can be improved, increasing their usefulness for sensing, energy storage, and catalytic applications. For instance, nitrogen-doped CQDs show enhanced catalytic activity and photoluminescence (Yadav et al., 2023).

E. **CQDs from Biomass:** Biomass waste can also be used to synthesize CQDs, which is an economical and environmentally beneficial method. Biomass-derived CQDs show strong fluorescence, good biocompatibility, and low toxicity while retaining the intrinsic features of the source material. Applications for these CQDs include pollution sensing, bioimaging, and renewable energy (Kong et al., 2014).

F. **CQDs from CNTs:** CNTs can be converted into CQDs using techniques like chemical oxidation and laser ablation. These CQDs are appropriate for use in electronics, sensors, and energy storage because they inherit the superior mechanical and electrical qualities of CNTs (Wang & Hu, 2014).

G. **CQDs from Fullerenes:** CQDs that are formed from fullerenes are produced by hydrothermally treating and chemically oxidizing fullerenes. These CQDs are utilized in medication delivery, bioimaging, and photovoltaic applications due to their distinctive optical characteristics (Magesh et al., 2022).

A versatile and promising family of nanomaterials with a broad variety of uses is CQDs. Each of the different kinds of CQDs—GQDs, CNDs, PDs, doped CQDs, biomass-derived CQDs, CNT-derived CQDs, and fullerene-derived CQDs—has special qualities that suit them for applications.

8.4.2 Synthesis

With distinctive optical and electrical characteristics, CQDs are an intriguing class of nanomaterials. Their prospective uses in medicine delivery, energy devices, bioimaging, and sensors have attracted a lot of attention. A brief overview of the development of CQDs, advantages, and limitations is discussed in Table 8.3.

Table 8.3 Summary of carbon quantum dot (CQD) synthesis techniques, along with their advantages and limitations.

Method	Description	Advantages	Challenges	References
Top-down methods				
Arc discharge	Uses a high-temperature plasma arc in an inert gas environment to vaporize graphite	Can modify size/characteristics by adjusting arc current/gas pressure	High energy consumption	Wang and Hu (2014)
Laser ablation	Targets a carbon surface in a liquid with a laser beam	Adjusts the CQDs' size by laser settings	Complex setup	Wu et al. (2023)
Electrochemical exfoliation	Applies voltage to graphite in an electrolyte solution	Simple, scalable	Graphene layers may vary in quality	Hu et al. (2011)
Chemical oxidation	Oxidizes carbon precursors with strong acids	Cost-effective, scalable	Alters electrical properties.	Molaei (2024)
Bottom-up methods				
Hydrothermal synthesis	Heats carbon precursors at high pressure and temperature in a sealed autoclave	Allows control over size/characteristics	High pressure/temperature required	Nammahachak et al. (2022)
Solvothermal synthesis	No	Produces different surface functionalities	Organic solvent handling	Omer et al. (2018)
Microwave-assisted synthesis	Uses microwaves to heat carbon precursors in a solvent	Quick, high yields	Potential for uneven heating	Romero et al. (2021)

Method	Description	Advantages	Challenges	References
Pyrolysis	Heats carbon precursors in an inert atmosphere	Regulates size and characteristics by temperature and duration	Energy intensive	Ahlawat et al. (2023)
Green synthesis				
Biomass-derived CQDs	Utilizes natural carbon sources and hydrothermal/ pyrolysis methods	Eco-friendly, cost-effective	Quality control of biomass sources	Wu et al. (2023)
Enzymatic synthesis	Uses enzymes to convert carbon precursors	High specificity, biocompatible	Limited scalability	Yadav et al. (2023)
Post-synthesis				
Surface functionalization and doping	Adds chemical groups or heteroatoms to modify properties	Enhances stability, biocompatibility, and solubility	Maintains or improves CQDs' functionality	Yan et al. (2018); Kandasamy (2019)

8.4.3 Applications of the CQDs

8.4.3.1 Dye Degradation

It has been demonstrated that CQDs can efficiently break down dyes like methyl orange, rhodamine B, and methylene blue. While CQDs' photoluminescence capabilities allow for efficient light absorption and the production of ROS, which destroy dyes, their high surface area and functional groups aid in the adsorption of dye molecules (Roopan et al., 2024).

8.4.3.2 Pesticide Degradation

Pesticides like atrazine and chlorpyrifos are also degraded by CQDs. The process entails the production of ROS, which target pesticide molecules and disassemble them into less toxic forms (Jung et al., 2022).

8.4.3.3 Pharmaceutical Degradation

Drugs like anti-inflammatories and antibiotics are becoming more polluting in water bodies. These substances have been broken down by CQDs, lessening their negative effects on the environment (Yadav et al., 2022).

8.4.3.4 Water Splitting

Water molecules can be divided into hydrogen and oxygen through a process called water splitting, which employs light energy. Because of their capacity to produce electron-hole pairs and absorb a wide range of light, CQDs have demonstrated promise as photocatalysts for water splitting. When exposed to light, CQDs can be utilized to create hydrogen from water. Protons are converted to hydrogen by the electrons produced by CQDs, while water is converted to oxygen by the holes. The large surface area and effective electron transport capabilities of CQDs improve this process (Li et al., 2020). In water splitting, the oxygen evolution reaction (OER) is a crucial stage. By offering active sites for the oxidation of water molecules, which produces oxygen, CQDs can help this reaction happen (Wadkar et al., 2024).

8.4.3.5 Carbon Dioxide Reduction

Another significant application of CQDs in photocatalysis is the conversion of carbon dioxide (CO_2) to useful compounds and fuels. Under light irradiation, CQDs can convert CO_2 into compounds such as methanol, formic acid, and methane. Methane, a useful fuel, can be produced by CQDs from CO_2. When CQDs are exposed to light, they produce electrons and holes, which go on to contribute to the process of reducing CO_2 to methane (Kong et al., 2024). The reduction of CO_2 can also yield methanol, another useful product. The process of CO_2 adsorption and reduction is facilitated by CQDs, which offer active sites for this purpose (Wu et al., 2023). Reduction of CO_2 can yield the valuable chemical formic acid. By producing ROS that cause CO_2 to be reduced to formic acid, CQDs can catalyze this reaction (Cheng et al., 2024).

8.4.4 Challenges of CQDs

A. **Synthesis Challenges**
   a. **Aggregation:** CQDs tend to aggregate during synthesis, which may have an impact on their size, homogeneity, and general characteristics.
   b. **Control Over Properties:** CQDs' effectiveness in a variety of applications may be impacted by the difficulty of precisely controlling their size, shape, and surface characteristics (Yadav et al., 2023).

B. **Stability Issues**
   a. **Photostability:** When exposed to light, CQDs may experience photobleaching, which causes a gradual decrease in fluorescence.
   b. **Chemical Stability:** The practical applications of CQDs depend on their long-term chemical stability in various conditions (Kong et al., 2024)

C. **Scalability**
   a. **Production Costs:** The high cost of large-scale production and the complexity of synthesis methods can limit the commercial viability of CQDs.
   b. **Consistency:** Producing CQDs with consistent quality and properties on a large scale remains a significant challenge (Kong et al., 2024).

D. **Performance Limitations**
   a. **Photostability:** When exposed to light, CQDs may experience photobleaching, which causes a gradual decrease in fluorescence.

b. **Chemical Stability:** The practical applications of CQDs depend on their long-term chemical stability in various (Kong et al., 2024).

E. **Environmental and Health Concerns**

a. **Toxicity:** Even though CQDs are thought to be low in toxicity, further research is needed to determine the long-term impacts on human health and the environment.

b. **Biocompatibility:** To prevent negative consequences, it is crucial to guarantee that CQDs are biocompatible for use in biological and medical applications (Kong et al., 2024).

These are the challenges and limitations to overcome in the field of CQDs to overcome (Kong et al., 2024; Chowmasundaram et al., 2023; Panachikkool & Pandiyarajan, 2024).

8.4.5 Prospects

Notwithstanding the encouraging future of CQDs, several obstacles still exist. Among them are

A. **Advanced Synthesis Techniques**

- The intent is to create novel synthesis methods that are economical, ecologically benign, and scalable.
- Research Insight: To create CQDs with a high yield and reliable quality, green synthesis techniques employing biomass and other renewable resources are being investigated (Bhattacharya et al., 2024).

B. **Surface Functionalization**

- The objective is to increase the stability, biocompatibility, and functional characteristics of CQDs by improving their surface functionalization.
- Research Insight: In some applications, functionalization with metals, polymers, and other nanomaterials can improve the performance of CQDs (Kong et al., 2024).

C. **Hybrid Nanomaterials**

- The goal is to combine CQDs with other nanomaterials to produce hybrid systems that have complementary qualities.
- Research Insight: Improved photocatalytic and electrical properties have been demonstrated by hybrid nanomaterials, such as CQD-metal oxide composites (Yadav et al., 2023).

D. **Application-specific Optimization**

- Through controlled synthesis and functionalization, the properties of CQDs can be tailored for specific applications.
- Research Insight: By optimizing the size, shape, and surface chemistry of CQDs, their performance in applications such as drug delivery, solar cells, and sensors can be improved (Wang & Hu, 2014).

E. Comprehensive Toxicity Studies

- To guarantee the safe deployment of CQDs in environmental and biomedical applications, extensive toxicity and biocompatibility investigations must be carried out.
- Research Insight: To completely comprehend the safety profile of CQDs, further in vivo research and environmental impact assessments are required (Chung et al., 2021).

8.5 Graphene Carbon Nitride (g-C_3N_4)

Graphene carbon nitride, known as graphitic carbon nitride (g-C_3N_4), is a two-dimensional conjugated polymer consisting of carbon and nitrogen. Although nitrogen atoms are integrated into the lattice, the structural analogy to graphene is notable. This substance is well-known for its excellent stability, nontoxicity, ease of manufacture, and semiconductor qualities (Huang et al., 2021; Capaz, 2022; Zurutuza & Marinelli, 2014).

8.5.1 Types of Graphene Carbon Nitride

Different forms of graphene carbon nitride (g-C_3N_4) can be distinguished by their structural variants and synthesis techniques. These are a few noteworthy types that are discussed here:

A. **Graphene-based g-C_3N_4:** This kind improves the electrical and mechanical characteristics of g-C_3N_4 by fusing the qualities of graphene and g-C_3N_4 (Akple et al., 2023).
B. **Triazine-based g-C_3N_4:** This structure has distinct electrical characteristics and strong thermal stability since it is made up of associated triazine units (C_3N_3) (Lin et al., 2022).
C. **Heptazine-based g-C_3N_4:** This form of g-C_3N_4 is distinguished by its exceptional photocatalytic activity and stability, as it is comprised of heptazine units (C_6N_7) (Lin et al., 2022).
D. **g-C_3N_4/Metal Oxide Composites:** These composites enhance photocatalytic activity and electron-hole separation efficiency by combining g-C_3N_4 with metal oxides (de Sousa et al., 2016).
E. **g-C_3N_4/Graphene Hybrids:** These compounds work together to enhance the applications of both compounds in energy conversion and catalysis (Wang et al., 2017a).

These variants demonstrate the adaptability and possible applications of g-C_3N_4.

8.5.2 Synthesis of Graphene Carbon Nitride

Several techniques are used to synthesize graphene carbon nitride (g-C_3N_4), each with pros and cons of its own. These are a few noteworthy synthesis techniques that are summarized in Table 8.4.

Table 8.4 Summary of graphitic nitrogen ($g-C_3N_4$) synthesis techniques, along with their advantages and limitations.

Method	Description	Advantages	Challenges	Reference
Thermal polymerization	Heating nitrogen-rich precursors in an inert environment to high temperatures	Scalable, simple	High temperature required	Ahmad et al. (2022)
CVD	Depositing $g-C_3N_4$ from a vapor phase onto substrates like silicon	Useful for thin film creation	Requires substrate preparation	Ahmad et al. (2022); Banciu et al. (2022); Saeed et al. (2020)
Electrochemical deposition	Depositing $g-C_3N_4$ on substrates via an electrochemical process	Good for films/ membranes	Involves a precise electrochemical setup	Ahmad et al. (2022)
Solvothermal/ hydrothermal	High-pressure, high-temperature reaction in a solvent	Can produce specific morphologies	Requires autoclave and product yield was comparatively low	
Microwave-assisted synthesis	Rapid heating with microwave radiation	Efficient, large surface areas	Requires a microwave reactor	Venkatesh et al. (2023)
Template-assisted synthesis	Using a template to control morphology	Produces specific structures (e.g., nanotubes)	Template removal required	Ahmad et al. (2022)
Exfoliation methods	Exfoliating bulk $g-C_3N_4$ into nanosheets chemically or mechanically	Can create nanosheets	Additional refinement needed	Ahmad et al. (2022)

8.5.3 Market Growth of $g-C_3N_4$

Due to its unique features, graphitic carbon nitride ($g-C_3N_4$) finds extensive use in both industrial and medical fields.

A. **Photocatalysis:** For environmental cleanup purposes, such as breaking down contaminants and creating hydrogen through water splitting, $g-C_3N_4$ is commonly employed in photocatalytic applications (Akple et al., 2023).

B. **Energy Conversion:** Its stability and appropriate bandgap make it useful in energy conversion systems such as solar cells (Akple et al., 2023).
C. **Sensors and Actuators:** Sensors and actuators that detect gases and other chemicals use materials based on g-C_3N_4 (Akple et al., 2023).
D. **Electrochemical Applications:** For enhanced performance, it is utilized in electrochemical devices such as supercapacitors and batteries (Han et al., 2017).
E. **Medical Applications:** The application of g-C_3N_4 in drug delivery systems and biocompatible coatings is being researched (Ajiboye et al., 2020). Because of its strong quantum yield, consistent fluorescence, and excellent biocompatibility, g-C_3N_4 quantum dots, or g-CNQDs, are employed in biosensors (Liu et al., 2019). Because g-CNQDs have brilliant, steady fluorescence, which makes biological processes easier to see, they are also used in bioimaging (Wang & Hu, 2014). Materials based on g-C_3N_4 can be employed in drug delivery systems to ensure the precise and regulated release of medicinal substances (Liu et al., 2019). g-C_3N_4 materials are being developed for theragnostic applications, which include simultaneous diagnosis and treatment of disorders. Theragnostic applications combine therapy and diagnostics (Liu et al., 2019).

8.5.4 Challenges for g-C_3N_4

Although g-C_3N_4 has potential, there are a few issues that need to be resolved before it can be widely used.

A. **Insufficient Light Absorption:** The photocatalytic effectiveness of g-C_3N_4 is limited due to its limited light absorption in the visible band (Bhanderi et al., 2024; Wang et al., 2023).
B. **Low Surface Area:** Applications requiring high surface interactions may find the material less effective due to its reduced surface area (Bhanderi et al., 2024; Wang et al., 2023).
C. **Poor Electronic Conductivity:** The low electronic conductivity of g-C_3N_4 compromises its usefulness in energy storage and electronic applications (Bhanderi et al., 2024; Wang et al., 2023).
D. **Fast Recombination of Charge Carriers:** The efficacy of g-C_3N_4 in photocatalytic processes is decreased by the quick recombination of photogenerated electron-hole pairs (Bhanderi et al., 2024; Wang et al., 2023).
E. **Scalability:** Ensuring consistent quality and large-scale production of g-C_3N_4 is still a challenging task (Bhanderi et al., 2024; Wang et al., 2023).

By addressing these obstacles via sophisticated synthesis techniques, doping, and composite construction, g-C_3N_4's performance can be improved, and its range of applications can be expanded.

8.5.5 Prospects

A. **Quantum Efficiency:** There is still much work to be done to increase these materials' quantum efficiency. To improve their efficiency by doping, surface modification, and hybridization, research is still being conducted (Bhanderi et al., 2024).

B. **Charge Carrier Recombination:** Improving charge carrier recombination is essential to raising photocatalyst efficiency. Techniques like forming heterojunctions with different materials are being investigated (Bhanderi et al., 2024).

C. **Scalability and Cost:** It will be necessary to create scalable and affordable synthesis techniques for these materials before they can be widely used (Li & Du, 2023).

With further research targeted at resolving present issues and realizing the full potential of graphene and g-C_3N_4 in a range of applications, the future of these materials appears bright.

8.6 Conclusion

From laboratory research to industrial application, carbon scaffold-based nano photocatalysts have seen a major evolution in terms of both scientific improvements and economic viability. These materials have profited from the ongoing innovation in synthesis techniques and the use of more reasonably priced raw materials. Initially, they were hampered by high manufacturing costs and scaling concerns. Their photocatalytic performance has been improved, and their increased viability for large-scale applications has been facilitated by this advancement. Carbon-based photocatalysts are currently gaining significant traction due to their superior qualities and affordability in fields like CO_2 reduction, hydrogen production, and wastewater treatment. The ability of these materials to replace more costly and less sustainable alternatives, which would lower operating costs and promote greener industrial practices, makes their economic benefits clear. Prospects for nano photocatalysts based on carbon scaffolds appear promising. It is anticipated that continued research into enhancing their effectiveness and cutting production costs will hasten their acceptance in several industries. These developments will probably have a significant positive economic impact, promote sustainable industrial growth and aid in the shift to a greener economy. The future of sustainable technology will be greatly influenced by the incorporation of carbon-based photocatalysts as the industry develops

References

Ahlawat, A., Dhiman, T. K., Solanki, P. R., & Rana, P. S. (2023). Facile synthesis of carbon dots via pyrolysis and their application in photocatalytic degradation of rhodamine B (RhB). *Environmental Science and Pollution Research*, 31, 46752–46759. https://doi.org/10.1007/s11356-023-25604-6

Ahmad, F., Khan, Z. H., & Singh, S. (2022). Graphitic carbon nitrides: Synthesis, properties, and applications in perovskite solar cells. In Z.H. Khan (Eds.), *Nanomaterials for Innovative Energy Systems and Devices. Materials Horizons: From Nature to Nanomaterials* (pp. 45–76). Springer, Singapore.

Ajiboye, T. O., Kuvarega, A. T., & Onwudiwe, D. C. (2020). Graphitic carbon nitride-based catalysts and their applications: A review. *Nano-Structures & Nano-Objects*, 24, 100577. https://doi.org/10.1016/j.nanoso.2020.100577

Akple, M. S., Takyi, G. K. S., & Chimmikuttanda, S. P. (2023). Graphitic carbon nitride (g-C_3N_4) based materials: current application trends in health and other multidisciplinary fields. *International Nano Letters*, 13, 223–234. https://doi.org/10.1007/s40089-023-00395-3

Al-Nuaim, M. A., Alwasiti, A. A., & Shnain, Z. Y. (2023). The photocatalytic process in the treatment of polluted water. *Chemical Papers*, 77, 677–701. https://doi.org/10.1007/s11696-022-02468-7

Alsafari, I. A., Fatima, R., Warsi, M. F. et al. (2022). Photocatalytic and antibacterial activity study of ternary oxide of Ni-Al-Cd and their nanocomposite with carbon nanotubes. *Journal of Taibah University for Science*, 16, 1016–1025. https://doi.org/10.1080/16583655.2022.2135812

Aqel, A., El-Nour, K. M. M. A., Ammar, R. A. A., & Al-Warthan, A. (2012). Carbon nanotubes, science and technology part (I) structure, synthesis and characterisation. *Arabian Journal of Chemistry*, 5, 1–23. https://doi.org/10.1016/j.arabjc.2010.08.022

Banciu, C. A., Nastase, F., Istrate, A.-I., & Veca, L. M. (2022). 3D graphene foam by chemical vapor deposition: Synthesis, properties, and energy-related applications. *Molecules*, 27, 3634. https://doi.org/10.3390/molecules27113634

Bhanderi, D., Lakhani, P., & Modim, C. K. (2024). Graphitic carbon nitride (g-C_3N_4) as an emerging photocatalyst for sustainable environmental applications: A comprehensive review. *RSC Sustainability*, 2, 265–287. https://doi.org/10.1039/D3SU00382E

Bhattacharya, T., Preetam, S., Mukherjee, S. et al. (2024). Anticancer activity of quantum size carbon dots: Opportunities and challenges. *Discover Nano*, 19, 122. https://doi.org/10.1186/s11671-024-04069-7

Bramhaiah, K. & Bhattacharyya, S. (2022). Challenges and future prospects of graphene-based hybrids for solar fuel generation: Moving towards next generation photocatalysts. *Materials Advances*, 3, 142–172. https://doi.org/10.1039/D1MA00748C

Budiarso, I. J., Dabur, V. A., Rachmantyo, R. et al. (2024). Carbon nitride- and graphene-based materials for the photocatalytic degradation of emerging water pollutants. *Materials Advances*, 5, 2668–2688. https://doi.org/10.1039/D3MA01078C

Capaz, R. B. (2022). Grand challenges in graphene and graphitc research. *Frontiers in Carbon*, 1, 1034557. https://doi.org/10.3389/frcrb.2022.1034557

Cheng, Y., Yang, R., Xia, L. et al. (2024). Graphene quantum dot-mediated anchoring of highly dispersed bismuth nanoparticles on porous graphene for enhanced electrocatalytic CO_2 reduction to formate. *Nanoscale*, 16, 2373–2381. https://doi.org/10.1039/D3NR05853K

Chowmasundaram, Y. A. P., Tan, T. L., Nulit, R., Jusoh, M., & Rashid, S. A. (2023). Recent developments, applications and challenges for carbon quantum dots as a photosynthesis enhancer in agriculture. *RSC Advances*, 13, 25093–25117. https://doi.org/10.1039/D3RA01217D

Chung, C.-Y., Chen, Y.-J., Kang, C.-H. et al. (2021). Toxic or not toxic, That is the carbon quantum dot's question: A comprehensive evaluation with zebrafish embryo, eleutheroembryo, and adult models. *Polymers (Basel)*, 13, 1598. https://doi.org/10.3390/polym13101598

Daneshvar, F., Chen, H., Noh, K., & Sue, H.-J. (2021). Critical challenges and advances in the carbon nanotube–metal interface for next-generation electronics. *Nanoscale Advances*, 3, 942–962. https://doi.org/10.1039/D0NA00822B

Das, R., Shahnavaz, Z., Ali, M. E. et al. (2016). Can we optimize Arc discharge and laser ablation for well-controlled carbon nanotube synthesis? *Nanoscale Research Letters*, 11, 510. https://doi.org/10.1186/s11671-016-1730-0

de Sousa, J. M., Botari, T., Perim, E. et al. (2016). Mechanical and structural properties of graphene-like carbon nitride sheets. *RSC Advances*, 6, 76915–76921. https://doi.org/10.1039/C6RA14273G

Di, L., Xian, T., Sun, X. et al. (2019). Facile Preparation of CNT/Ag_2S nanocomposites with improved visible and NIR light photocatalytic degradation activity and their catalytic mechanism. *Micromachines*, 10, 503. https://doi.org/10.3390/mi10080503

Djurišić, A. B., He, Y., & Ng, A. M. C. (2020). Visible-light photocatalysts: Prospects and challenges. *APL Materials*, 8. https://doi.org/10.1063/1.5140497

Douglas, A., Carter, R., Li, M., & Pint, C. L. (2018). Toward small-diameter carbon nanotubes synthesized from captured carbon dioxide: Critical role of catalyst coarsening. *ACS Applied Materials & Interfaces*, 10, 19010–19018. https://doi.org/10.1021/acsami.8b02834

Dubey, A., Dave, S., Lakhani, M., & Sharma, A. (2016). Applications of graphene for communication, electronics and medical fields: A review. In *2016 International Conference on Electrical, Electronics, and Optimization Techniques (ICEEOT)* (pp. 2435–2439). IEEE.

Dubey, P. K., Hong, J., Lee, K., & Singh, P. (2023). Graphene-based materials: Synthesis and applications. In D.K. Singh, S. Singh, & P. Singh (Eds.), *Nanomaterials* (pp. 59–84). Springer Nature Singapore, Singapore.

Eatemadi, A., Daraee, H., Karimkhanloo, H. et al. (2014). Carbon nanotubes: Properties, synthesis, purification, and medical applications. *Nanoscale Research Letters*, 9, 393. https://doi.org/10.1186/1556-276X-9-393

Fang, S., Rahaman, M., Bharti, J. et al. (2023). Photocatalytic CO_2 reduction. *Nature Reviews Methods Primers*, 3, 61. https://doi.org/10.1038/s43586-023-00243-w

Gore, J. P. & Sane, A. (2011). Flame synthesis of carbon nanotubes. In S. Yellampalli (Ed.), *Carbon Nanotubes - Synthesis, Characterization, Applications* (pp. 147–170). InTech.

Guo, S., Cai, Z., Yang, Y., & Ding, Q. (2024). Beware of graphene's huge and hidden environmental costs. *Nature*, 629, 286–286. https://doi.org/10.1038/d41586-024-01339-3

Gupta, N., Gupta, S. M., & Sharma, S. K. (2019). Carbon nanotubes: Synthesis, properties and engineering applications. *Carbon Letters*, 29, 419–447. https://doi.org/10.1007/s42823-019-00068-2

Hamzah, N., Yasin, M. F. M., Yusop, M. Z. M. et al. (2017). Rapid production of carbon nanotubes: A review on advancement in growth control and morphology manipulations of flame synthesis. *Journal of Materials Chemistry A*, 5, 25144–25170. https://doi.org/10.1039/C7TA08668G

Han, Q., Chen, N., Zhang, J., & Qu, L. (2017). Graphene/graphitic carbon nitride hybrids for catalysis. *Materials Horizons*, 4, 832–850. https://doi.org/10.1039/C7MH00379J

Hasnain, M. S. & Nayak, A. K. (2019). Classification of carbon nanotubes. In *Carbon Nanotubes for Targeted Drug Delivery* (pp. 11–15). Springer.

Hassaan, M. A., El-Nemr, M. A., Elkatory, M. R. et al. (2023). Principles of photocatalysts and their different applications: A review. *Topics in Current Chemistry*, 381, 31. https://doi.org/10.1007/s41061-023-00444-7

He, F., Jeon, W., & Choi, W. (2021). Photocatalytic air purification mimicking the self-cleaning process of the atmosphere. *Nature Communications*, 12, 2528. https://doi.org/10.1038/s41467-021-22839-0

Hu, S., Liu, J., Yang, J. et al. (2011). Laser synthesis and size tailor of carbon quantum dots. *Journal of Nanoparticle Research*, 13, 7247–7252. https://doi.org/10.1007/s11051-011-0638-y

Huang, R., Li, X., Gao, W. et al. (2021). Recent advances in photocatalytic nitrogen fixation: from active sites to ammonia quantification methods. *RSC Advances*, 11, 14844–14861. https://doi.org/10.1039/D0RA10439F

Ibrahim, A., Klopocinska, A., Horvat, K., & Abdel Hamid, Z. (2021). Graphene-based nanocomposites: Synthesis, mechanical properties, and characterizations. *Polymers (Basel)*, 13, 2869. https://doi.org/10.3390/polym13172869

Iijima, S. (1991). Helical microtubules of graphitic carbon. *Nature*, 354, 56–58. https://doi.org/10.1038/354056a0

Illarionov, G. A., Morozova, S. M., Chrishtop, V. V. et al. (2020). Memristive TiO_2: Synthesis, technologies, and applications. *Frontiers in Chemistry*, 8. https://doi.org/10.3389/fchem.2020.00724

Janas, D. (2020). Perfectly imperfect: A review of chemical tools for exciton engineering in single-walled carbon nanotubes. *Materials Horizons*, 7, 2860–2881. https://doi.org/10.1039/D0MH00845A

Jeon, I.Y., Wook, D., Ashok, N., & Baek, J.-B. (2011). Functionalization of carbon nanotubes. In S. Yellampalli (Ed.), *Carbon Nanotubes - Polymer Nanocomposites*. InTech.

Jiang, J., Feng, T., Cheng, X. et al. (2006). Synthesis and growth mechanism of Fe-catalyzed carbon nanotubes by plasma-enhanced chemical vapor deposition. *Nuclear Instruments and Methods in Physics Research Section B: Beam Interactions with Materials and Atoms*, 244, 327–332. https://doi.org/10.1016/j.nimb.2005.10.012

Jung, H., Sapner, V. S., Adhikari, A. et al. (2022). Recent progress on carbon quantum dots based photocatalysis. *Frontiers in Chemistry*, 10. https://doi.org/10.3389/fchem.2022.881495

Kandasamy, G. (2019). Recent advancements in doped/Co-doped carbon quantum dots for multi-potential applications. *C*, 5, 24. https://doi.org/10.3390/c5020024

Kandy, M. M. (2020). Carbon-based photocatalysts for enhanced photocatalytic reduction of CO_2 to solar fuels. *Sustainable Energy Fuels*, 4, 469–484. https://doi.org/10.1039/C9SE00827F

Karthikeyan, A. & Mariappan, R. (2023). Characterization and electrochemical performance of Mn-doped Co_3O_4 nanoparticles for supercapacitor applications. *Journal of Materials Science: Materials in Electronics*, 34, 2092. https://doi.org/10.1007/s10854-023-11493-5

Khan, M. E., Khan, M. M., & Cho, M. H. (2017). Ce^{3+}-ion, surface oxygen vacancy, and visible light-induced photocatalytic dye degradation and photocapacitive performance of CeO_2-graphene nanostructures. *Scientific Reports*, 7, 5928. https://doi.org/10.1038/s41598-017-06139-6

Kim, M. H., Lee, S. G., Kim, K. S. et al. (2018). Environmental disinfection with photocatalyst as an adjunctive measure to control transmission of methicillin-resistant Staphylococcus aureus: A prospective cohort study in a high-incidence setting. *BMC Infectious Diseases*, 18, 610. https://doi.org/10.1186/s12879-018-3555-1

Koe, W. S., Lee, J. W., Chong, W. C. et al. (2020). An overview of photocatalytic degradation: photocatalysts, mechanisms, and development of photocatalytic membrane. *Environmental Science and Pollution Research*, 27, 2522–2565. https://doi.org/10.1007/s11356-019-07193-5

Kong, J., Wei, Y., Zhou, F. et al. (2024). Carbon quantum dots: Properties, preparation, and applications. *Molecules*, 29, 2002. https://doi.org/10.3390/molecules29092002

Krishnan, A., Swarnalal, A., Das, D. et al. (2024). A review on transition metal oxides based photocatalysts for degradation of synthetic organic pollutants. *Journal of Environmental Sciences*, 139, 389–417. https://doi.org/10.1016/j.jes.2023.02.051

Kumar Jagadeesan, A., Thangavelu, K., & Dhananjeyan, V. (2020). Carbon nanotubes: Synthesis, properties and applications. In P. Pham, P. Goel, S. Kumar, & K. Yadav (Eds.), *21st Century Surface Science - A Handbook*. IntechOpen.

Kumar, M. (2011). Carbon nanotube synthesis and growth mechanism. In S. Yellampalli (Ed.), *Carbon Nanotubes - Synthesis, Characterization, Applications* (pp. 333–356). InTech.

Kuvarega, A. T., Selvaraj, R., & Mamba, B. B. (2021). Graphene-based photocatalytic materials: An overview. In S. Balakumar, V. Keller, & M. V. Shankar (Eds.), *Nanostructured Materials for Environmental Applications* (pp. 433–454). Springer International Publishing, Cham.

Lashgari, M. & Soodi, S. (2020). CO_2 conversion into methanol under ambient conditions using efficient nanocomposite photocatalyst/solar-energy materials in aqueous medium. *RSC Advances*, 10, 15072–15078. https://doi.org/10.1039/D0RA01733G

Lashgari, M., Elyas-Haghighi, P., & Takeguchi, M. (2017). A highly efficient pn junction nanocomposite solar-energy-material [nano-photovoltaic] for direct conversion of water molecules to hydrogen solar fuel. *Solar Energy Materials and Solar Cells*, 165, 9–16. https://doi.org/10.1016/j.solmat.2017.02.028

Li, J., Zeng, H., Zeng, Z. et al. (2021). Promising graphene-based nanomaterials and their biomedical applications and potential risks: A comprehensive review. *ACS Biomaterials Science & Engineering*, 7, 5363–5396. https://doi.org/10.1021/acsbiomaterials.1c00875

Li, W., Wei, Z., Wang, B. et al. (2020). Carbon quantum dots enhanced the activity for the hydrogen evolution reaction in ruthenium-based electrocatalysts. *Materials Chemistry Frontiers*, 4, 277–284. https://doi.org/10.1039/C9QM00618D

Li, Y. & Du, H. (2023). Engineering graphitic carbon nitride for next-generation photodetectors: A mini review. *RSC Advances*, 13, 25968–25977. https://doi.org/10.1039/D3RA04051H

Li, Y., Yin, K., Wang, L. et al. (2018). Engineering MoS_2 nanomesh with holes and lattice defects for highly active hydrogen evolution reaction. *Applied Catalysis B: Environmental*, 239, 537–544. https://doi.org/10.1016/j.apcatb.2018.05.080

Lin, H., Xiao, Y., Geng, A. et al. (2022). Research progress on graphitic carbon nitride/metal oxide composites: Synthesis and photocatalytic applications. *International Journal of Molecular Sciences (IJMS)*, 23, 12979. https://doi.org/10.3390/ijms232112979

Linsebigler, A. L, Lu, G., & Yates, J. T. (1995). Photocatalysis on TiO_2 surfaces: Principles, mechanisms, and selected results. *Chemical Reviews*, 95, 735–758. https://doi.org/10.1021/cr00035a013

Liu, H., Wang, X., Wang, H., & Nie, R. (2019). Synthesis and biomedical applications of graphitic carbon nitride quantum dots. *Journal of Materials Chemistry B*, 7, 5432–5448. https://doi.org/10.1039/C9TB01410A

Liu, N., Tang, Q., Huang, B., & Wang, Y. (2021). Graphene synthesis: Method, exfoliation mechanism and large-scale production. *Crystals*, 12, 25. https://doi.org/10.3390/cryst12010025

Lou, M., Bao, J. L., Zhou, L. et al. (2022). Direct H_2S decomposition by plasmonic photocatalysis: Efficient remediation plus sustainable hydrogen production. *ACS Energy Letters*, 7, 3666–3674. https://doi.org/10.1021/acsenergylett.2c01755

Ma, Y., Dichiara, A. B., He, D. et al. (2016). Control of product nature and morphology by adjusting the hydrogen content in a continuous chemical vapor deposition process for carbon nanotube synthesis. *Carbon N Y*, 107, 171–179. https://doi.org/10.1016/j.carbon.2016.05.060

Ma, Y., Lin, L., Takata, T. et al. (2023). A perspective on two pathways of photocatalytic water splitting and their practical application systems. *Physical Chemistry Chemical Physics*, 25, 6586–6601. https://doi.org/10.1039/D2CP05427B

Madurani, K. A., Suprapto, S., Machrita, N. I. et al. (2020). Progress in graphene synthesis and its application: History, challenge and the future outlook for research and industry. *ECS Journal of Solid State Science and Technology*, 9, 093013. https://doi.org/10.1149/2162-8777/abbb6f

Magesh, V., Sundramoorthy, A. K., & Ganapathy, D. (2022). Recent advances on synthesis and potential applications of carbon quantum dots. *Frontiers in Chemistry*, 9. https://doi.org/10.3389/fmats.2022.906838

Mathew, J., John, N., & Mathew, B. (2023). Graphene oxide-incorporated silver-based photocatalysts for enhanced degradation of organic toxins: A review. *Environmental Science and Pollution Research*, 30, 16817–16851. https://doi.org/10.1007/s11356-022-25026-w

Molaei, M. J. (2024). Synthesis and application of carbon quantum dots derived from carbon black in bioimaging. *Journal of Fluorescence*, 34, 213–226. https://doi.org/10.1007/s10895-023-03252-w

Mombeshora, E. T. & Muchuweni, E. (2023). The current impacts and future prospects of graphene derivatives in polymer-based supercapacitors. *Discover Materials*, 3, 32. https://doi.org/10.1007/s43939-023-00068-0

Mou, Z., Meng, T., Li, J. et al. (2024). Significantly enhanced photocatalytic hydrogen production performance of MoS_2/CNTs/CdS with carbon nanotubes as the charge mediators. *International Journal of Hydrogen Energy*, 51, 748–757. https://doi.org/10.1016/j.ijhydene.2023.07.060

Mubarak, N. M., Abdullah, E. C., Jayakumar, N. S., & Sahu, J. N. (2014). An overview on methods for the production of carbon nanotubes. *Journal of Industrial & Engineering Chemistry*, 20, 1186–1197. https://doi.org/10.1016/j.jiec.2013.09.001

Muley, S. & Ravindra, N. M. (2019). Graphene: Properties, synthesis, and applications. In M. I. Pech-Canul & N. M. Ravindra (Eds.), *Semiconductors* (pp. 219–332). Springer International Publishing, Cham.

Murjani, B. O., Kadu, P. S., Bansod, M. et al. (2022). Carbon nanotubes in biomedical applications: Current status, promises, and challenges. *Carbon Letters*, 32, 1207–1226. https://doi.org/10.1007/s42823-022-00364-4

Nammahachak, N., Aup-Ngoen, K. K., Asanithi, P. et al. (2022). Hydrothermal synthesis of carbon quantum dots with size tunability via heterogeneous nucleation. *RSC Advances*, 12, 31729–31733. https://doi.org/10.1039/D2RA05989D

Natarajan, T. S., Lee, J. Y., Bajaj, H. C. et al. (2017). Synthesis of multiwall carbon nanotubes/TiO_2 nanotube composites with enhanced photocatalytic decomposition efficiency. *Catalysis Today*, 282, 13–23. https://doi.org/10.1016/j.cattod.2016.03.018

Ng, S., Lau, M. Y. L., & Ong, W. (2021). Engineering layered double hydroxide-based photocatalysts toward artificial photosynthesis: State-of-the-art progress and prospects. *Solar RRL*, 5(6), 2000535. https://doi.org/10.1002/solr.202000535

Nguyen, A. L., Liu, W., Khor, K. A. et al. (2022). The emergence of graphene research topics through interactions within and beyond. *Quantitative Science Studies*, 3, 457–484. https://doi.org/10.1162/qss_a_00193

Omer, K. M., Tofiq, D. I., & Hassan, A. Q. (2018). Solvothermal synthesis of phosphorus and nitrogen doped carbon quantum dots as a fluorescent probe for iron(III). *Microchimica Acta*, 185, 466. https://doi.org/10.1007/s00604-018-3002-4

Ōsawa, E. (2015). Fullerenes, carbon nanotubes and carbon nano-onions—They could be plasma crystals. *Forma*, 30(2), S1–S5. https://doi.org/10.5047/forma.2015.s002

Panachikkool, M. & Pandiyarajan, T. (2024). Graphene quantum dots as game-changers in solar cell technology: A review of synthetic processes and performance enhancement. *Carbon Letters*, 34, 445–475. https://doi.org/10.1007/s42823-023-00617-w

Pandiyan, R., Mahalingam, S., & Ahn, Y.-H. (2019). Antibacterial and photocatalytic activity of hydrothermally synthesized SnO_2 doped GO and CNT under visible light irradiation. *Journal of Photochemistry and Photobiology B: Biology*, 191, 18–25. https://doi.org/10.1016/j.jphotobiol.2018.12.007

Pierson, H. O. (1999). The CVD of the allotropes of carbon. In *Handbook of Chemical Vapor Deposition (CVD)* (pp. 185–216). Elsevier.

Purabgola, A., Mayilswamy, N., & Kandasubramanian, B. (2022). Graphene-based TiO_2 composites for photocatalysis & environmental remediation: Synthesis and progress. *Environmental Science and Pollution Research*, 29, 32305–32325. https://doi.org/10.1007/s11356-022-18983-9

Qin, X., Ola, O., Zhao, J. et al. (2022). Recent progress in graphene-based electrocatalysts for hydrogen evolution reaction. *Nanomaterials*, 12, 1806. https://doi.org/10.3390/nano12111806

Ramos, P. G., Rivera, H., Sánchez, L. A. et al. (2024). Graphene-based semiconductors for photocatalytic degradation of organic dye from wastewater: A comprehensive review. *Water, Air, & Soil Pollution*, 235, 292. https://doi.org/10.1007/s11270-024-07111-7

Rasool, M. A., Sattar, R., Anum, A. et al. (2022). An insight into carbon nanomaterial-based photocatalytic water splitting for green hydrogen production. *Catalysts*, 13, 66. https://doi.org/10.3390/catal13010066

Rilda, Y., Arief, S., Agustien, A. et al. (2022). Growth inhibition of bacterial pathogens by photo-catalyst process of nano-alloys FeCuNi doped TiO_2 under ultraviolet irradiation. *Heliyon*, 8, e10611. https://doi.org/10.1016/j.heliyon.2022.e10611

Romero, M. P., Alves, F., Stringasci, M. D. et al. (2021). One-pot microwave-assisted synthesis of carbon dots and in vivo and in vitro antimicrobial photodynamic applications. *Frontiers in Microbiology*, 12. https://doi.org/10.3389/fmicb.2021.662149

Roopan, S. M., Prakash, S. H., Manjupriya, R. et al. (2024). Biomass-derived carbon quantum dots-supported metal oxide composite for the photocatalytic degradation of toxic pollutants. *Biomass Conversion and Biorefinery*. https://doi.org/10.1007/s13399-024-05422-7

Saeed, M., Alshammari, Y., Majeed, S. A., & Al-Nasrallah E (2020). Chemical vapour deposition of graphene—Synthesis, characterisation, and applications: A review. *Molecules*, 25, 3856. https://doi.org/10.3390/molecules25173856

Saifuddin, N., Raziah, A. Z., & Junizah, A. R. (2013). Carbon nanotubes: A review on structure and their interaction with proteins. *Journal of Chemistry*, 2013(1), 676815. https://doi.org/10.1155/2013/676815

Sajna, M. S., Simon, S. M., Unnikrishnan, N. V., & Sadasivuni, K. K. (2022). An overview of graphene-based 2D/3D nanostructures for photocatalytic applications. *Topics in Catalysis*, 68, 1498–1522. https://doi.org/10.1007/s11244-021-01539-5

Saleh, T. A. (2013). The role of carbon nanotubes in enhancement of photocatalysis. In S. Suzuki (Ed.), *Syntheses and Applications of Carbon Nanotubes and Their Composites* (pp. 479–493). InTech.

Santhiran, A., Iyngaran, P., Abiman, P., & Kuganathan, N. (2021). Graphene synthesis and its recent advances in applications—A review. *C*, 7, 76. https://doi.org/10.3390/c7040076

Sayapina, A. G., Yakusheva, A. S., & Kuznetsov, D. V. (2023). Investigation the functional properties of metal-containing carbon quantum dots. *Nanobiotechnology Reports*, 18, S17–S24. https://doi.org/10.1134/S2635167623600360

Schmaltz, T., Wormer, L., Schmoch, U., & Döscher, H. (2024). Graphene roadmap briefs (No. 3): Meta-market analysis 2023. *2D Materials*, 11, 022002. https://doi.org/10.1088/2053-1583/ad1e78

Sharma, R., Jaiswal, S., Chauhan, R. et al. (2024). Applications of graphene-based photocatalysts for efficient functionalized degradation of some common antibiotics. *Inorganic Chemistry Communications*, 168, 112941. https://doi.org/10.1016/j.inoche.2024.112941

Shoukat, R. & Khan, M. I. (2021). Carbon nanotubes: A review on properties, synthesis methods and applications in micro and nanotechnology. *Microsystem Technologies*, 27, 4183–4192. https://doi.org/10.1007/s00542-021-05211-6

Shoukat, R. & Khan, M. I. (2022). Carbon nanotubes/nanofibers (CNTs/CNFs): A review on state of the art synthesis methods. *Microsystem Technologies*, 28, 885–901. https://doi.org/10.1007/s00542-022-05263-2

Sobamowo, M. G., Akanmu, J. O., Adeleye, O. A. et al. (2021). Coupled effects of magnetic field, number of walls, geometric imperfection, temperature change, and boundary conditions on nonlocal nonlinear vibration of carbon nanotubes resting on elastic foundations. *Forces in Mechanics*, 3, 100010. https://doi.org/10.1016/j.finmec.2021.100010

Song, B., Xu, P., Zeng, G. et al. (2018). Carbon nanotube-based environmental technologies: the adopted properties, primary mechanisms, and challenges. *Reviews in Environmental Science and Bio/Technology*, 17, 571–590. https://doi.org/10.1007/s11157-018-9468-z

Sowmiya, K. C. & Vijayalakshmi, K. A. (2024). Enrich the competency of $MnCo_2O_4$ electrode exposed under air plasma to achieve superior coulombic efficiency in super capacitors. *Applied Physics A*, 130, 583. https://doi.org/10.1007/s00339-024-07738-7

Stroyuk, O., Raievska, O., & Zahn, D. R. T. (2020). Graphitic carbon nitride nanotubes: A new material for emerging applications. *RSC Advances*, 10, 34059–34087. https://doi.org/10.1039/D0RA05580H

Subramanian, V. R., Sarker, S., Yu, B. et al. (2013). TiO_2 nanotubes and its composites: Photocatalytic and other photo-driven applications. *Journal of Materials Research*, 28, 280–293. https://doi.org/10.1557/jmr.2012.392

Tao, Z., Zhao, Y., Wang, Y., & Zhang, G. (2024). Recent advances in carbon nanotube technology: Bridging the gap from fundamental science to wide applications. *C- Journal of Carbon Research*, 10, 69. https://doi.org/10.3390/c10030069

Tontapha, S., Uppachai, P., & Amornkitbamrung, V. (2021). Fabrication of functional materials for dye-sensitized solar cells. *Frontiers in Energy Research*, 9. https://doi.org/10.3389/fenrg.2021.641983

Trivedi, P. A., Naik, J. B., & Patil, P. B. (2024). Exploring graphene and its derivatives for various applications: Photocatalysis. *Chemical Papers*, 78, 5705–5722. https://doi.org/10.1007/s11696-024-03475-6

Upadhyay, R. K., Soin, N., & Roy, S. S. (2014). Role of graphene/metal oxide composites as photocatalysts, adsorbents and disinfectants in water treatment: A review. *RSC Advances*, 4, 3823–3851. https://doi.org/10.1039/C3RA45013A

Urade, A. R., Lahiri, I., & Suresh, K. S. (2023). Graphene properties, synthesis and applications: A review. *JOM: The journal of the Minerals, Metals & Materials Society*, 75, 614–630. https://doi.org/10.1007/s11837-022-05505-8

Venkatesh, D., Deepthi, G., Mangatayaru, K. G., & Noorjahan, M. (2023). Microwave-assisted synthesis, characterization, photocatalytic degradation of antibiotics, and fluorometric selective sensing activity of g-C_3N_4 supported CuO composites. *Journal of Fluorescence*, 33, 987–1002. https://doi.org/10.1007/s10895-022-03125-8

Voiry, D., Salehi, M., Silva, R. et al. (2013). Conducting MoS_2 nanosheets as catalysts for hydrogen evolution reaction. *Nano Letters*, 13, 6222–6227. https://doi.org/10.1021/nl403661s

Wadkar, D. A., Jadhav, K. B., Jituri, S. D., & Mujawar, S. H. (2024). Carbon quantum dots decorated NiCo layered double hydroxide for electrochemical oxygen evolution reaction and supercapacitor application. *Journal of Materials Science: Materials in Electronics*, 35, 1037. https://doi.org/10.1007/s10854-024-12793-0

Wang, A., Wang, C., Fu, L. et al. (2017a). Recent advances of graphitic carbon nitride-based structures and applications in catalyst, sensing, imaging, and LEDs. *Nano-Micro Letters*, 9, 47. https://doi.org/10.1007/s40820-017-0148-2

Wang, Q., Li, Y., Huang, F. et al. (2023). Recent advances in g-C_3N_4-based materials and their application in energy and environmental sustainability. *Molecules*, 28, 432. https://doi.org/10.3390/molecules28010432

Wang, X., Chen, Y., Zheng, B. et al. (2017b). Significant enhancement of photocatalytic activity of multi-walled carbon nanotubes modified WSe_2 composite. *Materials Letters*, 197, 67–70. https://doi.org/10.1016/j.matlet.2017.03.150

Wang, X., Feng, Y., Dong, P., & Huang, J. (2019). A mini review on carbon quantum dots: Preparation, properties, and electrocatalytic application. *Frontiers in Chemistry*, 7. https://doi.org/10.3389/fchem.2019.00671

Wang, Y. & Hu, A. (2014). Carbon quantum dots: Synthesis, properties and applications. *Journal of Materials Chemistry C*, 2, 6921. https://doi.org/10.1039/C4TC00988F

Wu, J., Chen, T., Ge, S. et al. (2023). Synthesis and applications of carbon quantum dots derived from biomass waste: A review. *Environmental Chemistry Letters*, 21, 3393–3424. https://doi.org/10.1007/s10311-023-01636-9

Yadav, N., Gaikwad, R. P., Mishra, V., & Gawande, M. B. (2022). Synthesis and photocatalytic applications of functionalized carbon quantum dots. *Bulletin of the Chemical Society of Japan*, 95, 1638–1679. https://doi.org/10.1246/bcsj.20220250

Yadav, P. K., Chandra, S., Kumar, V. et al. (2023). Carbon quantum dots: Synthesis, structure, properties, and catalytic applications for organic synthesis. *Catalysts*, 13, 422. https://doi.org/10.3390/catal13020422

Yahyazadeh, A., Nanda, S., & Dalai, A. K. (2024). Carbon nanotubes: A review of synthesis methods and applications. *Reactions*, 5, 429–451. https://doi.org/10.3390/reactions5030022

Yan, F., Jiang, Y., Sun, X. et al. (2018). Surface modification and chemical functionalization of carbon dots: A review. *Microchimica Acta*, 185, 424. https://doi.org/10.1007/s00604-018-2953-9

Yan, Y., Nashath, F. Z., Chen, S. et al. (2020). Synthesis of graphene: Potential carbon precursors and approaches. *Nanotechnology Reviews*, 9, 1284–1314. https://doi.org/10.1515/ntrev-2020-0100

Yang, M.Q. & Xu, Y.J. (2016). Photocatalytic conversion of CO_2 over graphene-based composites: Current status and future perspective. *Nanoscale Horizons*, 1, 185–200. https://doi.org/10.1039/C5NH00113G

Yang, R., Fan, Y., Zhang, Y. et al. (2023). 2D transition metal dichalcogenides for photocatalysis. *Angewandte Chemie International Edition*, 62. https://doi.org/10.1002/anie.202218016

Yang, S. (2023). Properties, applications, and prospects of carbon nanotubes in the construction industry. *Architecture, Structures and Construction*, 3, 289–298. https://doi.org/10.1007/s44150-023-00090-z

Yang, Y., Kim, N. D., Varshney, V. et al. (2017). In situ mechanical investigation of carbon nanotube–graphene junction in three-dimensional carbon nanostructures. *Nanoscale*, 9, 2916–2924. https://doi.org/10.1039/C6NR09897E

Yunus, R. M., Rosman, N. N., & Shah, N. R. A. M. (2023). 3D graphene as a photocatalyst for water splitting. In R. K. Gupta (Eds.), *3D Graphene. Carbon Nanostructures* (pp. 359–374). Springer, Cham.

Zafar, M., Iqbal, T., Fatima, S. et al. (2022). Carbon nanotubes for production and storage of hydrogen: Challenges and development. *Chemical Papers*, 76, 609–625. https://doi.org/10.1007/s11696-021-01922-2

Zare, P., Aleemardani, M., Seifalian, A. et al. (2021). Graphene oxide: Opportunities and challenges in biomedicine. *Nanomaterials*, 11, 1083. https://doi.org/10.3390/nano11051083

Zhang, J., Dai, M., Zhang, S. et al. (2022). Recent progress on carbon-nanotube-based materials for photocatalytic applications: A review. *Solar RRL*, 6(9), 2200243. https://doi.org/10.1002/solr.202200243

Zhou, D., Anoshkina, E. V., Chow, L., & Chai, G. (2006). Synthesis of carbon nanotubes by electrochemical deposition at room temperature. *Carbon N Y*, 44, 1013–1016. https://doi.org/10.1016/j.carbon.2005.11.006

Zurutuza, A. & Marinelli, C. (2014). Challenges and opportunities in graphene commercialization. *Nature Nanotechnology*, 9, 730–734.

9

Self-assembled Carbon Nanophotocatalysts: An Eco-friendly Strategy for Environment and Energy Conservation

Dharitri Borah[1], Abhilash Borgohain[1], Jayashree Rout[2], and Thajuddin Nooruddin[3]*

[1] *Department of Environmental Science, Arunachal University of Studies, Namsai, Arunachal Pradesh, India*
[2] *Department of Ecology and Environmental Science, Assam University, Silchar, Assam, India*
[3] *B.S. Abdur Rahman Crescent Institute of Science & Technology, Seethakathi Estate, Vandalur, Chennai, Tamil Nadu, India*

* Corresponding author: dhbtna@gmail.com

9.1 Introduction

Nanomaterials are of the size range from 1 to 100 nm and present at least one dimension with a wide range of applications such as catalysis, conductor, delivery vehicle, etc. (Tonelli et al., 2023). Nanoscience/nanotechnology deals with nanoparticles (NPs) of characteristic chemical composition, and dimension/sizes with controlled monodispersity (Rajeshkumar et al., 2013). NPs are constructed in several forms like carbon dots (CDs), carbon nanotubes (CNTs), nanocomposites (NCs), zerovalent NPs (ZNPs), and metal oxide NPs (MONPs) that are useful for energy generation, space, defense, agriculture, biomedical applications, communication systems, as well as environmental remediation (Usmani et al., 2017; Chaudhary et al., 2020; Abualnaja et al., 2021; Alprol et al., 2023a). Also, NPs are classified as carbon/lipid/polymeric, ceramic, and semiconductors. Pollution is caused by human intervention, and is largely contributed by rapid industrialization (Tonelli et al., 2023). Pollutants contaminating air, soil, and water are diverse in their chemical (organic dyes/herbicides; persistent), inorganic (heavy metals), and pathogenic characteristics. Persistent pollutants accumulate and biomagnify, causing toxic effects to plants, animals, and humans. NPs cause environmental remediation by dint of different chemical characteristics participating in pollutant removal, adsorption, and

Photocatalysts based on Carbon Scaffolds: Environmental and Industrial Applications, First Edition. Edited by Ramesh Oraon et al.
Published 2026 by Wiley-VCH GmbH.

converting into less harmful pollutants. NPs are also known to remove dyes like malachite green/crystal violet from industrial effluents, which are genotoxic, carcinogenic together with mutagenic where their chemical/physical synthesis results in persistent chemicals (reducing/stabilizing agents) harmful to the environment. For industrial effluents, photocatalysis-based remediation is preferred as per the ease of use, cost-effectiveness, quick oxidation, and less harmful emissions. Green synthesis decreases the formation of toxic byproducts and pollution while using nontoxic solvents and renewable feedstocks (Singh et al., 2018; Alprol et al., 2023b). Plasmonic photocatalysts have different synthetic routes and present different modes of interaction with pollutants, generating active radicals and identification (Leong et al., 2018). Often, plasmonic photocatalysts are developed by positioning noble metals onto the semiconductor surface during fabrication. Enhanced visible light absorption favors electron/hole generation. Also, it acts as a local electric field enhancer and prolongs the lifetime of the charge carrier providing a distinctive feature to plasmonic photocatalysts. C-based photocatalysts present remarkable physicochemical and electrochemical properties (Kandy, 2019). Photocatalytic activity is modulated in the visible light spectrum, which scavenges electrons from the photocatalyst surface by dint of its high electrical conductivity. Also, C-materials (CNT/graphene/graphite/carbon nitride/conducting polymers) are suitable for the photocatalyst dispersion appertaining to broad surface area enhancing catalytic reactions and CO_2 adsorption (Figure 9.1).

Solar energy-based water splitting photocatalysis for H_2-fuel is among the cost-effective, safe, sustainable, and eco-friendly strategies (Li et al., 2022a).

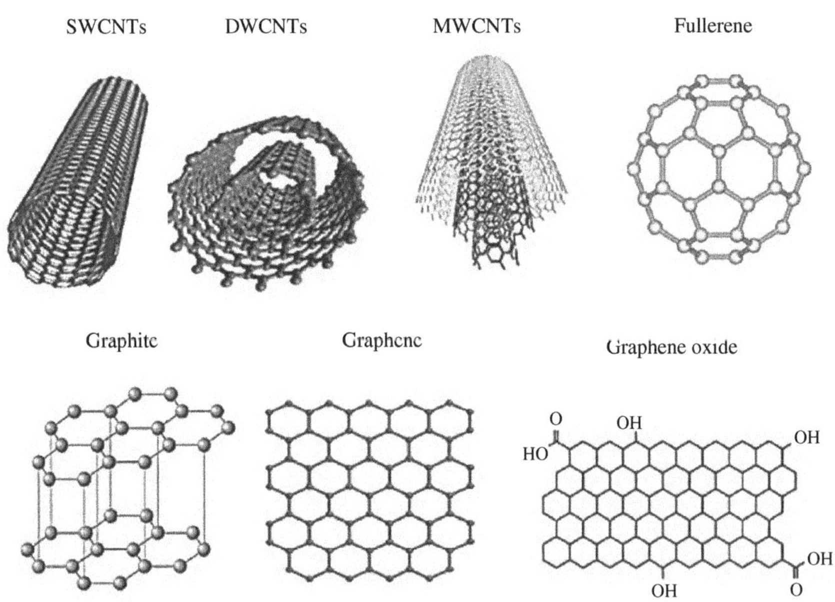

Figure 9.1 Different structures of C-nanomaterials. *Source*: Maryam Azizi-Lalabadi et al. (2020) / with permission of Elsevier.

C-nanostructures such as diamond, graphite, graphene, CNTs, and fullerene are applicable in generation of H_2 fuels. The C-based materials are highly available, possess appealing band structures, and are chemically stable. CDs are characterized by the advanced fluorescent C-nanomaterials with an ultrasmall size range of 2–10 nm. CDs are an alternative to the metal/semiconductor-based luminescent materials, presenting unique photophysical-chemical characteristics with luminescent activity, photostability, biocompatibility, and absorption spectra (continuous/broad). CDs have wide applications in bioimaging, optoelectronic applications, catalysis, and sensing (Li et al., 2022b). Technologies for storage of electrochemical energy and generation of energy from renewable resources (wind/solar/geothermal/tidal/hydro) have evolved into reliable, efficient, and practical technologies. Synthesis with contemporary technologies is of utmost importance for environmentally benign NPs (Rajeshkumar et al., 2013). Biogenic-green synthesis involves the use of biomass (algae/cyanobacteria/fungi/plants/animals), which is eco-friendly and sustainable. Algae/cyanobacteria present diverse applications (C-rich), such as in food, feeds, pharmaceuticals, fertilizer, biolubricants, and biofuel. Large numbers of algae/cyanobacteria-based NPs are synthesized as they are rich in secondary metabolites, fast-growing (higher growth rate), and industrially scalable (Chaudhary et al., 2020). Reportedly, NPs are developed from algae/cyanobacteria like *Padina* sp., *U. lactuca*, *Chlorella vulgaris*, *Anabaena variabilis*, *Spirulina platensis*, etc. (Bhuyar et al., 2020; Ismail et al., 2021; Aldayel et al., 2022; Bensy et al., 2022; Tonelli et al., 2023). Also, Au-NPs derived from the macroalgae *Sargassum horneri* reduced toxic dyes (methyl orange, methylene blue, rhodamine B) by catalysis (Waghchaure et al., 2022). Carcinogenic pollutant Pb was removed from pharmaceutical effluents by Au-NPs derived from *Nannochloropsis* sp. Algae-based synthesis of NPs is useful for a wide range of applications like wastewater remediation, mitigation of pollutants, antibacterial/fungal/cancer activity, electronics, cosmetics, optics, sensing, food packaging, chemical industries, mechanics, etc. (Khan et al., 2022). The contemporary C-based electrode materials are applied for renewable resource-based energy conversion. The C-based materials with a wide range of structures and morphologies are preferred owing to their huge abundance, cost-effectiveness, conductivity, high chemical/thermal stabilities, specific surface area, etc. Microalgae feedstocks for the synthesis of nanostructures are utilized as they have short growth phases with greater photosynthetic efficiency (Zhang et al., 2022a). Use of microalgae like *C. pyrenoidosa* avoids pre-crushing. It broadens the contact surface area and enables easy degradability by reason of the absence of lignin.

The elemental doping together with morphological adjustment as self-assembly of the NPs is quite advantageous in photocatalysis (Han et al., 2024). Doping increases the surface area, light utilization, and photocatalysis. Also, light absorption is largely dependent upon the morphology of the nanostructures. Dopants are useful for altering the crystallinity and surface defects, which ultimately provide active sites while modifying the electronic structures (Maliutina et al., 2023). Interestingly, NP-properties differ from bulk material. The synthesis of assembled NPs is dependent upon the controlling ability of interactions among the electronic/ magnetic/ optical properties (Nie et al., 2010). It augments energy availability, environmental friendliness together with high efficiency. Use of biomass as C-source is practically applicable, pertaining to its abundance, industrially cost-effective, nontoxic, biodegradable, and biocompatible characteristics (Cui et al., 2021). Often, lignin is incorporated as a hybrid composite material/C-based support for semiconductors, augmenting photoelectrochemical properties useful for pollutant degradation. The lignocellulosic biomass is among the most promising renewable resources, highly available from the wastes generated in the wood and agricultural sectors. Carbohydrates, lipids, proteins, and inorganic minerals are non-lignocellulosic resources obtained from algae, animal hair, feathers, bones, fish/seafood waste, manure, etc. (Li & Jiang, 2017). Facile synthesis of well-defined structured microcapsules/carbonized spheres was useful for photocatalytic degradation of pollutants. However, the functioning (sorption/photochemical reaction) of the hybrid material is dependent upon the porosity, surface area together with electrical conductivity. The starch extracted from the plant has diverse applications; however, it presents limitations due to its small pore volume together with specific surface area. The bio-assembly method increased its activity using quinoa starch as a C-source. An efficient H_2 evolving photocatalyst (CeO_2–MoxC/PSCA) was developed from starch and molybdenum salts (Yuan et al., 2025). Therefore, the present study aims to the construction of self-assembled C-nanophotocatalysts with emphasis on biogenic synthesis, including algae and cyanobacteria. Additionally, their applications in environmental remediation and generation of clean energy is also emphasized with their limitations and prospects.

9.2 Nanoparticles as Photocatalysts

Photocatalysts have originated from the term "photo"-light and "catalyst"-changes the kinetics but not the thermodynamics, which follows new reaction routes together with activation energy (Estrada et al., 2023). The C-nanomaterials have unique properties of hybridization allowing tailoring of morphology and structures (Mohapatra et al., 2023). The photocatalysis efficacy is improved by differently structured C-semiconductors. Among the photocatalysts, heterogenous forms have drawn considerable attention by dint of wide applications in energy generation together with environmental remediation (Estrada et al., 2023). Often, their light-absorbing capability is hindered by the faster recombination rate of electron-hole pairs. Also, a few photocatalysts are chemically unstable, and their activity is reduced in low light. The NC like graphene materials present more stability, mechanical strength, and surface area for reactivity. Composite materials like 2D–2D graphene-like g-C_3N_4/ultrathin $Bi_4O_5Br_2$ presented effective separation of charges with the required energy band (Ji et al., 2017). A maximum contact was generated between $Bi_4O_5Br_2$ (ultrathin) and C_3N_4 (graphene-like) to enhance the photocatalysis. The mechanically prepared CdS/g-C_3N_4 hybrid nano photocatalyst presented degradation (16–34%) of methyl orange (MO) dye in the initial step (Pourshirband et al., 2020). NC with graphene-like C_3N_4 nanosheets was produced to increase the photocatalysis of CDs under visible light. The conjugative π-structures (graphene-like carbon/C60) that could harvest natural/artificial radiations increased the photocatalytic efficacy. Also, metal sulfides can be combined with C-materials (Estrada et al., 2023). Graphitic nanostructures accept/shuttle the electrons (photogenerated) by the conjugative π-system that allows efficient separation of electron-hole pairs. Carbon nanotubes (CNTs-1D) of several micrometers/millimeters' length are constructed from graphene sheets by rolling on their axis (0.5–100 nm diameter). The dispersion of CNTs can be promoted by the aqueous surface modifiers like homopolymers, surfactants, block copolymers, etc. Surface oxidation generates functional groups like hydroxyl/carboxylic/carbonyl that enhance the dispersion of CNTs. The heterostructure photocatalysts like CDs/rGO and CDs/GO degrade pollutant organic dyes. These NCs could be recycled (four times) and are more effective than CDs alone. The CDs/graphene photocatalysts could degrade Cr (VI) and photo-oxidize alcohols (Estrada et al., 2023). Also, photocatalytic heterostructures could be constructed by coupling ZnS with C-nanomaterials. The specifically designed C-based photocatalysts improve CO_2 reduction, N_2-fixation, and organic synthesis reactions (Xu et al., 2022a). Carbon neutrality draws attention to reducing global climate change. The C-based photocatalysts are efficient materials for the generation of carbon-neutral H_2 energy (Li et al., 2022a). Oxygenated functional group associated graphene oxide (GO), graphene/graphene counter parts (GQDs, GO, rGO) are the potential NC by dint of their superior properties, matrix distribution, and electronic transmission efficiency. The photocatalytic devices are boosted to produce H_2 by absorbing more solar/artificial light. For increasing efficacy, different methods like ion doping, heterostructure construction, and functionalization of C-materials are accomplished (Li et al., 2022a). Numerous active/defective sites present on C-materials enable anchoring and generation of homogenous nanomaterials (Rasool et al., 2023). C-materials are abundant in nature. However, its lightweight nature necessitates the extraneous supportive materials. Hybrid semiconductors are more competent than the pure forms. The porous C-nanostructures (carbon doping/ nanotubes/diamond/ graphene/graphite/C60 fullerenes) are more stable, inert and adjustable in structure (Khalid et al., 2017; Pang et al., 2020). In visible light, reduced CDs are efficient photocatalysts for removing model dye pollutants viz. methylene blue (MB)/ methyl orange (MO)/their mixtures and other toxic substances (Duarah & Karak, 2017). Reduced CDs present better photocatalysis than CDs, and can be functionalized by various techniques like hydrothermal carbonization (Flores-Oña & Fullana, 2022). A remarkable photocatalysis activity degrading MB dye is reported from carbon quantum dots/nitride (CQDs/HpCN)-proton functionalized self-assembled composites under visible light (Jian et al., 2016). Also, graphitic carbon nitrides are useful for the generation of green energy together with environmental remediation (Kumar et al., 2020). The metal free S@g-C_3N_4/B@g-C_3N_4 synthesized from CDs/ rGO (self-assembled) could degrade chloramphenicol (10 mg/L) under visible light. g-C_3N_4 was dopped by different methods/band structures that generated junctions for photocatalysts with an effective Z-scheme. Photocatalyst-based conversion of CO_2 is sustainable and economical. rGO/pCN (metal-free) photocatalysts converted CO_2 into methane under reduced light (Ong et al., 2015). Biomass-based activated-C was loaded on TiO_2 for degrading reactive red 120/ofloxacin (Alghamdi et al., 2022).

The reusable nanocomposite of CZF@CQDs efficiently degraded pollutant dye in industrial effluents (Malitha et al., 2024). Biogenic-C is derived by pyrolysis (slow/fast), hydrothermal (high/low temperature; microwave), with disparate activation (physical/chemical) techniques (Son et al., 2021). The $Bi_4O_5Br_2$/g-C_3N_4 nanosheet-photocatalyst was stable and degraded RhB dye (Yi et al., 2020). The newly synthesized nanosheets performed better charge separation together with absorption (45%) of visible light. Semiconductor-based photocatalysis follows zero waste and acts as an industrially efficient technology for energy generation. Also, the efficacy of photocatalysts depends upon the time frame of e-h pair for reacting with the semiconductor surface. The nanoscale diameter enhances the transfer of e-h pair from the core to the surface by dint of its reduced diameter accelerating the redox rate (Hassaan et al., 2023).

9.3 Synthesis of Carbon Nanophotocatalysts

C-based nanophotocatalysts are derived by disparate approaches like chemical vapor deposition, pyrolysis, template-assisted, and supramolecular self-assembly methods (Figure 9.2). CDs are grouped into carbon quantum dots (CQDs; 1–5 nm), graphene quantum dots (GQDs; 1–20 nm), carbon nanodots (CNDs; 1–100 nm), polymer dots (PDs; 1–20 nm), and carbonized polymer dots (CPDs; 1–20 nm). CDs are synthesized by hydro/solvothermal, microwave-assisted, chemical oxidation/exfoliation, electrochemical oxidation/exfoliation, laser ablation/irradiation, ultrasonic-assisted, thermal pyrolysis, self-assembly, anchor/support-based techniques (Liu et al., 2021). However, every method used exhibits its own *pros* and *cons*. C/N can be adjusted in the high-temperature/pressure method with expensive equipment and energy consumption (Li et al., 2024). Equipment cost for the ionothermal method is low; however, it has intricate washing processes (downstream). Limitations in solvothermal method are by the feedstock/pressure conditions, despite its simple, requiring low-cost equipment. Microwave-assisted methods are time/energy efficient, limited by high temperature conditions.

Also, C-nanophotocatalysts were synthesized by the standard and solvent-assisted hydrothermal carbonization (Flores-Oña & Fullana, 2022). For the standard method, initially glucose (10 g) dissolved in distilled water (400 mL) was heated (200 °C) for one hour. The reactor was cooled down, and the biochar was separated by a membrane filter (0.45 μm). The ultracentrifugation was carried out at 20 000 rpm for 15 minutes. Then, the water was evaporated, and C-NPs were dried (two hours). For the solvent-assisted method, butyl acetate was used (100 mL) along with water (300 mL) and glucose (10 g). The blend was separated after ultrafiltration, and two phases were allowed to separate. The organic phase was ultracentrifuged (20 000 rpm) for 15 minutes. Finally, the C-NPs were dried. CDs were synthesized by oxidizing carbon fibers in nitric acid, followed by ultrafiltration (Li et al., 2022b). The CDs were added to nitric acid, heated (120 °C), and refluxed (48 hours). The CDs were separated and neutralized by NaOH. Membrane filters (0.22 μm) and centrifugal filters with cut-off sizes of 50, 30, 10, and 3 kDa were

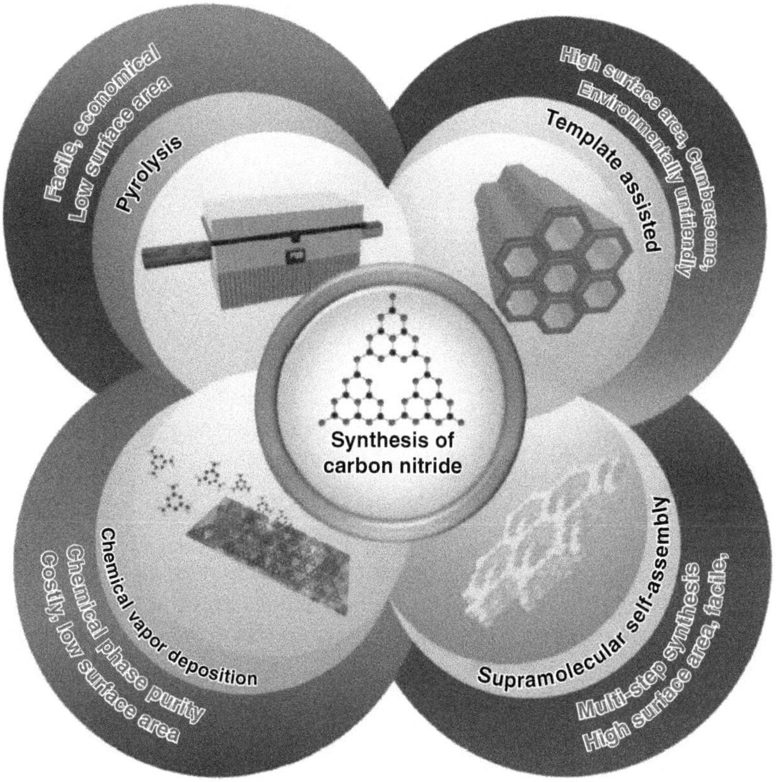

Figure 9.2 Comparative study of synthetic methods. *Source*: Dharmarajan et al. (2024) / John Wiley & Sons / CC BY-NC 4.0.

used for filtration. Thermally exfoliated graphitic carbon nitride (TE–g–C_3N_4) was synthesized using melamine (Ganesan et al., 2024). In brief, melamine (10 g) was placed in an alumina crucible, and raised to a temperature of 550 °C (5 °C/min) for four hours (static air), cooled down, and yellow-colored agglomerates were collected (Figure 9.3). TE–g–C_3N_4 was synthesized by heating pre-existing bulk g-C_3N_4 materials in an open alumina crucible employing disparate temperatures (450/500/550 °C) for two hours (5 °C/min).

9.3.1 Biogenic Synthesis of Nanophotocatalysts

C-NPs derived from different biogenic sources are used to construct nanostructures with disparate morphologies useful for environmental remediation and energy generation (Table 9.1). Disparate methods include self-assembly in solution/at interfaces, templating method, and assisted self-assembly. Biomass-derived C is evidenced for high-quality applications in energy conversion/storage together with environmental protection (Zhang et al., 2023). Biomass-derived NPs have different structures viz. zero-dimensional (spherical structures), 1D (nanotubes/

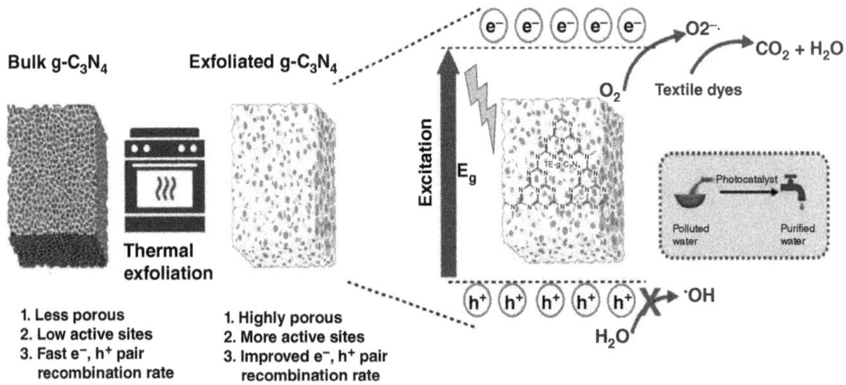

Figure 9.3 Preparation of TE-g-C_3N_4 and its photocatalytic mechanism. *Source*: Ganesan et al. (2024) / Springer Nature / CC BY 4.0.

nanofibers), 2D (nanosheets), and 3D (hierarchical structures). All these materials essentially have large surface area, activated ionic channels, and enhanced electronic conductivity (Yan et al., 2022). Tubular-shaped nanostructures were synthesized from hard carbon of cotton biomass. The electrochemical performance was enhanced, and the diffusion distance of Na^+ was reduced. Sustainable C could be obtained from biomaterials by hydrolysis of lignin at a temperature of 1000/1400 °C (Thomas et al., 2022). The lignin content and temperatures were responsible for the carbon quality/microstructure. Graphitization with a higher degree (1400 °C) and electrical conductivity (335 S/m) created a large surface area rooting for soda lignin. Wheat husk-porous C exhibited high specific surface area/mesopore volume obtained by chloride salt sealing and successive KOH activation (Wang et al., 2022a). Nitrogen-doped C-nanopaper can be obtained from chitin with nanoporous structures (Yeamsuksawat et al., 2023). Use of renewable raw materials/refined compounds is preferred since these methods do not use toxic chemicals and excessive energy (Chahal et al., 2021). Carbon from agro-wastes is useful for the synthesis of nanostructures with characteristic structural and functional properties (Fathy et al., 2020; Jirimali et al., 2022). The synthesis conditions regulate the chemical composition/functional groups, surface area, and pore size (Li et al., 2019; Ghosh et al., 2019; Xiong et al., 2020; Jirimali et al., 2022). The different methods followed for the construction of nanomaterials from the agro-wastes are pyrolysis, microwave, hydrothermal, and carbonization methods. Pyrolysis is a thermochemical method where C-materials are disintegrated into smaller components in the environment without oxygen. In microwave irradiation, irradiation frequency/temperature and time are finely controlled. The agricultural waste-derived microcellulose has wide applications in porous C-material synthesis by microwave irradiation. Hydrothermal processes are conducted in the presence of water and high pressure. In Microwave Hydrothermal Carbonization (MHC), microwaves are used to heat the hydrothermal unit for carbonization.

Table 9.1 Biogenic source-assisted generation and applications of NPs/photocatalysts.

Photocatalyst	Biogenic source	Synthesis method	Application	Reference
Modified g-C_3N_4	Biochar based (spent carbon-rich biomass)	Facile supramolecular self-assembly	Degradation of phenanthrene	Lin et al. (2022a)
TiO_2/potassium alginate–CNT photocatalysts (coating on cotton fabrics)	Alginate (brown algae-derived/cotton	Layered self-assembly technique	Degradation of gaseous/liquid pollutants/Formaldehyde/RhB	Wang et al. (2017b)
CDs	Kelp (brown algae)	Microwave irradiation	Selective detection of Co^{2+} (linear range:1–200 μmol/L; Limit: 0.39 μmol/L)	Zhao et al. (2019)
TiO_2/wood charcoal composite	Chinese honey locust spine wood	Bio-templated synthesis	Bisphenol removal	Luo et al. (2015)
AgI/UiO-66-NH @CW	Wood chip	Facile solvothermal and ion exchange deposition	Degradation of RhB (Model dye)	Liang et al. (2024)
Fe_3O_4@Dwood	Demethylated lignin	Polymerization	Organic pollutant-photochemical oxidation	Guo et al. (2023)
Fluorescent CDs	Onion waste	Autoclave	Selective/sensitive Fe^{3+} detection	Bandi et al. (2016)
CDs	*Chlorella pyrenoidosa* (microalga)	Hydrothermal	Iron detection	Zhang et al. (2022a)
CQDs	Waste lemon peel	Hydrothermal process-facile	Aqueous detection of Cr^{6+}/MB degradation (model pollutant)	Tyagi et al. (2016)
CQDs/nano-TiO_2 incorporated wood (QNIW)	*Cunninghamia lanceolata* (Lamb.) Hook	Green synthesis	Adsorption/photocatalysis for Cr^{6+} removal	Qi et al. (2023)
Nitrogen (N)/Phosphorus (P) co-doped CQDs	*Eleocharis dulcis*	Green hydrothermal strategy	Sensing applications	Bao et al. (2018)

(*Continued*)

Table 9.1 (Continued)

Photocatalyst	Biogenic source	Synthesis method	Application	Reference
TiO_2 based on wood templates	Poplar/Paulownia/Chinese fir/larch	Hydrothermal/ high-temperature calcining	Degradation of formaldehyde	Liu et al. (2020)
UiO-66/wood composite	Natural wood	Solvothermal method- (in-situ/ one step	Ofloxacin photodegradation	Shi et al. (2022)
Ternary Pt/Au/TiO_2-decorated plasmonic wood carbon	Basswood	Simple two-step calcination	Photodegradation of Tetracycline	Wang et al. (2019a)
CQDs@MIL-88B (Fe)	Poplar wood flour	Solvothermal method	Degradation of MB	Zhang et al. (2021)
Hybrid material Activated carbon (AC)-TiO_2 composite	Pistachio shells	Sol–gel method	Photodegradation of azo dye/RR 120/ofloxacin	Alghamdi et al. (2022)
Reduced biobased luminescent CDs	Leaves (*CIlocasia esculenta*, *Mesua ferrea*, tea) with *Syzygium aromaticum*-flower bud	Facile one-step hydrothermal synthetic route	MB, MO, MB/MO mixture, and organic contaminants (OCs)	Duarah and Karak (2017)
Cu_2O@carbon nanocapsules	Sodium alginate-Brown algae	Self-generated by calcination	Degradation of MR dye	Su et al. (2019b)

The mango peels were used to develop CQDs by the hydrothermal method (Malitha et al., 2024). Initially, peels were washed thoroughly and sun-dried (7 days). The dried peels were oven-heated for six hours at 80 °C, and crushed. Then 0.1 N H_2SO_4 (150 mL) was added for 10 minutes and washed. The sample was oven-dried at 100 °C for six hours. Then, sodium hypochlorite (100 mL) was added for four hours and repeatedly washed to reach pH 7. The sample was autoclaved (hydrothermal) for 10 hours at 180 °C after being washed with dichloromethane and then frozen for 48 hours. The frozen sample was lyophilized at −56 °C under a pressure of 0.01 mbar for 20 hours. Cobalt zinc ferrite-CQDs coated (CZF@CQDs) was synthesized by adding nano photocatalyst CZF to an aqueous NaOH (1 N) solution and stirred (30 minutes). Under cooling conditions (ice bath), CZF/CQDs were added with stirring (1:10 (w/v) dropwise, 20 minutes). To the mixture, an equal weight of ammonium peroxydisulfate dissolved in HCl (1 M) was added for five hours. The mixture was incubated overnight, filtered, washed (rectified spirit), and dried (70 °C) for a consistent mass. CDs are synthesized by top-down and bottom-up synthesis methods (Long et al., 2021). The top-down approach involves the conversion of large carbons (graphite/graphene/CNTs/activated carbon) into quantum-sized/fluorescent CQDs, laser ablation, and chemical/electrochemical oxidation. The bottom-up synthesis involves carbonization of small molecules by pyrolysis, hydrothermal, and microwave methods (Wang et al., 2017a; Long et al., 2021). In the oxidation process (top-down), single-walled CNTs were purified between nitric acid and arc-discharged soot under the oxidation process (Xu et al., 2004; Long et al., 2021). The electrochemical oxidation is a common avenue pertaining to the retention of high purity, yield, size modulation, reproducibility, and low cost (Hutton et al., 2017; Long et al., 2021). The hydrothermal method is widely used for bottom-up synthesis. This carbonization method is widely followed by dint of its low cost/abundant precursors. Although large instruments are not required, it requires a hydrothermal reactor maintaining high temperature/pressure for a prolonged reaction. Sufficient energy is accomplished by microwaves/electronic waves (300 MHz to 300 GHz) to break down the C-rich raw materials.

TiO_2-wood charcoal composite photocatalyst from the Chinese honey locust spine resource was (C-source) prepared by dip-sol-gel and calcination method (Luo et al., 2015). TiO_2 (10–20 nm), when dispersed in wood charcoal of the plant, presented wood-like characteristics. Zhang et al. (2024) reported C-based photocatalysts that are derived from woods like Balsa, Fir, Bass, Beech, etc. Although Biochar-C derived from pyrolysis has a strong affinity for adsorption, it cannot eliminate the pollutants alone in aqueous conditions. Often, chemical methods are followed to synthesize photocatalysts with low recyclability. Wood-biomass is cost-effective and accessible in huge abundance in nature. The composites derived from biochar and photocatalysts can improve the catalytic activity. The photodegradation/water treatment efficiency is higher compared to a single NP. A self-floating heterojunction was fabricated by the AgI/UiO-66-NH@carbonized wood (CW) (Liang et al., 2024). The CW increased the dispersibility and recyclability of photocatalysts. Also, disposal onion waste was utilized for the facile one-step synthesis of CDs with high fluorescent properties and a quantum yield of 28%. The simple autoclave technique was used and characterized by TEM, XRD, DLS, TGA, FTIR, UV-Visible, as well as fluorescent spectroscopy. The nanophotocatalysts exhibited aqueous dispersibility, fluorescence (excitation-dependent), and excellent stability in different pH values and ionic solutions (high strength) with continuous irradiation (Bandi et al., 2016). Photoluminescent (QY 14%) CDs (water-soluble) were generated from lemon peel waste by hydrothermal process (Tyagi et al., 2016). The synthesized particles were spherical in size (1–3 nm) and functionalized with oxygen. Compared with the photocatalytic activity of TiO_2 nanofibers, TiO_2–wsCQDs was more (2.5 times) efficient in photocatalysis of MB dye. The composite could be scaled up to a gram scale. In brief, lemon peels (5 g) were washed, dried (100 °C; 10 hours), treated with 100 mL H_2SO_4 (0.1 M), and washed. Then samples were filtered, dried again at 100 °C (four hours). Then sodium hypochlorite (150 mL) was added, incubated (four hours), and washed with water (pH 7). The peels were then autoclaved (Teflon-lined) for 200 °C (12 hours), cooled down, washed (dichloromethane), centrifuged (10 000 rpm for 30 minutes), and finally dried (100 °C). *Cunninghamia lanceolata* (raw) wood was used to prepare stable and separable (aphotolytic content) CQDs /nano-TiO incorporated wood (QNIW) with a reaction rate coefficient of 0.0661 minute (Qi et al., 2023).

Fluorescent high-QY CDs (water-soluble nitrogen/phosphorus [N/P] co-doped) were synthesized from *Eleocharis dulcis* following the hydrothermal method (Bao et al., 2018). Under hydrothermal conditions, the CDs presented remarkable photoluminescence without any additives or surface modifications. In brief, *Eleocharis dulcis* was washed repeatedly. The sample was peeled, sarcocarp was chopped and juiced. 40 mL of the juice was transferred to a Teflon-lined autoclave at a temperature 120 °C for five hours, and finally cooled down. The solution (dark brown) was centrifuged (12 000 rpm) for 10 minutes. The supernatant was collected by filtration with a microporous membrane (0.22 μm), dialyzed (1000 MWCO), and labeled as CQDs-120. CQDs-90 and CQDs-150 were generated by altering the temperature to 90 °C and 150 °C.

CDs are synthesized from the medicinal plant *Azadirachta indica* (Gedda et al., 2023). The collected leaves were sundried and powdered. Then the powder (2 g) was added to demineralized water (50 mL) for one hour while stirring. The large/undissolved particulates were separated from supernatant filtrate, added into a flask (round-bottomed), and placed on a sand bath for reflux at 180 °C for 12 hours under constant stirring. CDs were formed, signified by a reddish brown color. The CDs were further purified by centrifugation, filtration (0.22 μm), and finally stored at 4 °C. Also, Yadav et al. (2018) synthesized CQDs from *Azadirachta* leaves. In brief, leaves (10 g) were crushed and added to 100 mL of distilled water. Then the suspension was poured into a 250 mL Teflon-lined autoclave and kept in an oven (150 °C for 4 hours). On cooling, the carbonaceous (brown-colored) solution was centrifuged at 15 000 rpm for 10 minutes. Supernatant was dialyzed (24 hours) against distilled water, and the brown nanodots were kept at 4 °C. The microbe-rich sludge wastes were converted into CDs by microwave irradiation (Hu & Gao, 2020). The elemental composition of dried sewage was C (25.36 wt%), H (4.07 wt%), N (7.19 wt%), S (1.26 wt%), and O (15.11 wt%). The synthesis route requires less hazardous reactants, simple methodologies, and presents a high QY (21.7%). Omoriyekomwan et al. (2017) constructed hollow carbon nanofibers (HCNFs) on the biochar surface during microwave pyrolysis (500/600 °C) of palm kernel shell. Interestingly, HCNFs were detectable only by microwave heating. Conventional heating was not useful by means of differences in the heat transfer mechanism.

9.3.2 Synthesis of Self-assembled C-nanophotocatalysts

NPs are self-assembled via disparate methods like self-assembly in solution/at interfaces, templating methods, and assisted self-assembly. In the templating method, nanostructured objects like CNTs, block co-polymers, viruses, and DNA can be used as templates for defined structures/shapes. Although chemically synthesized hard templates provide a well-defined shape, they lack spacing between deposited NPs. Soft templates from synthetic polymers, proteins, viruses, DNA, etc., provide numerous well-defined binding sites for NP attachment. The hierarchically structured, organized NPs exhibit strategies similar to a natural system (Nie et al., 2010). The interfacial properties of epoxy composites (C-fiber reinforced) were improved by multilayered films of GO/SiO_2 (Fu et al., 2019). These films of GO/SiO_2 were deposited on the carbon fiber's surface (Figure 9.4). PEI acted as a connected bridge between the stacking layers and carbon fibers. Fiber wettability and roughness were enhanced by GO/SiO_2. The nanocomposite also exhibited highly significant interfacial properties.

Self-assembly connects top-down and bottom-up approaches. Designing is plausible in hierarchical order with complex formation resembling that of biological

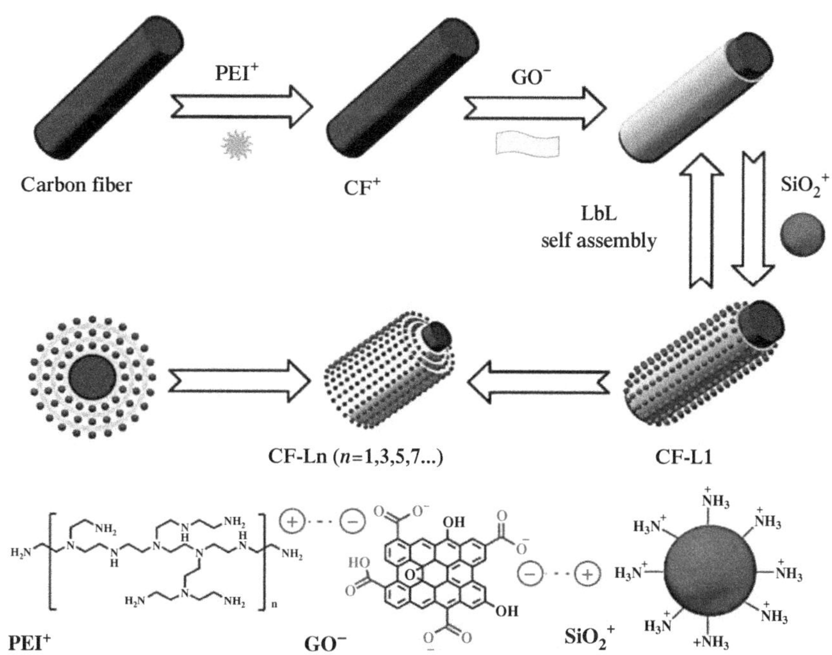

Figure 9.4 Schematic illustration of layer-by-layer self-assembly. *Source:* Fu et al. (2019) / with permission of Elsevier.

systems (Whitesides et al., 2005). This technique is low-cost and high-yielding, with a wide range of scientific and technological applications. Complexity increases with bottom-up self-assembly. Self-assembly is beneficial since it allows the aggregation of too small nanostructures which would otherwise be very difficult to manipulate individually, into an ordered functioning structure. Self-assembly is useful in generating scalable/inexpensive real 3D structures. Nanostructures present unique characteristics of quantum, microelectronics, and manufacturing with broad applications in fundamental science. Pattern synthesis is plausible in top-down 2D synthesis methods. However, the method is capital-intensive, and limited by scalability. When the top-down approach is not applicable, the self-assembly is the only technique used to fabricate the nanostructures.

Xu et al. (2021) synthesized biomass-based porous C (BPC)/graphene composite aerogels with excellent conductivity and high specific surface by π–π stacking/hydrophobic interactions. The efficacy of BPC was significantly enhanced with a diminutive graphene amount in maintaining/forming pore structure/size with effective surface area. The capacitance retention rate was 81.6% with a current density of 20 A/g. A supramolecular assembly route of synthesis produces stable aggregates by self-templating. The approach presents a well-organized architecture and morphology without interactions (noncovalent) with external template (Li et al., 2024). A self-assembled method was followed to synthesize S@g-C_3N_4/B@g-C_3N_4 (CRSB) junction modified by CQDs and rGO (Kumar et al., 2020). For the assembly, the pre-synthesized S@g-C_3N_4 (S@CN) and B@g-C_3N_4 (B@CN) were added to methanol under ultrasonication for three hours. The powder was dried (50 °C) overnight under a fume hood with vigorous stirring (six hours). Glucose was carbonized in alkaline conditions under ultrasonication (90 minutes) to initiate the CQDs synthesis. RGO (3–11%) was added to catalyst SB-40 (optimized), and the assembled RGO/S@C_3N_4/B@C_3N_4 was synthesized. To decorate, the assembly was dipped in a solution of CQDs, and autoclaved (180 °C, 24 hours). CRSB at 8% concentration was the best nanophotocatalyst. In an electrostatic self-assembly method, CQDs were attached onto proton-functionalized graphitic-C_3N_4 (Jian et al., 2016). Urea was pyrolyzed at 520 °C (2.5 hours) to fabricate the nanocomposite. Concentrated HNO_3 was used to reflux the sample (80 °C, 3 hours). A controlled nanomaterial with phase separation properties was developed by self-assembly method using a ternary solution (one hydrotrope/two immiscible fluids) (Li et al., 2022b). The micro-sized droplets were subjected to self-evaporation in a mixture of ethanol/C-dot water/oils on a surface (hydrophobic). The structures (porous/granular) of self-assembled CDs were dependent upon the proportion of in the droplet. Also, the parameters like solubility, surface tension, and vapor pressure were estimated. C-nanosheets (C8-structured) were developed from self-assembled CQDs synthesized from *Plectranthus amboinicus* leaves following three steps (Ann Mary et al., 2025). On the incitation of PVP, the growth of crystalline nanosheets and altered the morphology with cubic structure (C8), while addition of urea stopped the growth. Both NPs and nanosheets presented broad blue fluorescence. The stable cellulose nanocrystals (CNCs) were self-assembled into chiral nematic (ChN) liquid crystalline mesophases (Ackroyd et al., 2023). As confinement was increased,

ChNCNC could not bend at 90° of the square capillaries, and anchored on perpendicular surfaces. ChN creates radial layers shaped with concentric circles. A sol–gel route-self-assembly technique was followed for the synthesis of C-TiO_2 thin films using Tween80 (T80) as C-source (Hassan et al., 2014). Characterization by XRD showed that the calcination occurred at 400 °C. Initially, T80 (10 mL) was dissolved in isopropanol (40 mL) with the addition of titanium tetra-isopropoxide (6 mL) with vigorous stirring. After that acetic acid (3 mL) was added to the solution, stirred (24 hours), and a homogenous/stable solution was formed. For the construction of thin film, borosilicate glass was used as a substrate, washed with ethanol, dried at 60 °C (30 minutes), dipped in and pulled out at a rate of 12.5 cm/min (±0.3). The surface effective area was 10 cm^2. The coated glass was dried at 60 °C (10 minutes), calcined in a multisegmented furnace (400 °C), and cooled. The dip-coating and calcination were repeated for nine layers. A self-assembled 3D CNT-scaffolded $Li_4Ti_5O_{12}$@C microsphere (CNT@$Li_4Ti_5O_{12}$@C) was synthesized by a facile spray drying approach (Hu et al., 2019). CNTs were found to be uniformly dispersed in the microsphere. The scaffold was prepared initially by making CNT slurry. The slurry and acetic acid were added to deionized water with continuous stirring. Then the powders of lithium nitrate and titanium oxide (stoichiometric) were added to the solution. The well-dispersed mixture was formed in 2 hours, which was further dried (200 °C) to form spherical powders. Finally, the powders were calcinated at 800 °C for 15 hours under 5% Ar/H_2 (5%) conditions. A self-assembly method was followed to construct magnetic composites – Fe_3O_4@C@Cu_2O (420 nm) of bean-like nanostructure (Li et al., 2011). NPs presented a size of 5–10 nm. The C-layer acted as linker and stabilizer connecting Fe_3O_4 and Cu_2O. The composite presented ferromagnetism, photocatalytic activity (Rhodamine-B/methyl orange/alizarin red) and well dispersed. Self-assembled, one pot-based synthesis of CDs with high retention were derived from lecithin as C-source in presence of water (Padidam et al., 2024). The CDs (<50 nm) presented superior photothermal efficacy (red shifted fluorescence) even in a low concentration. Shen et al. (2023) constructed self-assembled hybrid CDs by altering spectral characters of H-/J-aggregation. Wood-based (Poplar wood flour) CQDs were synthesized by simple and low-cost solvothermal method. The method stabilized the MIL88B (Fe) and CQDs, increasing the photolytic activity in visible light (Zhang et al., 2021). CQDs@MIL-88B (Fe) utilized sunlight more efficiently.

CNT/ZnO/TiO_2 multifunctional nanotubes were synthesized by hydrothermal synthesis and ultrasonic-assisted acid treatment (Bai et al., 2015). Lignin contains a large amount of carbon (~60%) to be used for the synthesis of self-assembling C-spheres with self-propelling/photocatalytic properties (Cui et al., 2021). Hydrothermal carbonization was conducted to produce C-precursors from lignin (hydroxyl-abundant). A supramolecular approach was followed to synthesize g-C_3N_4 doped with Fe. This approach was followed by a thermal exfoliation method, and the nanostructure was used for photocatalysis to produce H_2 and N_2-fixation (Koli et al., 2023). The photocatalysts presented a surface area of 23.6 m^2/g. Biomass is rich in cellulose (60–90%). As crystalline cellulose is difficult to dissolve, ionic liquids are used as a dissolving solvent system for the synthesis of C-NPs (Mugadza et al., 2020). Carbonaceous-nitrogenous sensitization was followed to produce TiO_2 QDs (~4 nm) using oleic acid/oleylamine (Zhou et al., 2013). C/N-TiO_2 nanospheres were constructed by self-assembling QDs to a meso-structure (200 nm) by a bottom-up approach. In brief, oleic acid (5 mL), cyclohexane (25 mL), oleylamine (3 mL), and tetrabutyl titanate (1 mL) were mixed together for 5 minutes on stirring (ambient temperature). The solution (40 mL) was autoclaved (Teflon-lined) at 180 °C (18 hours). The QDs were ethanol precipitated, cooled down, and centrifuged for 2 minutes (3000 rpm). Cyclohexane (5 mL) and ethanol (40 mL) were used for washing QDs. The QDs were yellowish and derived on drying and milling. Nanospheres were assembled in an aqueous solution (100 mL) of sodium dodecylsulphate (4 mg/L) with vigorous stirring for 5 minutes and were ultrasonicated. Continuous stirring was performed (two hours) while the emulsion was heated (75 °C), and QDs were assembled into mesostructured nanospheres. The nanospheres were collected by centrifugation, washed (three times), and dried (60 °C, 10 hours). Also, activated carbon cloth (ACC)-eriochrome cyanine R-based photocatalysts (metal-free) were synthesized. Interestingly, the photocatalysts with a reasonable molar extinction coefficient presented a required bandgap under visible light (Gupta et al., 2023).

9.3.3 Synthesis of Nanophotocatalysts from Algae and Cyanobacteria

Many algae/cyanobacteria under diverse classes have been delved into the construction of NPs. Kelp is a macroalgae/seaweed under the class Phaeophyceae. CDs were synthesized by microwave irradiation using the macroalga C-source (Zhao et al., 2019). Ethylenediamine was used as a nitrogen dopant. In brief, initially kelp (5 g) was chopped and added to ethylenediamide (0.4 mL) deionized water (30 mL) in a microwave vessel of Teflon. The vessel was sealed, microwaved at 200 °C (10 minutes) with 800 W irradiation. The solution was cooled down to centrifuge at 10 000 rpm (20 minutes). The sample was filtered before dialysis (0.22 μm) against pure water (12 h) and stored (4 °C). A hydrothermal method was followed to synthesize microalga *Chlorella pyrenoidosa*-based CDs (20 nm) with a yield of 7.2% (Zhang et al., 2022a). The yield was increased to 7.9% when doped with N, S, and P. The proximate analysis revealed carbohydrate (32.2%), proteins (48.9%), and lipid (13.4%) of dry weight. A similar method was followed to synthesize fluorescent nitrogen-doped CDs (NCDs) of 2–8 nm. The C-source was the brown macroalga *Sargassum fluitans* (Godavarthi et al., 2017). The NCDs with remarkable fluorescent properties (quantum yield: 18.2%) were characterized. In brief, an extract was prepared from the dried and pulverized microalga (10 g) after repeated washing (six to eight times) with distilled water in 30 minutes interval. Reflux of the sample was conducted for 30 minutes with distilled water (100 mL). The extract (10 mL) thus prepared was diluted to 50 mL, autoclaved at 180 °C for five hours. The extract was cooled down, centrifuged at 10 000 rpm, filtered by a cellulose nitrate membrane (0.2 μm), and dialyzed. Brown macroalga-derived Na-alginates were used for synchronous in-situ reduction of GO to rGO (Su et al., 2019a). The hydrogel 3D modification enhanced photocatalytic efficacy (Figures 9.5 and 9.6).

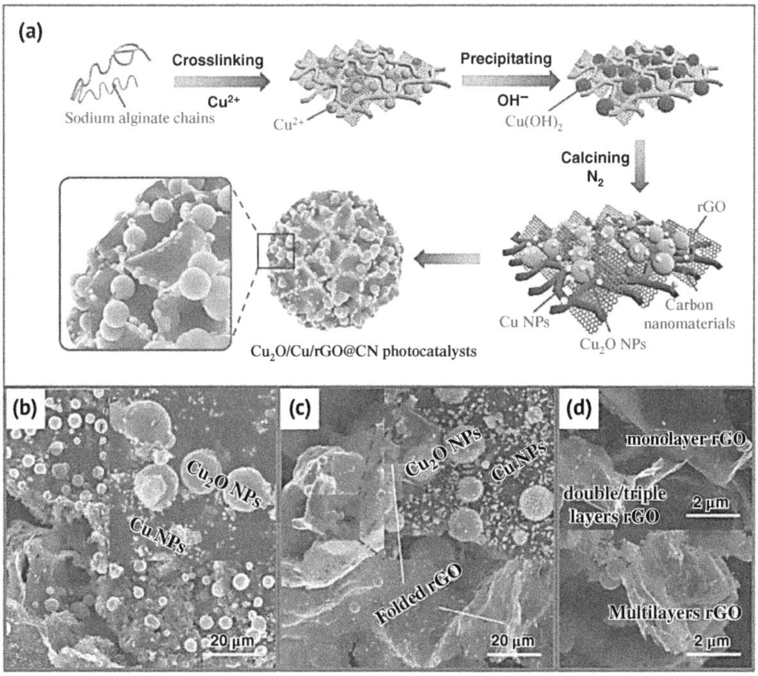

Figure 9.5 (a) Synthesis of Cu_2O/Cu/rGO@CN by sodium alginate hydrogel template, SEM microphotographs of (b) Cu_2O/Cu@CN, (c) Cu_2O/Cu/rGO@CN-4, and (d) rGO. *Source*: Reproduced with permission from Su et al. (2019a) / Elsevier.

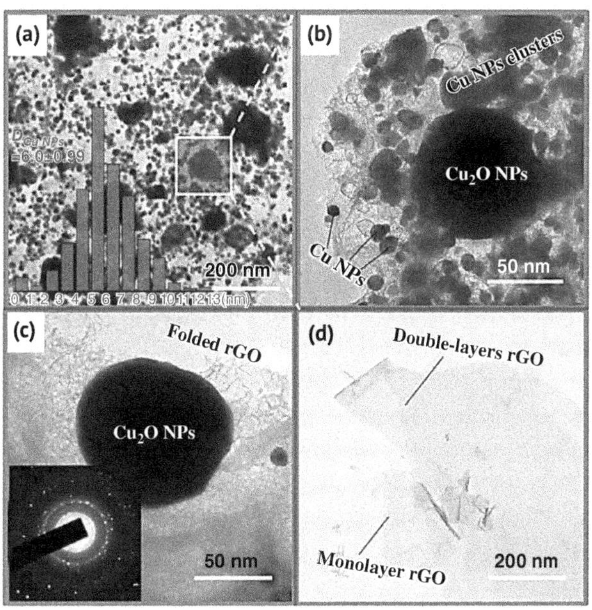

Figure 9.6 TEM image of (a–c) Cu_2O/Cu/rGO@CN-4 and (d) GO nanosheets. *Source*: Reproduced with permission from Su et al. (2019a) / Elsevier.

Algal blooms were collected under diverse classes (Cyanophyceae/Chlorophyceae/Bacillariophyceae/Euglenophyceae). Washed samples (running water) were subjected to drying under natural low-light conditions. Pulverized sample (1 g) was mixed with 20 mL phosphoric acid (40%), and microwaved [600 W (5 min 50 s)], raising a brownish-black solution signifying the formation of CDs. Then 10 mL water was added, neutralized with sodium hydroxide, filtered (Whatman 40), and centrifuged at 20 000 rpm for 20 minutes. The supernatant was dialyzed (cellulose ester membrane) for 48 hours to remove small molecules and excess NaOH from the solution. Its remarkable solubility in water/methanol/ethanol/isopropanol/acetone/DMF/DMSO, particularly in polar solvents indicated that CDs were wrapped by surface functional groups (multiple). CD yield was recorded to be 110 mg/gDW of algal powder (Ramanan et al., 2016).

Microorganism-based synthesis of NPs is performed by extracellular and intracellular methods (Singh, 2020). Mostly, extracellular syntheses of NPs are speedily scalable, comprehend unification, and follow simple downstream processing. These NPs are useful as a scaffold for molecular assembly in nanoscale devices pertaining to their size, monodisperses, and ability to modify chemical groups. Many genera under Cyanophyceae viz. *Anabaena* sp., *Calothrix* spp., *Spirulina platensis*, *Plectonema boryanum*, *Lyngbya majuscula*, *Phormidium valderianum*, etc., are used in synthesizing NPs. C-phycoerythrin is the pigment found in cyanobacteria and is extractable with water. CDs (5 nm) were synthesized from the cyanobacterium pigment extracted from *Phormidium tenue*. CDs were characterized by visible spectroscopy, FTIR, EDAX, and TEM (Mubarak Ali et al., 2012). The pigment was mixed with aqueous $CdCl_2$ (0.25 mM) and Na_2S (1 mM). Formation of CDs was indicated by changes in color. CDs were separated after 5 days and lyophilized. The carbon obtained from *Chlorella* (Chlorophyceae), *Enteromorpha* (Chlorophyceae), and carrageenan (Rhodophyceae) is hard and effective in constructing NC with disparate morphologies (Yan et al., 2022). Also, biomass-derived C from microalga *Chlorella* (N, P co-doped) and WS_2 nanocrystals were confined (facile fabrication), which was useful for inhibiting the stacking of WS_2 crystals (Kang et al., 2021). With the molten salt approach, carbons of reasonable surface area were achieved (N, S co-doped) from *Enteromorpha prolifera*, raising a hierarchical porous structure (Zhang et al., 2022b).

9.4 Self-assembled C-nanophotocatalysts for Bioremediation

Self-assembled nanostructures are useful in the remediation of industrial effluents, indoor respiratory/neurotoxic gases (nonaldehyde/p-xylene), and other wastewater treatments. The heterostructure g-C_3N_4/V_2O_5 (Z-scheme) constructed was efficient in continuous removal of nonaldehyde (97%) from the indoor environment (Bae & Jo, 2019). A melamine sponge was used to immobilize the NC with a 3D monolithic structure. The efficacy was found in the order g-C_3N_4/V_2O_5 > g-C_3N_4 > V_2O_5. Glucose was used in the hydrothermal method (one pot)-assisted Cu_2O/rGO composite fabrication as reducing agent/cross-linker (Cai et al., 2015). The aerogel was active in degrading MO (5 mg/L). A five hours irradiation could degrade about 70% of MO. The photocatalyst was reusable and efficient in the adsorption of model pollutant MO (44%). rGO benefitted charge transfer, augmenting the separation of charge carriers. Cu_2O electronic conductivity was enhanced by rGO. Reportedly, CeO_2/rGO composite aerogels-photocatalysts (self-assembled) degraded aqueous organic pollutants (Choi et al., 2016). The rGO–gel composites are lighter, exhibiting high surface area/absorbance rate while degrading the model pollutant RhB dye. The self-assembly was conducted in presence of a precursor (vitamin C). The efficacy was owed to the efficient charge transfer from interconnected CeO_2 (30/50–70/90 nm) to the rGO sheets. Organic dye-photocatalysis is dependent upon adsorption-desorption equilibrium and the occurrence of surface photocatalytic reactions. Intermolecular forces (non-covalent)/π–π stacking in RhB-rG nanosheets facilitated RhB adsorption (~62.76%). Photocatalytic efficiency could be improved by incorporating small molecules in g-C_3N_4 conjugated framework with doping/defect treatment. In a self-assembled structure, the presence of thiophene groups increased transfer rate of photoelectrons and efficacy under visible light, with a reusability of five cycles (Xu et al., 2022b). Dong et al. (2017) reported Bi_2WO_6/rGO as potential photocatalysts for removing dyes together with other pollutants (RhB/MO/phenol/sulfamonomethoxine/sulfanilamide). The self-assembled NC presented a shape of a hollow structure. Interestingly, the self-assembled composites of *Artemia* eggshell-ZnO could degrade dye (MB/RhB/neutral red) of industrial wastewater (Qian et al., 2021). A supramolecular self-assembly technique is useful for the construction of NC with a unique structure as well as optical properties (Lin et al., 2022b). CDs, rGO, CNTs, C-nanospheres are added to improve bulk g-C_3N_4 photocatalysis. KOH-modified char improved phenanthrene photocatalysis by addition of superoxide/hydroxyl radicals (76.72%). A superior photocatalysis was presented by C/g-C_3N_4 composites (hierarchical) by thermal diffusion (direct). Glucose (0.25%) increased photocatalysis rate (0.253/h) than bulk g-C_3N_4 (Huang et al., 2018). The photocatalysis was improved by hydroxyl radicals together with photoinduced holes. Interfacial coupling was conducted for $CuFe_2O_4$ (ferromagnetic) induced hotspots on nanosheets (feather-like-polymeric) of g-C_3N_4 (self-assembled). The photocatalysis was useful in disinfection and inactivation (0.157 log) of *E. coli* in wastewater in 120 minutes (Sabariselvan et al., 2024). A study by SEM confirmed the photocatalysis owing to hydroxyl/O_2 radicals. Recycling was facilitated by magnetic property, distinct separation, and insignificant leaching. Chloramphenicol is among the widely distributed antibiotics polluting water resources (ground/surface/drinking water). The metal-free S@C_3N_4/B@C_3N_4 (self-assembled) could degrade (0.0810/min) chloramphenicol 10 times higher than g-C_3N_4 (0.00802/min). The degradation was facilitated by increased interfacial contact, reduced recombination, pollutant adsorption, visible light absorption together with reactive oxygen. Alginates derived from brown macroalgae were used to construct recyclable TiO_2/potassium alginate-CNTs using cotton fabric (layer by layer-LBL) as a supportive material (Wang et al., 2017b). The pollutants like formaldehyde/RhB were efficiently removed by the nanocoating. The maximum photocatalysis was presented by 7BL for formaldehyde and RhB (92.7%). LBL increased contact between TiO_2 and CNTs. On the contrary, the porous structure improved the adsorption of pollutants on cotton fabrics. Tetracycline dye was degraded by Ag@AgI/BiOI/g-C_3N_4, which was fabricated by Z-scheme following a direct/indirect approach (Liang et al., 2018). Acetaminophen, methylparaben, and phenolics were photocatalyzed by WO_3/g-C_3N_4 (Meng et al., 2021). The nanophotocatalysts exhibited remarkable charge transfer while retaining electron redox ability on g-C_3N_4 conduction band/holes (WO_3-valence band). It provides active sites, reduces the distance to carry charge, and light harvesting. Hole-oxidation/superoxide/hydroxyl radicals mediated degradation was confirmed. The photocatalytic efficacy, stability, and recyclability are the important factors for a potential photocatalyst (Sun et al., 2019). The double-shelled hollow rods (DHR) of N, S-C/In_2O_3 created a microenvironment with greater optical performance, holes, and electrons with surface area. B-doped-3D flower-like g-C_3N_4 mesoporous nanospheres offered a reduced bandgap with more carrier movement (Han et al., 2024). The nanostructure improved light absorption/reflection/utilization rate and degraded RhB (97%) at a rate constant of 0.125/min (30 minutes). H_2O_2 production rate (15.9 times) was higher than g-C_3N_4. A template-free self-assembly method was followed to develop Co-doped tubular C_3N_4 (1D). The tabularization/metal doping improved the photocatalysis with reduced Co-leaching (Lin et al., 2023). The assembly degraded sulfisoxazole (100% in 40 minutes), produced H_2O_2 with robust applicability/stability/environmental resistance. The photocatalysis could occur in a broad pH range and in the presence of contaminants (anions/nutrients). A solvothermal method (one-pot facile) was followed to develop a lamellar BiOBr nanocomposite (Tie et al., 2020). The nanocomposite presented a conductive network that contained 1D-CNTs and reduced graphene-like flakes (2D-GF). The self-assemblies resulted in a 3D structure, which was useful for the degradation of model pollutant RhB (97.9%; 100 minutes) under visible light. Electron-hole recombination was repressed with sufficient surface contact and an electron-collecting/transfer system. The continuous generation of reactive oxygen/hydroxyl groups degraded RhB. Self-assembled C-nanosheets (mesoporous)-N-doped were synthesized from melamine and gallic acid (He et al., 2023). The nanostructure presented active sites of CO/graphitic N and pyridine N. The singlet oxygen (non-radical) actively participated in the degradation of sulfamethoxazole (90.3%) with a rate constant of 0.1705 minute. Single melamine was used as a fabrication precursor for synthesis of tubular g-C_3N_4 by supramolecular approach (Wang et al., 2018). The g-C_3N_4 (B-doped) was an efficient photocatalyst for removing NO. The heterogenous catalyst g-Fe_2O_3/GO presented reproducible photocatalysis of degrading model MB dye completely (50 mg/L). The catalyst was stable (pH 2–10.2), and could be separated to be reused for catalytic activity (Wang et al., 2019b).

9.5 Carbon Nanophotocatalysts for Energy Generation

Nanotechnology has evolved with each generation and is capable of developing efficient products (Ghasemzadeh & Shayan, 2020). Renewable energy is clean. However, it requires a faster rate of production. An energy perspective, NPs are efficient in

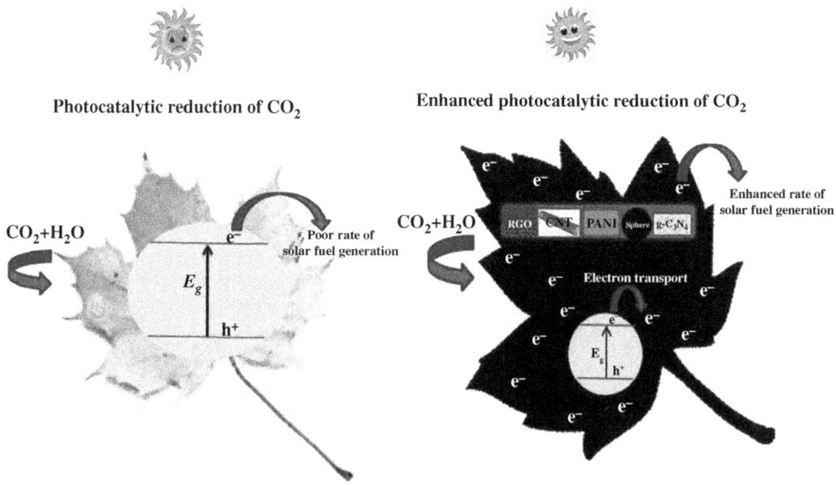

Figure 9.7 Schematic illustration of C-based materials increasing photocatalysis. *Source*: Reproduced with permission from Kandy (2019) / The Royal Society of Chemistry.

light absorption/conversion and storage. C-materials enhance the surface area and improve the photocatalysis (Figure 9.7). Reportedly, nanofluids are economical in power generation, promoting fluid displacement/heat transfer, improving power output. Generation of green energy is a must to reduce environmental pollution and climate change. Among the most efficient green technologies is the photocatalytic generation of H_2-fuel. H_2 is generally produced by methanol, which is energy-intensive and releases pollutants like CO. The mechanism of H_2 evolution by g-C_3N_4 is illustrated in Figure 9.8. On illumination, the electrons in the valence band (VB) are excited to the conduction band (CB). Simultaneously, the holes generated in g-C_3N_4-VB are quenched by triethanolamine. Electrons are further transferred to Ni $(dmgH)_2$ and initiate the catalytic reactions to release H_2 (Cao et al., 2014). The HER was 13 499 μmol/h/g, which was higher than g-C_3N_4 (16 times).

A simple method to produce H_2 is splitting the molecules of water Dharmarajan et al. (2024). Also, C_3N_4 is widely used owing to its characteristic molecular structure with a useful bandgap (visible region) among all other photocatalysts. Morphological/structural modifications can be attained by techniques like doping/hybridization. Polyhedral oligosilsesquioxanes (C-source) were assembled into hierarchical C-structures (porous). The structures presented a surface area of 2000 m^2/g with uniform micropores(~1 nm)/mesopores (~4 nm) and pore volume (1.19 cm^3/g). Structures with more mesopores presented the highest rate capability (94%) and capacitance retention (97%). Surface area and charge transfer were improved in self-assembled tubular g-C_3N_4 (C-doped), generating H_2 (3888.9 μmol/h) at a higher rate compare to pure g-C_3N_4 (Zhao et al., 2022). The band structure was adjustable, while bandgap could be narrowed with rapid photocatalysis. Barbituric acid in the supramolecular preorganization of asymmetric polymerization, fabricated g-C_3N_4 improved visible light harvesting, electron hole separation with redox capability for

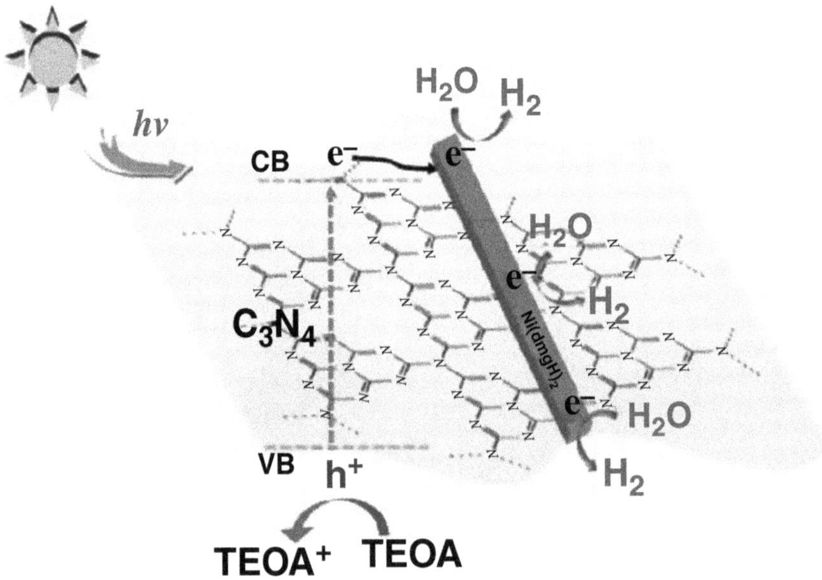

Figure 9.8 Illustration of H_2 evolution by g-C_3N_4/Ni$(dmgH)_2$. *Source*: Cao et al. (2014) / with permission of Elsevier.

H_2 generation (Li et al., 2020a). The H_2 – evolution rate (HER) increased (2.5%, three times) than symmetrically polymerized (melamine/cyanuric acid) calcined g-C_3N_4.

The photocatalytic HER (7460 μmol/g/h) was increased up to 53 times by the presence of a heptazine ring in g-C_3N_4 synthesized by supramolecular self-assembly method (Tang et al., 2024). The heptazine structure caused lone-pair electrons to be excited on N-atom and enhanced visible light utilization. Also, the shallow energy defect caused by the structure resulted in trapping of electrons and carrier recombination. The self-assembly exhibited $n \rightarrow \pi^*$ electron transition. TiO_2 is a semiconductor-photocatalyst useful for the cost-effective production of H_2 and solar cells (Colmenares et al., 2016). The C-material added to the semiconductor is responsible for moving away the electrons from e^-/h^+ pairs. The C-materials are derived from sources like lignin, STARBORN, cellulose/cellulose acetate/bacterial cellulose, chitosan, agro-wastes, wood, etc. The coupled g-C_3N_4 (5 mg) and Ni $(dmgH)_2$-sub-microwires (3.5 wt%) in the presence of triethanolamine is an effective/stable cocatalyst with HER of 1.18 μml/h under visible light (Cao et al., 2014). Wadhai and Thakur (2024) developed a 3D-nanowire photocatalyst (Ag/Cu-NWT). The HER was dependent upon the morphology of the bimetallic cost-effective method-decorated nanostructures. The HER ranged from 410 μmol/g (5 M NaOH) to 678 μmol/g (10 M NaOH). The co-production of CDs/biofuel was achieved by a hydrothermal approach (Zhang et al., 2022c). The size of the nanostructures was dependent upon the type of macromolecules. The largest size was presented by carbohydrate (2–59 nm), followed by protein (2.7–19.1 nm) and lignin (2–10 nm). Lin et al. (2022b) developed electrostatically self-assembled carbon nanofiber (CNF) and crystalline g-C_3N_4 (CCN) following microwave approach. The HER of the nanocomposite was 6398 μmol/g/h. The crystallinity, visible light harvest, surface area, and charge transfer were improved by the NC. $CoWO_4$/g-C_3N_4- a 2D heterojunction presented excellent heterointerface, charge density/separation under visible light generating H_2 (Wang et al., 2022b). Chen et al. (2013) developed multiwalled-CNTs by self-assembling Pt-nanochains onto CNTs for effective oxygen reduction. Also, HER is dependent upon the type of cocatalysts like CQDs (2199.2 μmol/h/g), nanotubes (11 850 μmol/h/g), nanosheets (3140 μmol/h/g), P dopant (1596 μmol/h/g), Co dopant (560 μmol/h/g), CeO_2 (1240.9 μmol/h/g), crystalline g-C_3N_4 (5534 μmol/h/g) and method of synthesis (Gao et al., 2022). A significant increase in HER (~25.6-fold) was exhibited by louver-like P-doped C_3N_4 (L-PCN) synthesized by supramolecular self-assembly (Li et al., 2020a). The significant increase in HER (1872.9 μmol/h/g) with photostability was observed at >420 nm with a quantum efficiency of 6.93%. The well-arrayed and appropriate separation of nanowires provided efficiency to the nanostructure. An et al. (2022) developed 3D-red crystalline C_3N_4 nanosheets (Red PCN) with characteristics optical properties adopting self-assembly strategy. The nanosheets were small curved subunits with minimum defects and ample reactive sites. Light absorption followed by charge transfer with surface reactions were synergistically optimized with HER 14 665 μmol/g/h. A photocatalyst must present broad-spectrum absorption potential, proper energy band positioning with electrochemical/thermodynamic/photoelectrochemical stabilities Feliczak-Guzik (2022). HER could be increased by introducing metal cocatalysts, doping metals, developing different structures by nonmetals (mesoporous/hollow structure/composites) with semiconductors (Hong et al., 2013). HER was increased to 48.2 μmol/h in the presence of photocatalyst NiS/C_3N_4 and triethanolamine under visible light. The optimal loading of NiS (1.1 wt%) on C_3N_4 enhanced HER (250 times) with a quantum efficiency of 1.9% (440 nm). The platinum-oxide cocatalyst increased HER and inhibited the pathways for reversible reactions (Dang et al., 2023). The monolayered nanosheets of g-C_3N_4 (grid-like) was derived by facile solvothermal self-assembled approach. HER increased 7.5 times reaching up to 16832.9 μmol/g/h at stationary phase. The C_3N_4 and layered double hydroxide (LDH) were self-assembled to generate CH_4 by photocatalytic reduction (electrostatic interaction) of CO_2 (Hong et al., 2014). The reduction efficiency was increased by the CO_3-anions with a remarkable increase in CH_4 production (3.7 μmol; 2.6 times higher) than without LDH. In carbon@g-C_3N_4, HER increased with amount of g-C_3N_4, reaching to 129 mol/h in the presence of visible light (>420 nm) with thickness 3 nm (Ma et al., 2017). This was eight times higher than g-C_3N_4 pertaining to the nanosheets that emerged in the structure of core–shell. The matrix presented high mechanical stability, reusability, and recyclability. Carbon nitride (mesoporous) exhibited HER of 120.3 μmol/h that was 1.7 times higher than C in bulk (25.6 μmol/h) in visible light (Zhao et al., 2018). The HER of g-C_3N_4 (S-doped) was 14.7 mmol/g/h, and presented active stability (95 hours). H_2 production was facilitated by C-S-H (active species) without CO emission (Wang et al., 2022c). HER of self-assembled g-C_3N_4 nanotubes (benzene ring grafted)-AHCN was 2780 μmol/h/g, which was higher than g-CNTSs-HCN (512 μmol/h/g) and g-C_3N_4 (187 μmol/h/g). The assembled nanostructure (AHCN)

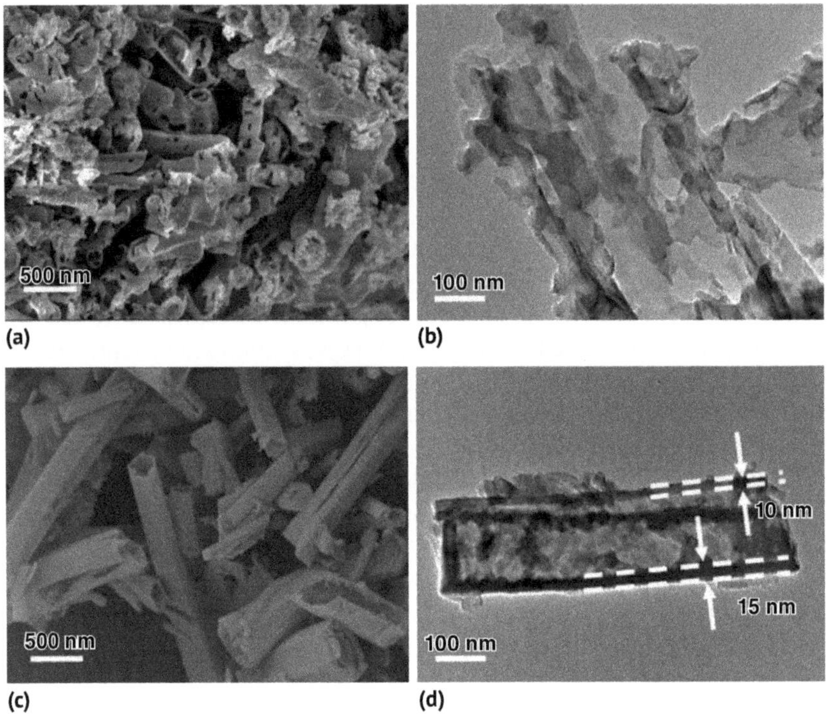

Figure 9.9 SEM/TEM images of HCN (A-B) and AHCN (C-D). *Source*: Reproduced with permission from Li et al. (2020b) / Elsevier.

presented higher charge separation and visible light (> 420 nm) absorption (Li et al., 2020b). AHCN was assembled into regular nanotubes, while HCNs were porous broken structures (Figure 9.9).

9.6 Limitations and Future Prospects

Many biogenic sources are exploited for green synthesis of nanostructures with their plausible application in environmental remediation (Tonelli et al., 2023). To deal with environmental remediation, NPs are synthesized from nonfood-producing biomass like algae/cyanobacteria. Algae/cyanobacteria are remarkable candidates for the construction of nanostructures (Chaudhury et al., 2020; Borah et al., 2022, 2023). They are sources of secondary metabolites and reducing/capping agents. Industrially scalable nanostructures can be generated only when adequate biomass is available presenting optimum growth. Although cultivation of microalgae is plausible round the year under optimized physicochemical parameters, the cultivation of macroalgae (seaweed) is limited to seasons. Biomaterial availability also depends upon location and cultivation techniques. A few limitations of algal NP syntheses include poor synthesis kinetics, morphological characteristics, strain specificity, poorly optimized conditions (pH, contact time, temperature, concentration, etc.), and low yields. Also, colloidal stability is important to reduce the agglomeration. Industrially, evaluation of growth kinetics together with cell viability is required to increase algal/cyanobacterial biomass with qualitative refining. However, cultivation of algae for NPs with desired structure is advantageous due to their low cost/eco-friendly approach, fast growth, with less energy intensity. For realistic applications, cost/scalability together with sustainability are required approaches for future perspectives (Leong et al., 2018). Also, it requires theoretical simulations wherever applicable at the electronic level. Algae-based nanostructure-syntheses are still burgeoning (Khan et al., 2022). These cellular factories are useful for reducing cost of cultivation by enabling simultaneous production/harvesting of NPs with bioremediation. An optimal C-content is essential in C-scaffold-based nanostructures to occur photocatalysis (Kandy, 2019). Photoactivity is retarded by means of over-addition of carbons in photocatalysts, as it hinders light absorption. Interestingly, atmospheric-C can also be deposited onto the large surface of photocatalysts making them available during reduction process. Hence, prior removal of impurities is required by annealing process. The possibilities for removal decrease when photocatalysts are structured with C-materials. Photostability/photo-corrosion, reusability, and recyclability are to be analyzed prior to the applications. With their unique characteristics, nanostructures such as GO, CNTs, CNFs, C_3N_4, CDs, etc., are relatively nontoxic, hold large surface area, pore volume/density and are useful for generation of fuels. However, more qualitative resources are to be exploited since C-materials enhance the photocatalysis with light absorption/utilization, charge separation/electron transportation, and QY (Kandy, 2019; Li et al., 2022a). The knowledge gap can be reduced with information on NPs such as CD precursors, synthesis/purification methods, physicochemical properties with their performance for diverse applications. A correlation must also be established among physicochemical attributes of NPs to develop products with required optical properties like peak excitation/emission wavelength/QY/electrochemical/luminescence sensing, etc. (Lin et al., 2020; Chahal et al., 2021). In-cage recombination that occurs within a very low time (sub-nanosecond) should be dealt with as QY is often not proportional to high quenching. It arises discrepancies because of unproductive recombination. Also, the limitations can be minimized by optimizing processes (light intensity/wavelength/type of reactor/temperature, etc.) and comparing/analyzing the reproducibility of different synthesis methods. Lack of homogeneity in reactor type/design, together with light irradiation, reduces the photocatalytic performance. In self-assembled luminescent nanostructures, π–π stacking is the most common interaction pertaining to organic fluorophores. Light-harvesting properties are to be studied, which increase with aggregation (Bhattacharyya et al., 2021). Visible light-based photoredox catalysis is an efficient alternative to conventional synthesis method. It is inspired by natural self-assembled machinery of photosynthesis. Supramolecular synthesis improves photocatalysis of known reactions. The properties of photoredox catalysts are customized by self-assembled supramolecular engineering. For photoredox catalysis, supramolecular gels (bottom-up) are promising and eco-friendly. C_3N_4-based supramolecular self-assembly is bioinspired approach that signifies photocatalysts with disparate morphologies and properties like water-splitting (Dharmarajan et al., 2024). However, a significant knowledge gap prevails in the detail interactions of reactants and gelator molecules. A detailed understanding of the C-based photocatalysts with high activity, CO_2-reduction and fuel conversion efficiency is required Kandy (2019). The uncontrolled aggregation of C-materials limits photocatalysis, which is required to be overcome by different approaches like integration with polymers, hard films (organic-inorganic matrices) or inducing self-assembly of CDs with precise control (Li et al., 2022b). A multidisciplinary approach from existing subfields is required to understand the underlying processes for quality use/reuse and recycle (Beil et al., 2024). New routes of synthesis are required for a sustainable and eco-friendly approach.

9.7 Conclusion

Self-assembled nanostructures are useful in the remediation of industrial effluents, indoor/outdoor air pollutants, and clean/sustainable energy generation. Assisted self-assembly, templating method, and self-assembly in solution/at interfaces are a few of the self-assembly approaches. Owing to the parallel processes, the approach offers order to the nanostructures without additional processing. C-materials improve the surface area of NC and photocatalysis. Biogenic-C is derived from plants (agricultural wastes, wood, leaves, biochar), algae/cyanobacteria (alginate/carrageenan/polysaccharide) by disparate activation, hydrothermal, and pyrolysis approaches. Among all these biogenic sources, algae/cyanobacteria are preferred as they are not foremost food crops or timbers. Optimized conditions for cultivation can avail the raw material (C-rich biomass) round the year irrespective of seasons and locations. Biomass tissues offer unique nanostructures with heteroatom doping sites/porosity and surface area. Photocatalysts harvest sunlight in a broad electromagnetic spectrum that is renewable and ample in nature. Also, nanostructures are constructed so as to absorb more solar/artificial light. Photocatalysis can split water molecules, generating H_2-fuel, which is cost-effective, safe, and eco-friendly. Effective C-NC are generated by doping/heterostructure construction/surface functionalization of the photocatalysts with metal ions/molecules/surfactants-polymers, etc. Green self-assembled synthesis is preferred since the approach is energy efficient, nontoxic, sustainable, and eco-friendly.

References

Abualnaja, K. M., Alprol, A. E., Abu-Saied, M. A. et al. (2021). Studying the adsorptive behavior of poly(Acrylonitrile-co-Styrene) and carbon nanotubes (Nanocomposites) impregnated with adsorbent materials towards methyl orange dye. *Nanomaterials*, 11, 1144.

Ackroyd, A. J., Paolis, A. D., Xu, Y. T. et al. (2023). Self-assembly of cellulose nanocrystals confined to square capillaries. *Nanoscale*, 15, 14388–14398.

Aldayel, M. F., Al Kuwayti, M. A., El Semary, N. A. H. (2022). Investigating the production of antimicrobial nanoparticles by *Chlorella vulgaris* and the link to its loss of viability. *Microorganisms*, 10, 145.

Alghamdi, Y. G., Krishnakumar, B., Malik, M. A. et al. (2022). Design and preparation of biomass-derived activated carbon loaded TiO_2 photocatalyst for photocatalytic degradation of reactive red 120 and ofloxacin. *Polymers*, 14, 880.

Alprol, A. E., Mansour, A. T., Abdelwahab, A. M. et al. (2023b). Advances in green synthesis of metal oxide nanoparticles by marine algae for wastewater treatment by adsorption and photocatalysis techniques. *Catalysts*, 13, 888.

Alprol, A. E., Mansour, A. T., El-Beltagi, H. S. et al. (2023a). Algal extracts for green synthesis of zinc oxide nanoparticles: Promising approach for algae bioremediation. *Materials*, 16, 2819.

An, S., Guo, Y., He, X. et al. (2022). Intermediate-induced repolymerization for constructing self-assembly architecture: Red crystalline carbon nitride nanosheets for notable hydrogen evolution. *Applied Catalysis B: Environmental*, 310, 121323.

Ann Mary, K. A., Palakulam, J. J., & Philip, R. (2025). Self-assembly of carbon quantum dots into C_8-structured nanosheets. *Carbon*, 234, 119991.

Bae, M. H. & Jo, W. K. (2019). Continuous photocatalytic mitigation of indoor noxious gases over a Z-scheme g-C_3N_4/V_2O_5 monolithic structure. *Building and Environment*, 161, 106235.

Bai, H., Zan, X., Zhang, L. et al. (2015). Multi-functional CNT/ZnO/TiO_2 nanocomposite membrane for concurrent filtration and photocatalytic degradation. *Separation and Purification Technology*, 156, 922–930.

Bandi, R., Gangapuram, B. R., Dadigala, R. et al. (2016). Facile and green synthesis of fluorescent carbon dots from onion waste and their potential applications as sensor and multicolour imaging agents. *RSC Advances*, 6, 28633–28639.

Bao, R., Chen, Z., Zhao, Z. et al. (2018). Green and facile synthesis of nitrogen and phosphorus co-doped carbon quantum dots towards fluorescent ink and sensing applications. *Nanomaterials*, 8, 386.

Beil, S. B., Bonnet, S., Casadevall, C. et al. (2024). Challenges and future perspectives in photocatalysis: Conclusions from an interdisciplinary workshop. *JACS Au*, 4, 2746–2766.

Bensy, A. D. V., Christobel, G. J., Muthusamy, K. et al. (2022). Green synthesis of iron nanoparticles from *Ulva lactuca* and bactericidal activity against enteropathogens. *Journal of King Saud University – Science*, 34, 101888.

Bhattacharyya, A., Sarkar, S. D., & Das. A. (2021). Supramolecular engineering and self-assembly strategies in photoredox catalysis. *ACS Catalysis*, 11, 710–733.

Bhuyar, P., Rahim, M. H. A., Sundararaju, S. et al. (2020). Synthesis of silver nanoparticles using marine macroalgae *Padina* sp. and its antibacterial activity towards pathogenic bacteria. *Beni-Suef University Journal of Basic and Applied Sciences*, 9, 3.

Borah, D., Rout, J., & Thajuddin, N. (2022). Bioenergy from algae -A biorefinery approach. In A. K. Kashyap, N. Behar, S. Sharma, et al. (Ed.), *Trends in Environmental Biology* (pp. 18–26). Kripa Drishti Publications.

Borah, D., Rout, J., & Thajuddin, N. (2023). Application of nanotechnology in bioenergy production from algae and cyanobacteria. In J. A. Malik & M. J. S. Mohamed (Eds.), *Modern Nanotechnology* (pp. 267–291). Springer, Cham.

Cai, J., Liu, W., & Li, Z. (2015). One-pot self-assembly of Cu_2O/RGO composite aerogel for aqueous photocatalysis. *Applied Surface Science*, 358, 146–151.

Cao, S.W., Yuan, Y. P., Barber, J. et al. (2014). Noble-metal-free g-C_3N_4/Ni(dmgH)$_2$ composite for efficient photocatalytic hydrogen evolution under visible light irradiation. *Applied Surface Science*, 319, 344–349.

Chahal, S., Macairan, J. R., Yousefi, N. et al. (2021). Green synthesis of carbon dots and their applications. *RSC Advances*, 11, 25354–25363.

Chaudhary, R., Nawaz, K., Khan, A. K. et al. (2020). An overview of the algae-mediated biosynthesis of nanoparticles and their biomedical applications. *Biomolecules*, 10, 1–36.

Chen, Y., Xu, J., Liu, X. et al. (2013). Electrostatic self-assembly of platinum nanochains on carbon nanotubes: A highly active electrocatalyst for the oxygen reduction reaction. *Applied Catalysis B: Environmental*, 140–141, 552–558.

Choi, J., Reddy, D. A., Islam, M. J. et al. (2016). Self-assembly of CeO_2 nanostructures/reduced graphene oxide composite aerogels for efficient photocatalytic degradation of organic pollutants in water. *Journal of Alloys and Compounds*, 688, 527–536.

Colmenares, J. C., Varma, R. S., & Lisowsk, P. (2016). Sustainable hybrid photocatalysts: Titania immobilized on carbon materials derived from renewable and biodegradable resources. *Green Chemistry*, 18, 5736–5750.

Cui, Y., Sheng, X., Anusuyadevi, P. R. et al. (2021). Self-assembled carbon spheres prepared from abundant lignin and urea for photocatalytic and self-propelling applications. *Carbon Trends*, 3, 100040.

Dang, T. T., Bhamu, K. C., Mahvelati-Shamsabadi, T. et al. (2023). Oxidized platinum cocatalyst and self-assembled graphene over graphitic carbon nitride for photocatalytic hydrogen evolution. *ACS Applied Nano Materials*, 6, 9825–9838.

Dharmarajan, N. P., Vidyasagar, D., Yang, J. H. et al. (2024). Bio-inspired supramolecular self-assembled carbon nitride nanostructures for photocatalytic water splitting. *Advanced Materials*, 36, 2306895.

Dong, S., Ding, X., Guo, T. et al. (2017). Self-assembled hollow sphere shaped Bi_2WO_6/RGO composites for efficient sunlight-driven photocatalytic degradation of organic pollutants. *Chemical Engineering Journal*, 316, 778–789.

Duarah, R. & Karak, N. (2017). Facile and ultrafast green approach to synthesize biobased luminescent reduced carbon nanodot: An efficient photocatalyst. *ACS Sustainable Chemistry & Engineering*, 5, 9454–9466.

Estrada, A. C., Lúcia Lopes, J., & Trindade, T. (2023). *Nanomaterials of Carbon and Metal Sulfides in Photocatalysis*. IntechOpen. https://doi.org/10.5772/intechopen.109658

Fathy, N. A., Basta, A. H., & Lotfy, V. F. (2020). 4 - Novel trends for synthesis of carbon nanostructures from agricultural wastes. In K. A. Abd-Elsalam (Ed.), *Micro*

and Nano Technologies, Carbon Nanomaterials for Agri-Food and Environmental Applications (pp. 59–74). Elsevier.
Feliczak-Guzik, A. (2022). Nanomaterials as photocatalysts-synthesis and their potential applications. *Materials (Basel)*, 25, 193.
Flores-Oña, D. & Fullana, A. (2022). High efficient solar photocatalytic carbon nanoparticles. *Frontiers in Catalysis*, 2, 1006564.
Fu, J., Zhang, M., & Jin, L. (2019). Enhancing interfacial properties of carbon fibers reinforced epoxy composites via Layer-by-Layer self assembly GO/SiO_2 multilayers films on carbon fibers surface. *Applied Surface Science*, 470, 543–554.
Ganesan, S., Kokulnathan, T., Sumathi, S. et al. (2024). Efficient photocatalytic degradation of textile dye pollutants using thermally exfoliated graphitic carbon nitride (TE-g-C_3N_4). *Scientific Reports*, 14:2284.
Gao, R. H., Ge, Q., Ge, Q. et al. (2022). Graphitic carbon nitride (g-C_3N_4)-based photocatalytic materials for hydrogen evolution. *Frontiers in Chemistry*, 10, 1048504.
Gedda, G., Sankaranarayanan, S. A., Putta, C. L. et al. (2023). Green synthesis of multi-functional carbon dots from medicinal plant leaves for antimicrobial, antioxidant, and bioimaging applications. *Scientific Report*, 13, 6371.
Ghasemzadeh, F. & Esmaeili Shayan, M. (2020). *Nanotechnology in the Service of Solar Energy Systems*. IntechOpen. https://doi.org/10.5772/intechopen.93014
Ghosh, S., Santhosh, R., Jeniffer, S. et al. (2019). Natural biomass derived hard carbon and activated carbons as electrochemical supercapacitor electrodes. *Scientific Report*, 9, 16315.
Godavarthi, S., Mohan Kumar, K., Vázquez Vélez, E. et al. (2017). Nitrogen doped carbon dots derived from *Sargassum fluitans* as fluorophore for DNA detection. *Journal of Photochemistry and Photobiology B: Biology*, 172, 36–41.
Guo, X., Zhao, X., Luo, X. et al. (2023). Sustainable wood-based iron photocatalyst for multiple uses with sunlight: Water treatment and radical photopolymerization. *Angewandte Chemie*, 62, e202301242.
Gupta, V., Yadav, R. K., Umar, A. et al. (2023). Highly efficient self-assembled activated carbon cloth-templated photocatalyst for NADH regeneration and photocatalytic reduction of 4-nitro benzyl alcohol. *Catalysts*, 13, 666.
Han, Y., Chao, M., Luo, C. et al. (2024). Self-assembled B-doped flower-like graphitic carbon nitride with high specific surface area for enhanced photocatalytic performance. *Journal of Colloid and Interface Science*, 657, 309–319.
Hassaan, M. A., El-Nemr, M. A., Elkatory, M. R. et al. (2023). Principles of photocatalysts and their different applications: A review. *Topics in Current Chemistry*, 381, 31.
Hassan, M. E., Cong, L., Liu, G. et al. (2014). Synthesis and characterization of C-doped TiO_2 thin films for visible-light-induced photocatalytic degradation of methyl orange. *Applied Surface Science*, 294, 89–94.
He, H., Ai, Y., He, Y. et al. (2023). Synthesis of metal-free nitrogen-doped mesoporous carbon nanosheets with reference to molecular self-assembly for efficient peroxymonosulfate activation. *Journal of Water Process Engineering*, 55, 104096.
Hong, J., Wang, Y., Wang, Y. et al. (2013). Noble-Metal-Free NiS/C_3N_4 for Efficient Photocatalytic Hydrogen Evolution from Water. *ChemSusChem*, 6, 2263-2268.

Hong, J., Zhang, W., Wang, Y. et al. (2014). Photocatalytic reduction of carbon dioxide over self-assembled carbon nitride and layered double hydroxide: The role of carbon dioxide enrichment. *ChemCatChem*, 6, 2315–2321.

Hu, Y. & Gao, Z. (2020). Sewage sludge in microwave oven: A sustainable synthetic approach toward carbon dots for fluorescent sensing of para-Nitrophenol. *Journal of Hazardous Materials*, 382, 121048.

Hu, Y., Lin, F., & Liu, Z. (2019). Three-dimensional carbon nanotube-scaffolded $Li_4Ti_5O_{12}$@C microsphere with enhanced rate property for electrochemical energy storage. *Ceramics International*, 45, 10976–10981.

Huang, R. L., Huang, W. Q., Li, D. F. et al. (2018). Self-assembled hierarchical carbon/g-C_3N_4 composite with high photocatalytic activity. *Journal of Physics D: Applied Physics*, 51, 135501.

Hutton, G. A., Martindale, B. C., & Reisner, E. (2017). Carbon dots as photosensitisers for solar driven catalysis. *Chemical Society Reviews*, 46, 6111–6123.

Ismail, G. A., Allam, N. G., El-Gemizy, W. M. et al. (2021). The role of silver nanoparticles biosynthesized by Anabaena variabilis and *Spirulina platensis* cyanobacteria for malachite green removal from wastewater. *Environmental Technology*, 42, 4475–4489.

Ji, M., Di, J., Ge, Y. et al. (2017). 2D-2D stacking of graphene-like g-C_3N_4/Ultrathin $Bi_4O_5Br_2$ with matched energy band structure towards antibiotic removal. *Applied Surface Science*, 413, 372–380.

Jian, X., Liu, X., Yang, H. et al. (2016). Construction of carbon quantum dots/proton-functionalized graphitic carbon nitride nanocomposite via electrostatic self-assembly strategy and its application. *Applied Surface Science*, 370, 514–521.

Jirimali, H., Singh, J., Boddula, R. et al. (2022). Nano-structured carbon: Its synthesis from renewable agricultural sources and important applications. *Materials*, 15, 3969.

Kandy, M. M. (2019). Carbon-based photocatalysts for enhanced photocatalytic reduction of CO_2 to solar fuels. *Sustainable Energy & Fuels*, 4, 469–484.

Kang, B., Wang, Y., He, X. et al. (2021). Facile fabrication of WS_2 nanocrystals confined in *Chlorella*-derived N, P co-doped bio-carbon for sodium-ion batteries with ultra-long lifespan. *Dalton Transactions*, 50, 14745–14752.

Khalid, N. R., Majid, A., Tahir, M. B. et al. (2017). Carbonaceous-TiO_2 nanomaterials for photocatalytic degradation of pollutants: A review. *Ceramics International*, 43:14552–14571.

Khan, F., Shahid, A., Zhu, H. et al. (2022). Prospects of algae-based green synthesis of nanoparticles for environmental applications. *Chemosphere*, 293, 133571.

Koli, V. B., Murugan, G., & Ke, S. C. (2023). Self-assembled synthesis of porous iron-doped graphitic carbon nitride nanostructures for efficient photocatalytic hydrogen evolution and nitrogen fixation. *Nanomaterials*, 13, 275.

Kumar, A., Kumari A., Sharma, G. et al. (2020). Carbon quantum dots and reduced graphene oxide modified self-assembled S@C_3N_4/B@C_3N_4 metal-free nano-photocatalyst for high performance degradation of chloramphenicol. *Journal of Molecular Liquids*, 300, 112356.

Leong, K. H., Aziz, A. A., Sim, L. C. Saravanan, P., Jang, M., & Bahnemann, D. (2018). Mechanistic insights into plasmonic photocatalysts in utilizing visible light. *The Beilstein Journal of Nanotechnology*, 9, 628–648.

Li, D. C. & Jiang, H. (2017). The thermochemical conversion of non-lignocellulosic biomass to form biochar: A review on characterizations and mechanism elucidation. *Bioresource Technology*, 246, 57–68.

Li, H., Cheng, B., Xu, J. et al. (2024). Crystalline carbon nitrides for photocatalysis. *EES Catalysis*, 2, 411–447.

Li, M., Wei, X. & Bao, L. (2022b). Self-assembly of carbon nanodots induced by liquid–liquid phase separation in a surface microdroplet. *Soft Matter*, 18, 6517–6528.

Li, S. K., Huang, F. Z., Wang, Y. et al. (2011). Magnetic Fe_3O_4@C@Cu_2O composites with bean-like core/shell nanostructures: Synthesis, properties and application in recyclable photocatalytic degradation of dye pollutants. *Journal of Materials Chemistry*, 21, 7459–7466.

Li, S., Peng, Y., Hu, C. et al. (2020b). Self-assembled synthesis of benzene-ring-grafted g-C_3N_4 nanotubes for enhanced photocatalytic H_2 evolution. *Applied Catalysis B: Environmental*, 279, 119401.

Li, X., Ding, C., Zhao, C. et al. (2020a). Activation of graphitic carbon nitride by solvent-mediated supramolecular assembly for enhanced hydrogen evolution. *Applied Surface Science*, 525, 146444.

Li, Z., Bai, Z., Mi, H. et al. (2019). Biowaste-derived porous carbon with tuned microstructure for high-energy quasi-solid-state supercapacitors. *ACS Sustainable Chemistry & Engineering*, 7, 13127–13135.

Li, Z., Li, K., Du, P. et al. (2022a) Carbon-based photocatalysts for hydrogen production: A review. *Chemosphere*, 308, 135998.

Liang, C., Niu, C.G., Guo, H. et al. (2018). Combination of efficient charge separation with the assistance of novel dual Z-scheme system: self-assembly photocatalyst Ag@AgI/BiOI modified oxygen-doped carbon nitride nanosheet with enhanced photocatalytic performance. *Catalysis Science & Technology*, 8, 1161–1175.

Liang, Z., Hao, L., Yu, Y. et al. (2024). AgI/UiO-66-NH_2@carbonized wood fabricated as self-floating heterojunction for enhancing visible light-driving photocatalytic degradation of Rhodamine B. *Industrial Crops and Products*, 210, 118125.

Lin, F., Zhou, S., Wang, G. et al. (2022b). Electrostatic self-assembly combined with microwave hydrothermal strategy: construction of 1D/1D carbon nanofibers/crystalline g-C_3N_4 heterojunction for boosting photocatalytic hydrogen production. *Nano Energy*, 99, 107432.

Lin, M., Li, F., Cheng, W. et al. (2022a). Facile preparation of a novel modified biochar-based supramolecular self-assembled g-C_3N_4 for enhanced visible light photocatalytic degradation of phenanthrene. *Chemosphere*, 288, 132620.

Lin, X., Xiong, M., Zhang, J. et al. (2020). Carbon dots based on natural resources: Synthesis and applications in sensors. *Microchemical Journal*, 106, 105604.

Lin, Z., Wang, Z., Xu, Z. et al. (2023). Self-assembly construction of 1D carbon nitride nanotubes and cobalt-modified for superior photocatalytic degradation of sulfonamide antibiotics. *Chemosphere*, 343, 140299.

Liu, Y., Roy, S., Sarkar, S. et al. (2021). A review of carbon dots and their composite materials for electrochemical energy technologies. *Carbon Energy*, 3, 795–826.

Liu, Y., Zhu, X., Yuan, D. et al. (2020). Preparation and characterization of TiO_2 based on wood templates. *Scientific Reports*, 10, 12444.

Long, C., Jiang, Z., Shangguan, J. et al. (2021). Applications of carbon dots in environmental pollution control: A review. *Chemical Engineering Journal*, 406, 126848.

Luo, L., Yang, Y., Xiao, M. et al. (2015). A novel biotemplated synthesis of TiO_2/wood charcoal composites for synergistic removal of bisphenol A by adsorption and photocatalytic degradation. *Chemical Engineering Journal*, 262, 1275–1283.

Ma, L., Fan, H., Fu, K. et al. (2017). Protonation of graphitic carbon nitride (g-C_3N_4) for an electrostatically self-assembling carbon@g-C_3N_4 core–shell nanostructure toward high hydrogen evolution. *ACS Sustainable Chemistry & Engineering*, 5, 7093–7103.

Malitha, M. D., Moll, M. T. H., Bashar, M. A. et al. (2024). Fabrication of a reusable carbon quantum dots (CQDs) modified nanocomposite with enhanced visible light photocatalytic activity. *Scientific Reports*, 14, 17976.

Maliutina, K. M., Omoriyekomwan, J. E., He, C. et al. (2023). Biomass-derived carbon nanostructures and their applications as electrocatalysts for hydrogen evolution and oxygen reduction/evolution. *Frontiers of Environmental Science & Engineering*, 2. https://doi.org/10.3389/fenve.2023.1228992.

Meng, J., Wang, X., Liu, Y. et al. (2021). Acid-induced molecule self-assembly synthesis of Z-scheme WO_3/g-C_3N_4 heterojunctions for robust photocatalysis against phenolic pollutant. *Chemical Engineering Journal*, 403, 126354.

Mohapatra, L., Cheon, D., & Yoo, S.H. (2023). Carbon-based nanomaterials for catalytic wastewater treatment: A review. *Molecules*, 28, 1805.

Mubarak Ali, D., Gopinath, V., Rameshbabu, N., & Thajuddin, N. (2012). Synthesis and characterization of CdS nanoparticles using C-phycoerythrin from the marine cyanobacteria. *Materials Letters*, 74, 8–11.

Mugadza, K., Stark, A., & Ndungu P. G. (2020). Synthesis of carbon nanomaterials from biomass utilizing ionic liquids for potential application in solar energy conversion and storage. *Materials (Basel)*, 13, 3945.

Nie, Z., Petukhova, A., & Kumacheva, E. (2010). Properties and emerging applications of self-assembled structures made from inorganic nanoparticles. *Nature Nanotechnology*, 5, 15–25.

Omoriyekomwan, J. E., Tahmasebi, A., Zhang, J. et al. (2017). Formation of hollow carbon nanofibers on bio-char during microwave pyrolysis of palm kernel shell. *Energy Conversion and Management*, 148, 583–592.

Ong, W. J., Tana, L. L., & Chaia, S. P. (2015). Surface charge modification via protonation of graphitic carbon nitride(g-C_3N4) for electrostatic self-assembly construction of 2D/2Dreducedgrapheneoxide (rGO)/g-C_3N_4 nanostructures toward enhanced photocatalytic reduction of carbon dioxide to methane. *Nano Energy*, 13, 757–770.

Padidam, S. R., Kadam, D. B., Thakkellapati, S. et al. (2024). Single-step synthesis of self-assembled carbon dots for enhanced cancer cell retention and theranostics applications. *Microchemical Journal*, 198, 110144.

Pang, Y. L., Lim, C. W., Shak, K. P. Y. et al. (2020). Biomass-based photocatalysts for environmental applications. In Inamuddin, A. M. Asiri, & E. Lichtfouse (Eds.), *Nanophotocatalysis and Environmental Applications* (pp. 55–86). Springer, Cham.

Pourshirband, N., Nezamzadeh-Ejhieh, A., & Mirsattari, S. N. (2020). The CdS/g-C_3N_4 nano-photocatalyst: Brief characterization and kinetic study of photodegradation and mineralization of methyl orange. *Spectrochimica Acta Part A: Molecular and Biomolecular Spectroscopy*, 248, 119110.

Qi, X., Cai, H., Zhang, X. et al. (2023). CdS quantum dots/nano-TiO_2 incorporated wood as a long-term stable and easily separable photocatalytic adsorbent for efficient Cr(VI) removal. *Chemical Engineering Journal*, 475, 146320.

Qian, C., Yin, J., Zhao, J. et al. (2021). Facile preparation and highly efficient photodegradation performances of self-assembled Artemia eggshell-ZnO nanocomposites for wastewater treatment. *Colloids and Surfaces A: Physicochemical and Engineering Aspects*, 610, 125752.

Rajeshkumar, S., Malarkodi, C., Gnanajobitha, G. et al. (2013). Seaweed-mediated synthesis of gold nanoparticles using *Turbinaria conoides* and its characterization. *Journal of Nanostructure in Chemistry*, 3, 44.

Ramanan, V., Thiyagarajan, S. K., Raji, K. et al. (2016). Outright green synthesis of fluorescent carbon dots from eutrophic algal blooms for in vitro imaging. *ACS Sustainable Chemistry & Engineering*, 4, 4724–4731.

Rasool, M. A., Sattar, R., Anum, A. et al. (2023). An insight into carbon nanomaterial-based photocatalytic water splitting for green hydrogen production. *Catalysts*, 13, 66.

Sabariselvan, L., Okla, M. K., Brindha, B. et al. (2024) Interfacial coupling of $CuFe_2O_4$ induced hotspots over self-assembled g-C_3N_4 nanosheets as an efficient photocatalytic bacterial disinfectant. *Environmental Pollution*, 342, 123076.

Shen, C. L., Lou, Q., Liu, K. K. et al. (2023). Supramolecular self-assembly of carbon nanodots through edge functionalized interaction. *Carbon*, 213, 118217.

Shi, L., Zou, X., Wang, T. et al. (2022). Sunlight photocatalytic degradation of ofloxacin using UiO-66/wood composite photocatalysts. *Chinese Chemical Letters*, 33, 442–446.

Singh, D. (2020). Cyanobacteria as a source of nanoparticle: Application and future projections. In P. K. Singh, A. Kumar, V. K. Singh, et al (Eds.), *Advances in Cyanobacterial Biology* (pp. 319–331). Academic Press.

Singh, J., Dutta, T., Kim, K.-H. et al. (2018). 'Green' synthesis of metals and their oxide nanoparticles: Applications for environmental remediation. *Journal of Nanobiotechnology*, 16, 1–24.

Son, B. T., Long, N. V., & Hang, N. T. N. (2021). The development of biomass-derived carbon-based photocatalysts for the visible-light-driven photodegradation of pollutants: A comprehensive review. *RSC Advances*, 11, 30574–30596.

Su, R., Ge, S., Li, H. et al. (2019a). Synchronous synthesis of Cu_2O/Cu/rGO@carbon nanomaterials photocatalysts via the sodium alginate hydrogel template method for visible light photocatalytic degradation. *Science of The Total Environment*, 693, 133657.

Su, R., Li, Q., Chen, Y. et al. (2019b). One-step synthesis of Cu_2O@carbon nanocapsules composites using sodium alginate as template and characterization of their visible light photocatalytic properties. *Journal of Cleaner Production*, 209, 220–229.

Sun, L., Li, R., Zhan, W. et al. (2019). Double-shelled hollow rods assembled from nitrogen/sulfur-codoped carbon coated indium oxide nanoparticles as excellent photocatalysts. *Nature Communications*, 10, 2270.

Tang, S., Wang, W., Ansari, H. M. et al. (2024). Atypical supramolecular self-assembly derived graphitic carbon nitride with $n \rightarrow \pi^*$ electron transition capable of efficient visible light hydrogen production. *Applied Surface Science*, 674, 160917.

Thomas, B., Sain, M., & Oksman, K. (2022). Sustainable carbon derived from sulfur-free lignins for functional electrical and electrochemical devices. *Nanomaterials*, 12, 3630.

Tie, W., Du, Z., Yue, H. et al. (2020). Self-assembly of carbon nanotube/graphitic-like flake/BiOBr nanocomposite with 1D/2D/3D heterojunctions for enhanced photocatalytic activity. *Journal of Colloid and Interface Science*, 579, 862–871.

Tonelli, F. M. P., Silva C. S., Delgado V. M. S. et al. (2023). Algae-based green AgNPs, AuNPs, and FeNPs as potential nanoremediators. *Green Processing and Synthesis*, 12, 20230008.

Tyagi, A., Tripathi, K.M., Singh, N. et al. (2016). Green synthesis of carbon quantum dots from lemon peel waste: applications in sensing and photocatalysis. *RSC Advances*, 6, 72423–72432.

Usmani, M. A., Khan, I., Bhat, A.H. et al. (2017). Current trend in the application of nanoparticles for waste water treatment and purification: A review. *Current Organic Synthesis*, 14, 206–226.

Wadhai, S. & Thakur, P. (2024). Synthesis of Ag/Cu decorated 3D self-assembled nanowire TiO_2 photocatalyst for hydrogen production: a promising pathway towards sustainable energy generation. *Environmental Science and Pollution Research*, 31, 60836–60851.

Waghchaure, R. H., Adole, V. A., & Jagdale, B. S. (2022). Photocatalytic degradation of methylene blue, rhodamine B, methyl orange and Eriochrome black T dyes by modified ZnO nanocatalysts: A concise review. *Inorganic Chemistry Communications*, 143, 109764.

Wang, F., Yu, X., Ge, M. et al. (2019b). Facile self-assembly synthesis of γ-Fe_2O_3 / graphene oxide for enhanced photo-Fenton reaction. *Environmental Pollution*, 248, 229–237.

Wang, H., Niu, R., Liu, J. et al. (2022b). Electrostatic self-assembly of 2D/2D $CoWO_4$/g-C_3N_4 p—n heterojunction for improved photocatalytic hydrogen evolution: Built-in electric field modulated charge separation and mechanism unveiling. *Nano Research*, 15, 6987–6998.

Wang, H., Thangamuthu, M., Wu, Z. et al. (2022c). Self-assembled sulphur doped carbon nitride for photocatalytic water reforming of methanol. *Chemical Engineering Journal*, 445, 136790.

Wang, M., Wang, P., Zhang, J. et al. (2019a). A Ternary Pt/Au/TiO_2 -Decorated plasmonic wood carbon for high-efficiency interfacial solar steam generation and photodegradation of tetracycline. *ChemSusChem*, 12, 467–472.

Wang, R., Lu, K. Q., Tang, Z. R. et al. (2017a). Recent progress in carbon quantum dots: Synthesis, properties and applications in photocatalysis. *Journal of Materials Chemistry*, 5, 3717–3734.

Wang, Y., Chen, Y., Zhao, H. et al. (2022a). Biomass-derived porous carbon with a good balance between high specific surface area and mesopore volume for supercapacitors. *Nanomaterials*, 12, 3804.

Wang, Y., Ding, X., Chen, X. et al. (2017b). Layer-by-layer self-assembly photocatalytic nanocoating on cotton fabrics as easily recycled photocatalyst for degrading gas and liquid pollutants. *Cellulose*, 24, 4569–4580.

Wang, Z., Chen, M., Huang, Y. et al. (2018). Self-assembly synthesis of boron-doped graphitic carbon nitride hollow tubes for enhanced photocatalytic NOx removal under visible light. *Applied Catalysis B: Environmental*, 239, 352–361.

Whitesides, G. M., Kriebel, J. K., & Mayers, B. T. (2005). Self-assembly and nanostructured materials. In W. T. S. Huck (Ed.), *Nanoscale Assembly. Nanostructure Science and Technology* (pp. 217–239). Springer, Boston.

Xiong, B., Li, H., & Xiao, Y. (2020) Progress on synthesis and applications of porous carbon Materials. *International Journal of Electrochemical Science*, 15, 1363–1377.

Xu, M., Wang, A., & Xiang, Y. (2021). Biomass-based porous carbon/graphene self-assembled composite aerogels for high-rate performance supercapacitor. *Journal of Cleaner Production*, 315, 128110.

Xu, R., Li, J., & Sui, G. (2022b). Constructing supramolecular self-assembled porous g-C_3N_4 nanosheets containing thiophene-groups for excellent photocatalytic performance under visible light. *Applied Surface Science*, 578, 152064.

Xu, X., Ray, R., Gu, Y. et al. (2004). Electrophoretic analysis and purification of fluorescent single-walled carbon nanotube fragments. *Journal of the American Chemical Society*, 126, 12736–12737.

Xu, Y., Li, S., Chen, M. et al. (2022a). Carbon-based nanostructures for emerging photocatalysis: CO reduction, N fixation, and organic conversion. *Trends in Chemistry*, 4, 984–1004.

Yadav, P. K., Singh, V. K., Chandra, S. et al. (2018). Green synthesis of fluorescent carbon quantum dots from *Azadirachta indica* leaves and their peroxidase-mimetic activity for the detection of H_2O_2 and ascorbic acid in common fresh fruits. *ACS Biomaterials Science & Engineering*, 5, 623–632.

Yan, H., Dong, Y., Cai, L. et al. (2022). Construction of 1D biomass-derived tubular carbon fiber/Ni nanoparticles composite for broadband and lightweight microwave absorbers. *Carbon*, 200, 317–326.

Yeamsuksawat, T., Zhu, L. T., Kasuga, T. et al. (2023). Chitin-derived nitrogen-doped carbon nanopaper with subwavelength nanoporous structures for solar thermal heating. *Nanomaterials*, 13, 1480.

Yi, F., Ma, J., Lin, C. et al. (2020). Insights into the enhanced adsorption/photocatalysis mechanism of a $Bi_4O_5Br_2$/g-C_3N_4 nanosheet. *Journal of Alloys and Compounds*, 821, 153557.

Yuan, M., Liu, C., Liu, C. et al. (2025). Hydrogen evolution using a self-supported starch carbon aerogel electrocatalyst loaded with CeO_2–MoxC in alkaline saline water. *New Journal of Chemistry*, 49, 2782–2790.

Zhang, C., Ge-Zhang, S., Wang, Y. et al. (2024). Wooden carbon-based photocatalyst for water treatment. *International Journal of Molecular Sciences*, 25, 4743.

Zhang, H., Gong, X., Song, Z. et al. (2021). Wood-based carbon quantum dots for enhanced photocatalysis of MIL-88B(Fe). *Optical Materials*, 113, 110865.

Zhang, M., Huang, K., Ding, Y. et al. (2022b). N, S Co-doped carbons derived from *Enteromorpha prolifera* by a molten salt approach: Antibiotics removal performance and techno-economic analysis. *Nanomaterials*, 12, 4289.

Zhang, M., Li, P., & Wu, D. (2023). Editorial for special issue: "Synthesis and application of biomass-derived carbon-based nanomaterial." *Nanomaterials*, 13, 2020.

Zhang, J., Xia, A., Chen, H. et al. (2022a). Biobased carbon dots production via hydrothermal conversion of microalgae *Chlorella pyrenoidosa. Science of The Total Environment*, 839, 156144.

Zhang, J., Xia, A., Zhu, X. et al. (2022c). Co-production of carbon quantum dots and biofuels via hydrothermal conversion of biomass. *Fuel Processing Technology*, 232, 107276.

Zhao, C., Li, X., Cheng, C., & Yang, Y. (2019). Green and microwave-assisted synthesis of carbon dots and application for visual detection of cobalt (II) ions and pH sensing. *Microchemical Journal*, 147, 183–190.

Zhao, S., Liu, Y., Wang, Y. et al. (2022). A self-assembly strategy to synthesize carbon doped carbon nitride microtubes with a large π-electron conjugated system for efficient H_2 evolution. *Chemical Engineering Journal*, 447, 137436.

Zhao, S., Zhang, Y., Fang, J. et al. (2018). Self-assembled mesoporous carbon nitride with tunable texture for enhanced visible-light photocatalytic hydrogen evolution. *ACS Sustainable Chemistry & Engineering*, 6, 8291–8299.

Zhou, G., Liu, X., Nan, C. et al. (2013). C/N-sensitized self-assembly of mesostructured TiO_2 nanospheres with significantly enhanced photocatalytic activity. *New Journal of Chemistry*, 37, 2582.

Index

Note: Page numbers in *italics* and **bold** refers to figures and tables respectively.

Photocatalysts based on Carbon Scaffolds: Environmental and Industrial Applications, First Edition. Edited by Ramesh Oraon et al.
Published 2026 by Wiley-VCH GmbH.

r

s

t

u